# A Practical Introduction to Hardware/Software Codesign

Patrick R. Schaumont

# A Practical Introduction to Hardware/Software Codesign

Second Edition

Springer

Patrick R. Schaumont
Bradley Department of Electrical
    and Computer Engineering
Virginia Tech
Whittemore Hall 302
Blacksburg, VA 24061
USA

Additional material to this book can be downloaded from http://extras.springer.com

ISBN 978-1-4899-9060-0          ISBN 978-1-4614-3737-6 (eBook)
DOI 10.1007/978-1-4614-3737-6
Springer New York Heidelberg Dordrecht London

Printed on acid-free paper

Springer is part of Springer Science+Business Media (www.springer.com)

*The most important day is today*

# Preface

How do we design efficient digital machines? Software programmers would say "by writing better code". Hardware designers would say "by building faster hardware". This book is on codesign – the practice of taking the best from software design and the best from hardware design to solve design problems. Hardware/software codesign can help a designer to make trade-offs between the flexibility and the performance of a digital system. Using hardware/software codesign, designers are able to combine two radically different ways of design: the sequential way of decomposition in time, using software, with the parallel way of decomposition in space, using hardware.

## About the Picture

The picture on the next page is a drawing by a famous Belgian artist, Panamarenko. It shows a human-powered flying machine called the Meganeudon II. He created it in 1973. While, in my understanding, noone has built a working Meganeudon, I believe this piece of art captures the essence of design. Design is not about complexity, and it is not about low-level details. Design is about ideas, concepts, and vision. Design is a fundamentally creative process.

But to realize a design, we need technology. We need to map ideas and drawings into implementations. Computer engineers are in a privileged position. They have the background to convert design ideas into practical realizations. They can turn dreams into reality.

## Intended Audience

This book assumes that you have a basic understanding of hardware, that you are familiar with standard digital hardware components such as registers, logic gates,

vii

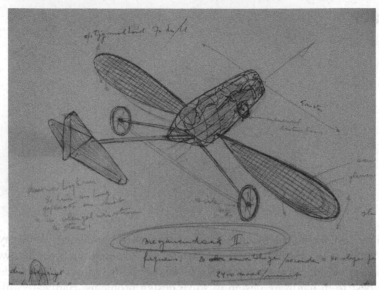

Panamarenko's Meganeudon II ((c) Panamarenko)

and components such as multiplexers, and arithmetic operators. The book also
assumes that you know how to write a program in C. These topics are usually
covered in an introductory course on computer engineering, or in a combination
of courses on digital design and software engineering.

The book is suited for advanced undergraduate students and beginning graduate
students, as well as researchers from other (non-computer engineering) fields.
For example, I often work with cryptographers who have no formal training in
hardware design but still are interested in creating dedicated architectures for highly
specialized algorithms. This book is also for them.

## Organization

The book puts equal emphasis on design methods, and modeling (design languages).
Design modeling helps a designer to think about a design problem, and to capture
a solution for the problem. Design methods are systematic transformations that
convert design models into implementations.

There are four parts in this book: Basic Concepts, the Design Space of Custom
Architectures, Hardware/Software Interfaces, and Applications.

## Part I: Basic Concepts

Chapter 1 covers the fundamental properties of hardware and software, and discusses the motivation for hardware/software codesign. Chapters 2 and 3 describe data-flow modeling and implementation. Data-flow modeling is a system-level specification technique, and a very useful one. Data-flow models are implementation-agnostic: they map into software as well as into hardware. They also support high-level performance analysis and optimization. Chapter 2 in particular discusses stability analysis, and optimizations such as pipelining and retiming. Chapter 3 shows how dataflow models can be realized in hardware and software. Chapter 4 introduces control-flow and data-flow analysis of C programs. By analyzing the control dependencies and the data dependencies of a C program, a designer obtains insight into possible hardware implementations of that C program.

## Part II: The Design Space of Custom Architectures

The second part is a tour along the vast design space of flexible, customized architectures. A review of four digital architectures shows how hardware gradually evolves into software. The Finite State Machine with Datapath (FSMD) discussed in Chap. 5 is the starting point. FSMD models are the equivalent of hardware modeling at the register-transfer level (RTL). Chapter 6 introduces micro-programmed architectures. These are still very much like RTL machines, but they have a flexible controller, which allows them to be reprogrammed with software. Chapter 7 reviews general-purpose embedded RISC cores. These processors are the heart of typical contemporary hardware/software systems. Finally, Chap. 8 ties the general-purpose embedded core back to the FSMD in the context of a System-on-Chip architecture (SoC). The SoC sets the stage for the hardware/software codesign problems that are addressed in the third part.

## Part III: Hardware/Software Interfaces

The third part describes the link between hardware and software in the SoC architecture, in four chapters. Chapter 9 introduces the key concepts of hardware/software communication. It explains the concept of synchronization schemes and the difference between communication-constrained design and computation-constrained design. Chapter 10 discusses on-chip bus structures and the techniques they use to move information efficiently between hardware and software. Chapter 11 describes micro-processor interfaces. These interfaces are the locations in a processor-based design where custom-hardware modules can be attached. The chapter describes a memory-mapped interface, the coprocessor interface, and a custom-instruction interface. Chapter 12 shows how hardware modules need to be encapsulated in order to "fit" into a micro-processor interface. This requires the design of a programmer's model for the custom hardware module.

## Part IV: Applications

The final part describes three in-depth applications of hardware-software codesign. Chapter 13 presents the design of a coprocessor for the Trivium stream cipher algorithm. Chapter 14 presents a coprocessor for the Advanced Encryption Standard. Chapter 15 presents a coprocessor to compute CORDIC rotations. Each of these designs uses different processors and microprocessor interfaces. Chapter 13 uses an 8051 microcontroller and an ARM, Chap. 14 uses an ARM and a Nios-II, and Chap. 15 uses a Microblaze.

Many of the examples in this book can be downloaded. This supports the reader in experiments beyond the text. The Appendix contains a guideline to the installation of the GEZEL tools and the examples.

Each of the chapters includes a Problem Section and a Further Reading Section. The Problem Section helps the reader to build a deeper understanding of the material. Solutions for selected problems can be requested online through Springerextras (http://extras.springer.com).

There are several subjects which are *not* mentioned or discussed in this book. As an introductory discussion on a complex subject, I tried to find a balance between detail and complexity. For example, I did not include a discussion of advanced concepts in software concurrency, such as threads, and software architectures, such as operating systems and drivers. I also did not discuss software interrupts, or advanced system operation concepts such as Direct Memory Access.

I assume that the reader will go through all the chapters in sequence. A minimal introduction to hardware-software codesign should include Chaps. 1, 4, 5, 7–12.

## A Note on the Second Edition

This book is the second edition of *A Practical Introduction to Hardware/Software Codesign*. The book was thoroughly revised over the first edition. Several chapters were rewritten, and new material was added. I focused on improving the overall structure, making it more logical and smooth. I also added more examples. Although the book grew in size, I did not extend its scope .

Here are some of the specific changes:

- The chapter on dataflow was split in two: one chapter on dataflow analysis and transformations and a second chapter on dataflow implementation. The discussion on transformations offers the opportunity to introduce performance analysis and optimization early on in the book.

- Chapter 6 includes a new example on microcontroller-based microprogramming, using an 8051.
- Chapter 7, on RISC processors, was reorganized with additional emphasis on the use of the GNU Compiler Toolchain, inspection of object code, and analysis of assembly code.
- Chapter 8, on SoC, includes a new example using an AVR microcontroller. Support for the AVR instruction-set simulator was recently added to GEZEL.
- Part III, on Hardware/Software Interfaces, was reorganized. Chapter 9 explains the generic concepts in hardware/software interface design. In the first edition, these were scattered across several chapters. By bringing them together in a single chapter, I hope to give a more concise definition of the problem.
- Part III makes a thorough discussion of three components in a hardware/software interface. The three components are on-chip buses (Chap. 10), Microprocessor Interfaces (Chap. 11), and Hardware Interfaces (Chap. 12). "Hardware Interface" was called "Control Shell" in the first edition. The new term seems more logical considering the overall discussion of the Hardware/Software Interface.
- Chapter 10, On-chip Busses, now also includes a discussion on the Avalon on-chip bus by Alterea. The material on AMBA was upgraded to the latest AMBA specification (v4).
- Chapter 11, Microprocessor Interfaces, now includes a discussion of the NiosII custom-instruction interface, as well as an example of it.
- Part IV, Applications, was extended with a new chapter on the design of an AES coprocessor. The Applications now include three different chapters: Trivium, AES, and CORDIC.
- A new Appendix discusses the installation and use of GEZEL tools. The examples from Chapters 5, 6, 8, 11, 13–15 are now available in source code distribution, and they can be compiled and run using the GEZEL tools. The Appendix shows how.
- The extras section of Springer includes the solution for selected Problems.
- I did a thorough revision of grammar and correction of typos. I am grateful for the errata pointed out on the first edition by Gilberta Fernandes Marchioro, Ingrid Verbauwhede, Soyfan, and Li Xin.

## Making it Practical

This book emphasizes ideas and design methods, in combination with hands-on, practical experiments. The book therefore discusses detailed examples throughout the chapters, and a separate part (*Applications*) discusses the overall design process.

The hardware descriptions are made in GEZEL, an open-source cycle-accurate hardware modeling language. The GEZEL website, which distributes the tools, examples, and other documentation, is at http://rijndael.ece.vt.edu/gezel2

Refer to Appendix A for download and installation instructions.

There are several reasons why I chose not to use a mainstream HDL such as VHDL, Verilog, or SystemC.

- A first reason is *reduced modeling overhead*. Although models are crucial for embedded system construction, detailed modeling issues often distract the readers' attention from the key issues. For example, modeling the clock signal in hardware requires a lot of additional effort and it is not essential when doing single-clock synchronous design (which covers the majority of digital hardware design today).
- A second reason is that GEZEL comes with *support for cosimulation* built in. GEZEL models can be cosimulated with different processor simulation models, including ARM, 8051, and AVR, among others. GEZEL includes a library-block modeling mechanism that enables one to define new cosimulation interfaces with other simulation engines.
- A third reason is *conciseness*. This is a practical book with many design examples. Listings are unavoidable, but they need to be short. Chapter 5 further illustrates the point of conciseness with a single design example each in GEZEL, VHDL, Verilog, and SystemC side-by-side.
- A fourth reason is the path to implementation. GEZEL models can be translated (automatically) to VHDL. These models can be synthesized using standard HDL logic synthesis tools.

I use the material in this book in a class on hardware/software codesign. The class hosts senior-level undergraduate students, as well as first-year graduate-level students. For the seniors, this class ties many different elements of computer engineering together: computer architectures, software engineering, hardware design, debugging, and testing. For the graduate students, it is a refresher and a starting point of their graduate researcher careers in computer engineering.

In the class on codesign, the GEZEL experiments connect to an FPGA backend (based on Xilinx/EDK or Altera/Quartus) and an FPGA prototyping kit. These experiments are implemented as homework. Modeling assignments in GEZEL alternate with integration assignments on FPGA. Through the use of the GEZEL backend support, students can even avoid writing VHDL code. At the end of the course, there is a "contest". The students receive a reference implementation in C that runs on their FPGA prototyping kit. They need to accelerate this reference as much as possible using codesign techniques.

## Acknowledgments

I would like to express my sincere thanks to the many people that have contributed to this effort.

My family is the one constant in my life that makes the true difference. I am more than grateful for their patience, encouragement, and enthusiasm. I remain inspired by their values and their sincerity.

The ideas in this book were shaped by interacting with many outstanding engineers. Ingrid Verbauwhede, my Ph.D. advisor and professor at Katholieke Universiteit Leuven, has supported GEZEL, for research and education, from its very start. She was also among the first adopters of the book as a textbook. Jan Madsen, professor at Denmark Technical University, and Frank Vahid, professor at University of California, Irvine, have been exemplary educators to me for many years. Every discussion with them has been an inspiration to me.

After the first edition of this book, I exchanged ideas with many other people on teaching codesign. I would like to thank Jim Plusquellic (University of New Mexico), Edward Lee (University of California at Berkeley), Anand Raghunathan (Purdue University), and Axel Jantsch (Royal Institute of Technology Sweden). I thank Springer's Chuck Glaser for encouraging me to write the second edition of the book and for his many useful and insightful suggestions. I thank Grant Martin (Tensilica) for writing a review on this book.

Throughout the years of GEZEL development, there have been many users of the tool. These people have had the patience to carry on, and bring their codesign projects to a good end despite the many bugs they encountered. Here are just a few of them, in alphabetical order and based on my publication list at http://rijndael. ece.vt.edu/gezel2/publications.html: Aske Brekling, Herwin Chan, Doris Ching, Zhimin Chen, Junfeng Fan, Xu Guo, Srikrishna Iyer, Miroslav Knezevic, Boris Koepf, Bocheng Lai, Yusuke Matsuoka, Kazuo Sakiyama, Eric Simpson, Oreste Villa, Shenling Yang, Jingyao Zhang. I'm sure there are others, and I apologize if I missed anyone.

Finally, I wish to thank the students at Virginia Tech that took the codesign class (ECE 4530). Each year, I find myself learning so much from them. They continue to impress me with their results, their questions, their ideas. They're all engineers, but, I can tell, some of them are really artists.

I hope you enjoy this book and I truly wish this material helps you to go out and do some real design. I apologize for any mistakes left in the book – and of course I appreciate your feedback.

Blacksburg, VA, USA                                                      Patrick R. Schaumont

# Contents

# Part I
# Basic Concepts

The first part of this book introduces important concepts in hardware-software codesign. We compare and contrast two schools of thought in electronic design: the mindset used by the hardware designer, as opposed to the mindset used by the software designer. We will demonstrate that hardware/software codesign is not just gluing together hardware and software components; instead, it's about finding the correct balance between flexibility and performance during design.

The trade-off between parallel and sequential implementations is another fundamental issue for the hardware/software co-designer; we will discuss a concurrent system model (data-flow), that can be converted into either a hardware (parallel) or else into a software (sequential) implementation.

Finally, we will show how a program in C can be analyzed and decomposed into control-flow and data-flow. This analysis is crucial to understand how a C program can be migrated into hardware. As we will discuss, a common approach to hardware/software codesign is to carry functionality from software into hardware, thereby improving the overall performance of the application.

# Part I
# Basic Concepts

# Chapter 1
# The Nature of Hardware and Software

## 1.1 Introducing Hardware/Software Codesign

Hardware/software codesign is a broad term to capture many different things in electronic system design. We start by providing a simple definition of *software* and *hardware*. It's by no means a universal definition, but it will help to put readers from different backgrounds on the same line. This section will also provide a small example of hardware/software codesign, and concludes with a definition of hardware/software codesign.

### *1.1.1 Hardware*

In this book, we will model hardware by means of single-clock synchronous digital circuits, created using combinational logic and flip-flops.

Such circuits can be modeled with building blocks such as for example registers, adders, and multiplexers. Cycle-based hardware modeling is often called register-transfer-level (RTL) modeling, because the behavior of a circuit can be thought of as a sequence of transfers between registers, with logic and arithmetic operations performed on the signals during the transfers.

Figure 1.1a gives an example of a hardware module captured in RTL. A register can be incremented or cleared depending on the value of the control signal rst. The register is updated on the up-going edges of a clock signal clk. The wordlength of the register is 8 bit. Even though the connections in this figure are drawn as single lines, each line represents a bundle of eight wires. Figure 1.1a uses graphics to capture the circuit; in this book, we will be using a hardware description language called GEZEL. Figure 1.1b shows the equivalent description of this circuit in GEZEL language. Chapter 5 will describe GEZEL modeling in detail.

Figure 1.1c illustrates the behavior of the circuit using a timing diagram. In such a diagram, time runs from left to right, and the rows of the diagram represent different

P.R. Schaumont, *A Practical Introduction to Hardware/Software Codesign,* 3
DOI 10.1007/978-1-4614-3737-6_1, © Springer Science+Business Media New York 2013

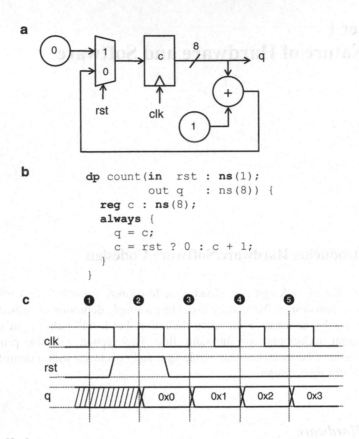

**Fig. 1.1** Hardware components

signals in the circuit. In this diagram, the register is cleared on clock edge 2, and it is incremented on clock edge 3, 4, and 5. Before clock edge 2, the value of the register is unknown, and the timing diagram indicates q's value as a shaded area. We will be using timing diagrams to describe the low-level behavior of hardware-software interfaces, and to describe events on on-chip buses.

The single-clock model is a very convenient abstraction for a designer who maps behavior (e.g. an algorithm) into discrete steps of one clock cycle. It enables this designer to envision how the hardware implementation of a particular algorithm should look like. The single-clock synchronous model cannot express every possible hardware circuit. For example, it cannot model events at a time resolution smaller than a clock cycle. As a result, some styles of hardware design cannot be captured with a single-clock synchronous model, including asynchronous hardware, dynamic logic, multi-phase clocked hardware, and hardware with latches. However, single-clock synchronous hardware is adequate to explain the key concepts of hardware-software co-design in this book.

**Listing 1.1** C example

```
1  int max;
2
3  int findmax(int a[10]) {
4    unsigned i;
5    max = a[0];
6    for (i=1; i<10; i++)
7      if (a[i] > max) max = a[i];
8  }
```

## 1.1.2 Software

Hardware/software codesign deals with hardware/software interfaces. The low-level construction details of software are important, because they directly affect the performance and the implementation cost of the hardware/software interface. This book will discuss important implementation aspects of software, such as the organization of variables into memory, and the techniques to control this from within a high-level programming language such as C.

We will model software as single-thread sequential programs, written in C or assembly. Programs will be illustrated using listings, for example Listings 1.1 and 1.2. Most of the discussions in this book will be processor-independent. In some cases, we will assume a 32-bit architecture (e.g. ARM) or an 8-bit architecture (e.g. 8051).

A single-thread sequential C program has a surprisingly good match to the actual execution of that program on a typical micro-processor. For example, the sequential execution of C programs matches the sequential instruction fetch-and-execute cycle of micro-processors. The variables of C are stored in a single, shared-memory space, corresponding to the memory attached to the micro-processor. There is a close correspondence between the storage concepts of a micro-processor (registers, stack) and the storage types supported in C (register int, local variables). Furthermore, common datatypes in C (char, int) directly map into units of micro-processor storage (byte, word). Consequently, a detailed understanding of C execution is closely related to a detailed understanding of the microprocessor activity at a lower abstraction level.

Of course, there are many forms of software that do not fit the model of a single-thread sequential C program. Multi-threaded software, for example, creates the illusion of *concurrency* and lets users execute multiple programs at once. Other forms of software, such as object-oriented software and functional programming, substitute the simple machine model of the micro-processor with a more sophisticated one. Such more advanced forms of software are crucial to master the complexity of large software applications. However, they make abstraction of (i.e. hide) the activities within a micro-processor. For this reason, we will concentrate on simple, single-thread C.

**Listing 1.2** ARM assembly example

```
        .text
findmax:
        ldr     r2, .L10
        ldr     r3, [r0, #0]
        str     r3, [r2, #0]
        mov     ip, #1
.L7:
        ldr     r1, [r0, ip, asl #2]
        ldr     r3, [r2, #0]
        add     ip, ip, #1
        cmp     r1, r3
        strgt   r1, [r2, #0]
        cmp     ip, #9
        movhi   pc, lr
        b       .L7
.L11:
        .align  2
.L10:
        .word   max
```

The material in this book does not follow any specific micro-processor, and is agnostic of any particular type of assembly language. The book emphasizes the relationship between C and assembly code, and assumes that the reader is familiar with the concept of assembly code. Some optimization problems in hardware/software codesign can only be handled at the level of assembly coding. In that case, the designer needs to be able to link the software, as captured in C, with the program executing on the processor, as represented by assembly code. Most C compilers offer the possibility to generate an assembly listing of the generated code, and we will make use of that feature. Listing 1.2 for example, was generated out of Listing 1.1.

Linking the statements of a C program to the assembly instructions, is easier than you would think, even if you don't know the microprocessor targeted by the assembly program. As an example, compare Listings 1.1 and 1.2. An ideal starting point when matching a C program to an assembly program, is to look for similar structures: loops in C will be reflected through their branch statements in assembly; if-then-else statements in C will be reflected, in assembly language, as conditional branches, and labels. Even if you're unfamiliar with the assembly format of a particular micro-processor, you can often derive such structures easily.

Figure 1.2 gives an example for the programs in Listings 1.1 and 1.2. The for-loop in C is marked with a label and a branch instruction. All the assembly instructions in between the branch and the label are part of the body of the loop. Once the loop structure is identified, it is easy to derive the rest of the code, as the following examples show.

- The if-statement in C requires the evaluation of a greater-then condition. In assembly, an equivalent cmp (compare) instruction can be found.

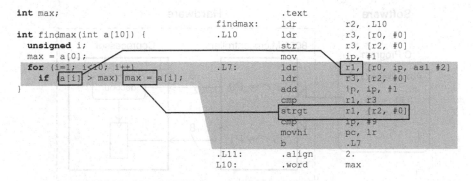

**Fig. 1.2** Mapping C to assembly

This shows that the operands r1 and r3 of the compare instruction must contain a[i] and max of the C program. Both of these variables are stored in memory; a[i] because it's an indexed variable, and max because it's a global variable. Indeed, looking at the preceding instruction in the C program, you can see that both r1 and r3 are defined through ldr (load-register) instructions, which require an address.

- The address for the load of r1 equals [r0, ip, asl #2], which stands for the expression r0 + (ip << 2). This may not be obvious if this is the first time you are looking at ARM assembly; but it's something you will remember quickly. In fact, the format of the expression is easy to explain. The register ip contains the loop counter, since ip is incremented once within the loop body, and the value of ip is compared with the loop boundary value of 9. The register r0 is the *base address* of a[], the location in memory where a[0] is stored. The shift-over-2 is needed because a[] is an array of integers. Microprocessors use byte-addressable memory, and integers are stored 4 byte-locations part.
- Finally, the conditional assignment of the max variable in C is not implemented using conditional branch instructions in assembly. Instead, a strgt (store-if-greater) instruction is used. This is a *predicated* instruction, an instruction that only executes when a given conditional flag is true.

The bottom line of this analysis is that, with a minimal amount of effort, you are able to understand a great deal on the behavior of a microprocessor simply by comparing C programs with equivalent assembly programs. In Chap. 7, you will use the same approach to analyze the quality of the assembly code generated by a compiler out of C code.

### 1.1.3  Hardware and Software

The objective of this book is to discuss the combination of hardware design and software design in all its forms. Hardware as well as software can be modeled using

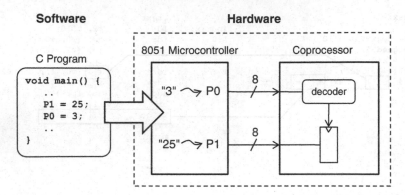

**Fig. 1.3** A codesign model

RTL programs and C programs respectively. A term *model* merely indicates they are not the actual implementation, but only a representation of it. An RTL program is a *model* of a network of logic gates; a C program is a *model* of a binary image of micro-processor instructions. It is not common to talk about C programs as *models*; in fact, software designers think of C programs as actual implementations. In this book, we will therefore refer to hardware *models* and C or assembly *programs*.

Models are an essential part of the design process. They are a formal representation of a designers' intent and they are used as input for simulation tools and implementation tools. In hardware/software codesign, we are working with models that are partly written as C programs, and partly as RTL programs. We will discuss this idea by means of a simple example.

Figure 1.3 shows an 8051 micro-controller and an attached coprocessor. The coprocessor is attached to the 8051 micro-controller through two 8-bit ports P0 and P1. A C program executes on the 8051 microcontroller, and this program contains instructions to write data to these two ports. When a given, pre-defined value appears on port P0, the coprocessor will make a copy of the value present on port P1 into an internal register.

This very simple design can be addressed using hardware/software codesign; it includes the design of a hardware model and the design of a C program. The hardware model contains the 8051 processor, the coprocessor, and the connections between them. During execution, the 8051 processor will execute a software program written in C. Listing 1.3 shows that C program. Listing 1.4 shows an RTL hardware model for this design, written in the GEZEL language.

The C driver sends three values to port P1, by calling a function sayhello. That function also cycles the value on port P0 between ins_hello and ins_idle, which are encoded as value 1 and 0 respectively.

The hardware model includes both the microcontroller and the coprocessor. The overall coprocessor behavior in Listing 1.4 is like this: when the ins input changes from 0 to 1, then the din input will be printed in the next clock cycle.

**Listing 1.3** 8051 driver program

```
1   #include <8051.h>
2
3   enum {ins_idle, ins_hello};
4
5   void sayhello(char d) {
6       P1 = d;
7       P0 = ins_hello;
8       P0 = ins_idle;
9   }
10
11  void terminate() {
12      // special command to stop simulator
13      P3 = 0x55;
14  }
15
16  void main() {
17      sayhello(3);
18      sayhello(2);
19      sayhello(1);
20      terminate();
21  }
```

The coprocessor is on lines 1–19. This particular hardware model is a combination of a finite state machine (lines 10–18) and a datapath (lines 1–9), a modeling method known as FSMD (for finite-state-machine with datapath). We will discuss FSMD in detail in Chap. 5. The FSMD is quite easy to understand. The datapath contains several instructions: decode and hello. The FSM controller selects, each clock cycle, which of those instructions to execute. For example, lines 15–16 shows the following control statement.

```
@s1 if (insreg == 1) then (hello, decode)  -> s2;
                     else (decode)          -> s1;
```

This means: when the value of insreg is 1, and the current state of the FSM controller is s1, then the datapath will execute instructions  hello and decode, and the FSM controller next-state will become s2. When the value of insreg would be 0, the datapath would execute only instruction decode and the next-state of the FSM controller would be  s1.

The 8051 microcontroller is captured in Listing 1.4 as well. However, the internals of the microcontroller are not shown; only the hardware interfaces relevant to the coprocessor are included. The 8051 microcontroller is captured with three ipblock (GEZEL library modules), on lines 21–38. The first ipblock is an i801system. It represents the 8051 microcontroller core, and it indicates the name of the compiled C program that will execute on this core (driver.ihx on line 22). The other two ipblock (lines 28–38) are two 8051 output ports, one to model port P0, and the second to model port P1.

**Listing 1.4** GEZEL model for 8051 platform

```
1  dp hello_decoder(in    ins : ns(8);
2                   in    din : ns(8)) {
3    reg insreg : ns(8);
4    reg dinreg : ns(8);
5    sfg decode    { insreg = ins;
6                    dinreg = din; }
7    sfg hello     { $display($cycle, " Hello! You gave me ",
       dinreg); }
8  }
9
10 fsm fhello_decoder(hello_decoder) {
11   initial s0;
12   state s1, s2;
13   @s0 (decode) -> s1;
14   @s1 if (insreg == 1) then (hello, decode) -> s2;
15                        else (decode)         -> s1;
16   @s2 if (insreg == 0) then (decode)         -> s1;
17                        else (decode)         -> s2;
18 }
19
20 ipblock my8051 {
21   iptype "i8051system";
22   ipparm "exec=driver.ihx";
23   ipparm "verbose=1";
24   ipparm "period=1";
25 }
26
27 ipblock my8051_ins(out data : ns(8)) {
28   iptype "i8051systemsource";
29   ipparm "core=my8051";
30   ipparm "port=P0";
31 }
32
33 ipblock my8051_datain(out data : ns(8)) {
34   iptype "i8051systemsource";
35   ipparm "core=my8051";
36   ipparm "port=P1";
37 }
38
39 dp sys {
40   sig ins, din : ns(8);
41   use my8051;
42   use my8051_ins(ins);
43   use my8051_datain(din);
44   use hello_decoder(ins, din);
45 }
46
47 system S {
48   sys;
49 }
```

Finally, the coprocessor and the 8051 ports are wired together in a top-level module, shown in lines 40–46. We can now simulate the entire model, including hardware and software, as follows. First, the 8051 C program is compiled to a binary. Next, the GEZEL simulator will combine the hardware model and the 8051 binary executable in a *cosimulation*. The output of the simulation model is shown below.

```
> sdcc driver.c
> /opt/gezel/bin/gplatform hello.fdl
i8051system: loading executable [driver.ihx]
9662 Hello! You gave me 3/3
9806 Hello! You gave me 2/2
9950 Hello! You gave me 1/1
Total Cycles: 10044
```

You can notice that the model produces output on cycles 9,662, 9,806, and 9,950, while the complete C program executes in 10,044 cycles. The evaluation and analysis of cycle-accurate behavior is a very important aspect of codesign, and we will address it throughout the book.

## *1.1.4 Defining Hardware/Software Codesign*

The previous example motivates the following traditional definition of hardware/-software codesign.

> Hardware/Software Codesign is the design of cooperating hardware components and software components in a single design effort.

For example, if you would design the architecture of a processor and at the same time develop a program that could run on that processor, then you would be using hardware/software codesign. However, this definition does not tell precisely what *software* and *hardware* mean. In the previous example, the software was a C program, and the hardware was an 8051 microcontroller with a coprocessor. In reality, there are many forms of hardware and software, and the distinction between them easily becomes blurred. Consider the following examples.

- A Field Programmable gate Array (FPGA) is a hardware circuit that can be reconfigured to a user-specified netlist of digital gates. The program for an FPGA is a 'bitstream', and it is used to configure the netlist topology. Writing 'software' for an FPGA really looks like hardware development – even though it is software.
- A soft-core is a processor implemented in the bitstream of an FPGA. However, the soft-core itself can execute a C program as well. Thus, software can execute on top of other 'software'.

- A Digital-Signal Processor (DSP) is a processor with a specialized instruction-set, optimized for signal-processing applications. Writing efficient programs for a DSP requires detailed knowledge of these specialized instructions. Very often, this means writing assembly code, or making use of a specialized software library. Hence, there is a strong connection between the efficiency of the software and the capabilities of the hardware.
- An Application-Specific Instruction-set Processor (ASIP) is a processor with a customizable instruction set. The hardware of such a processor can be extended, and these hardware extensions can be encapsulated as new instructions for the processor. Thus, an ASIP designer will develop a hardware implementation for these custom instructions, and subsequently write software that uses those instructions.
- The CELL processor, used in the Playstation-3, contains one control processor and eight slave-processors, interconnected through a high-speed on-chip network. The software for a CELL is a set of nine concurrent communicating programs, along with configuration instructions for the on-chip network. To maximize performance, programmers have to develop CELL software by describing simultaneously the computations and the communication activities in each processor.

These examples illustrate a few of the many forms of hardware and software that designers use today. A common characteristic of all these examples is that creating the 'software' requires intimate familiarity with the 'hardware'. In addition, hardware covers much more than RTL models: it also includes specialized processor datapaths, the FPGA fabric, multi-core architectures, and more.

Let us define the *application* as the overall function of a design, covering its implementation in hardware as well as in software. We can define hardware/software codesign as follows.

Hardware/Software Codesign is the partitioning and design of an application in terms of fixed and flexible components.

We used the term 'fixed component' instead of hardware component, and 'flexible component' instead of software component. A fixed component is often hardware, but it is not restricted to it. Similarly, a flexible component is often software, but it is not restricted to it. In the next section, we clarify the balance between fixed and flexible implementations.

## 1.2 The Quest for Energy Efficiency

Choosing between implementing a design in hardware or in software may seem trivial, if you only consider design effort. Indeed, from a designers' point-of-view, the easiest approach is to write software, for example in C. Software is easy and flexible, software compilers are fast, there are many libraries available with source code, and development systems (personal computers and laptops) are plentiful and cheap. Furthermore, why go through the effort of designing a new hardware architecture when there is already one available that will do the job for your implementation (namely, the RISC processor)?

In reality, choosing between a hardware implementation and a software implementation is much more subtle, and it is driven by both technological as well as by economical reasons. We start with two technological arguments (performance and energy efficiency), and next provide a more balanced view on the trade-off between hardware and software.

### 1.2.1 Performance

Another way to compare hardware and software is to compare them in terms of their performance. Performance could be expressed as the amount of work done per unit of time. Let's define a unit of work as the processing of 1 bit of data. The unit of time can be expressed in clock cycles or in seconds.

Figure 1.4 illustrates various cryptographic implementations in software and hardware that have been proposed over the past few years (2003–2008). All of them are designs that have been proposed for embedded applications, where the trade-off between hardware and software is crucial. The graph shows performance in bits per cycle, and demonstrates that hardware crypto-architectures have, on the average, a higher performance compared to embedded processors. Of course, the clock frequency should be taken into account. A hardware implementation may execute many operations per clock cycle, but a processor may run at a much higher clock frequency. The faster processor may outdo the advantage of the parallel hardware implementation in terms of performance.

### 1.2.2 Energy Efficiency

A second important factor in the selection between hardware and software is the energy needed for computations. This is especially important for portable, battery-operated applications. The *energy-efficiency* is the amount of useful work done per unit of energy. A better energy-efficiency clearly implies longer battery-life.

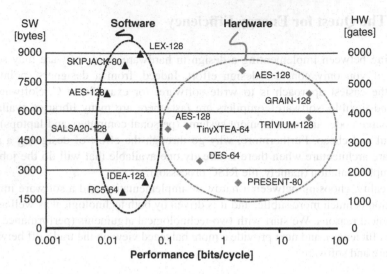

**Fig. 1.4** Cryptography on small embedded platforms

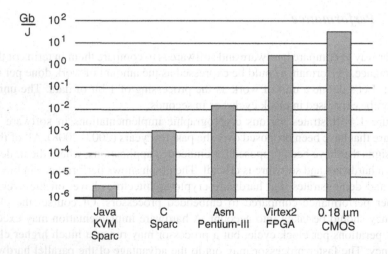

**Fig. 1.5** Energy efficiency

Figure 1.5 shows the example of a particular encryption application (AES) for different target platforms. The platforms include: Java on top of a Java Virtual machine on top of an embedded processor; C on top of an embedded processor; optimized assembly-code on top of a Pentium-III processor; Verilog code on top of a Virtex-II FPGA; and an ASIC implementation using 0.18 μm CMOS standard cells. The logarithmic Y-axis shows the amount of Gigabits that can be encrypted on each of these platforms with a single Joule of energy. Keep in mind that the application is the same for all these architectures, and consists of encrypting bits. As indicated by the figure, the energy-efficiency varies over many *orders of*

**Fig. 1.6** Driving factors in hardware/software codesign

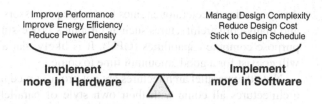

magnitude. If these architectures are being used in hand-held devices, where energy is a scarce resource, obviously there is a strong motivation to use a less flexible, more specialized architecture. For the same reason, you will never find a high-end workstation processor in a cell phone.

## 1.3   The Driving Factors in Hardware/Software Codesign

As pointed out in the previous section, energy-efficiency and relative performance are two important factors to prefer a (fixed, parallel) hardware implementation over a (flexible, sequential) software implementation. The complete picture, however, is more complicated. In the design of modern electronic systems, many trade-offs have to be made, often between conflicting objectives. Figure 1.6 shows that some factors argue for *more* software, while other factors argue for *more* hardware. The following are arguments in favor of increasing the amount of on-chip dedicated hardware.

- **Performance:** The classic argument in favor of dedicated hardware design has been increased performance: more work done per clock cycle. That is still one of the major factors. Increased performance is obtained by specializing the architecture used to implement the application. This can be done, for example, by introducing dedicated hardware components to accelerate part of the application.
- **Energy Efficiency:** Almost every electronic consumer product today carries a battery (iPod, PDA, mobile phone, Bluetooth device, ...). Batteries have limited energy storage. On the other hand, these consumer devices are used for similar services and applications as traditional high-performance personal computers. By moving (part of) the flexible software of a design into fixed hardware, the energy-efficiency of the overall application can increase (See Fig. 1.5).
- **Power Density:** The power dissipation of a circuit, and the associated thermal profile, is directly proportional to their clock frequency. In modern processors, the power density is at the limits of cost-effective cooling technology; further improvement of performance can therefore not be achieved by increasing the clock frequency even more. Instead, there is a broad and fundamental shift occurring towards parallel computer architectures. At this moment, there is no dominant parallel computer architecture that has demonstrated to be effective for all applications.

  Some of the current candidates for parallel computing include symmetric multiprocessors attached to the same memory (SMP); Field Programmable Gate

Arrays used as accelerator engines for classic processors (FPGA); and multi-core and many-core architectures such as Graphics Processing Engines with general-purpose compute capabilities (GPU). It is likely that all of these architectures will coexist for a good amount of time to come.

Note that parallel architectures are not programmed in C. The current parallel architectures all come with their own style of parallel programming, and the search for a good, universal, parallel programming model is still on.

The following arguments, on the other hand, argue for flexibility and thus for increasing the amount of on-chip software.

- **Design Complexity:** Modern electronic systems are so complex, that it's not a good idea to hard-code all design decisions in fixed hardware. Instead, a common approach has been to keep the implementation as flexible as possible, typically by using programmable processors that run software. Software, and the flexibility it provides, is then used as a mechanism to develop the application at a higher level of abstraction (in software), as a mechanism to cope with future needs, and as a mechanism to resolve bugs. Thus, the flexibility of software is used to cope with the complexity of the design.
- **Design Cost:** New chips are very expensive to design. As a result, hardware designers make chips programmable so that these chips can be reused over multiple products or product generations. The SoC is a good example of this trend. However, programmability can be found in many different forms other than embedded processors: reconfigurable systems are based on the same idea of reuse-through- reprogramming.
- **Shrinking Design Schedules:** Each new generation of technology tends to replace the older one more quickly. In addition, each of these new technologies is more complex compared to the previous generation. For a design engineer, this means that each new product generation brings more work, and that the work needs to be completed in a shorter amount of time.

  Shrinking design schedules require engineering teams to work on multiple tasks at the same time: hardware and software are developed concurrently. A software development team will start software development as soon as the characteristics of the hardware platform are established, even *before* an actual hardware prototype is available.

Finding the correct balance between all these factors is obviously a very complex problem. In this book, we will restrict the optimization space to performance versus resource cost.

Adding hardware to a software solution may increase the performance of the overall application, but it will also require more resources. In terms of the balance of Fig. 1.6, this means that we will balance *Design Cost* versus *Performance*.

## 1.4 The Hardware-Software Codesign Space

The trade-offs discussed in the previous section need to be made in the context of a *design space*. For a given application, there are many different possible solutions. The collection of all these implementations is called the hardware-software codesign space. Figure 1.7 gives a symbolic representation of this design space, and it indicates the main design activities in this design space.

### 1.4.1 The Platform Design Space

The objective of the design process is to implement a specification onto a target platform. In hardware-software codesign, we are interested in programmable components. Figure 1.7 illustrates several examples: A RISC microprocessor, a Field Programmable Gate Array (FPGA), a Digital Signal Processor (DSP), an Application-Specific Instruction-set Processor (ASIP) and finally an Application-Specific Integrated Circuit (ASIC).

Mapping an application onto a platform means writing software for that platform, and, if needed, customizing the hardware of the platform. Software as well as hardware have a very different meaning depending on the platform.

- In the case of a RISC processor, software is written in C, while the hardware is a general-purpose processor.

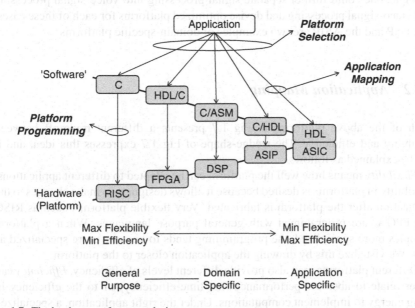

**Fig. 1.7** The hardware-software codesign space

- In the case of a Field Programmable Gate Array, software is written in a Hardware Description Language (HDL). When the FPGA contains a soft-core processor, as discussed above, we will also write additional platform software in C.
- A Digital Signal Processor uses a combination of C and assembly code for software. The hardware is a specialized processor architecture, adapted to signal processing operations.
- An Application-Specific Instruction-set Processor is a processor that can be specialized to a particular application domain, for example by adding new instructions and by extending the processor datapath. The 'software' of an ASIP thus can contain C code as well as a hardware description of the processor extensions.
- Finally, in the case of an ASIC, the application is written in HDL, which is then converted into a hardcoded netlist. In contrast to other platforms, ASICs are typically non-programmable. In an ASIC, the application and the platform have merged to a single entity.

The platforms in Fig. 1.7 are organized, from left to right, according to their flexibility. General-purpose platforms, such as RISC and FPGA, are able to support a broad range of applications. Application-specific platforms, such as the ASIC, are optimized to execute a single application. In between the general purpose platform and the application-specific platform, is a third class of architectures called the *domain-specific* platform. Domain-specific platforms are optimized to execute applications from a particular domain. Signal-processing, cryptography, networking, are all examples of domains. A domain may have sub-domains. For example, one could further separate signal processing into voice-signal processing and video-signal processing and devise optimized platforms for each of these cases. The DSP and the ASIP are two examples of domain-specific platforms.

## 1.4.2   Application Mapping

Each of the above platforms in Fig. 1.7 presents a different trade-off between flexibility and efficiency. The wedge-shape of Fig. 1.7 expresses this idea, and it can be explained as follows.

*Flexibility* means how well the platform can be adapted to different applications. Flexibility in platforms is desired because it allows designers to make changes to the application after the platform is fabricated. Very flexible platforms, such as RISC and FPGA, are programmed with general purpose languages. When a platform becomes more specialized, the programming tends to become more specialized as well. We visualize this by drawing the application closer to the platform.

Different platforms may also provide different levels of efficiency. *Efficiency* can either relate to absolute performance (i.e. time-efficiency) or to the efficiency in using energy to implement computations. Under the right application, a specialized

**Listing 1.5**  dot product in C64x DSP processor

```
        LDDW    .D2T2    *B_n++,B_reg1:B_reg0
| |     LDDW    .D1T1    *A_m++,A_reg1:A_reg0

        DOTP2 .M2X       A_reg0,B_reg0,B_prod
| |     DOTP2 .M1X       A_reg1,B_reg1,A_prod

        SPKERNEL 4, 0
| |     ADD     .L2      B_sum,B_prod,B_sum
| |     ADD     .L1      A_sum,A_prod,A_sum
```

platform will be more efficient than a general platform, because its hardware
components are optimized for that application. We can visualize this by moving
the platform closer to the application in the case of specialized platforms.

The effect of the flexibility-efficiency trade-off on the source code of software
can be illustrated with the following example. Consider the execution of the dot-
product on a DSP processor such as TI's C64x. In C, the dot-product is a vector
operation that can be expressed in single compact loop:

```
sum=0;
for (i=0; i<N; i++)
    sum += m[i]*n[i];
```

Listing 1.5 shows the body of the loop, optimized as assembly code for the
TI C64x DSP processor. The TI C64x is a highly parallel processor that has two
multiply-accumulate units. It can compute *two* loop iterations of the C loop at
the same time. Several instructions are preceded by | |. Those instructions will be
executing in parallel with the previous instructions. Even though Listing 1.5 spans
nine lines, it consists of only three instructions. Thus, Listing 1.5 has more efficiency
than the original C program, but the TI assembly software is specific to the TI
processor. A gain in efficiency was obtained at the cost of flexibility (or portability).

An interesting, but very difficult question is *how* one can select a platform for a
given specification, and *how* one can map an application onto a selected platform.
Of these two questions, the first one is the hardest. Designers typically answer it
based on their previous experience with similar applications. The second question
is also very challenging, but it is possible to answer it in a more systematic fasion,
using a design methodology. A *design method* is a systematic sequence of steps to
convert a specification into an implementation. Design methods cover many aspects
of application mapping, such as optimization of memory usage, design performance,
resource usage, precision and resolution of data types, and so on. A design method
is a canned sequence of design steps. You can learn it in the context of one design,
and next apply this design knowledge in the context of another design.

## 1.5   The Dualism of Hardware Design and Software Design

In the previous sections, we discussed the driving forces in hardware/software codesign, as well as its design space. Clearly, there are compelling reasons for hardware-software codesign, and there is a significant design space to explore. A key challenge in hardware-software codesign is that a designer needs to combine two radically different design paradigms. In fact, hardware and software are the dual of one another in many respects. In this section, we examine these fundamental differences. Table 1.1 provides a synopsis.

- **Design Paradigm:** In a hardware model, circuit elements operate in parallel. Thus, by using more circuit elements, more work can be done within a single clock cycle. Software, on the other hand, operates sequentially. By using more operations, a software program will take more time to complete. Designing requires the decomposition of a specification in lower level primitives, such as gates (in hardware) and instructions (in software). A hardware designer solves problems by decomposition in space, while a software designer solves problems by decomposition in time.
- **Resource Cost:** Resource cost is subject to a similar dualism between hardware and software. Decomposition in space, as used by a hardware designer, means that a more complex design requires more gates. Decomposition in time, as used by a software designer, implies that a more complex design requires more instructions to complete. Therefore, the resource cost for hardware is circuit area, while the resource cost for software is execution time.
- **Flexibility:** Software excels over hardware in the support of flexibility. Flexibility is the ease by which the application can be modified or adapted. In software, flexibility is essentially for free. In hardware on the other hand, flexibility is not trivial. Hardware flexibility requires that circuit elements can be easily reused for different activities or subfunctions in a design. A hardware designer has to think carefully about such reuse: flexibility needs to be designed into the circuit.
- **Parallelism:** A dual of flexibility can be found in the ease with which parallel implementations can be created. For hardware, parallelism comes for free as part of the design paradigm. For software, on the other hand, parallelism is a major challenge. If only a single processor is available, software can only implement concurrency, which requires the use of special programming constructs such as

**Table 1.1**  The dualism of hardware and software design

|                 | Hardware                  | Software                    |
|-----------------|---------------------------|-----------------------------|
| Design Paradigm | Decomposition in space    | Decomposition in time       |
| Resource cost   | Area (# of gates)         | Time (# of instructions)    |
| Flexibility     | Must be designed-in       | Implicit                    |
| Parallelism     | Implicit                  | Must be designed-in         |
| Modeling        | Model $\neq$ implementation | Model $\sim$ implementation |
| Reuse           | Uncommon                  | Common                      |

threads. When multiple processors are available, a truly parallel software implementation can be made, but inter-processor communication and synchronization becomes a challenge.

- **Modeling:** In software, modeling and implementation are very close. Indeed, when a designer writes a C program, the compilation of that program for the appropriate target processor will also result in the implementation of the program. In hardware, the model and the implementation of a design are distinct concepts. Initially, a hardware design is modeled using a Hardware Description Language. Such a hardware description can be simulated, but it is not an implementation of the actual circuit. Hardware designers use a hardware *description* language, and their programs are models which are later transformed to implementation. Software designers use a software *programming* language, and their programs are an implementation by itself.
- **Reuse:** Finally, hardware and software are also quite different when it comes to Intellectual Property Reuse or IP-reuse. The idea of IP-reuse is that a component of a larger circuit or a program can be packaged, and later reused in the context of a different design. In software, IP-reuse has known dramatic changes in recent years due to open source software and the proliferation of open platforms. When designing a complex program these days, designers will start from a set of standard libraries that are well documented and that are available on a wide range of platforms. For hardware design, IP-reuse still in its infancy. Compared to software, IP-reuse in hardware has a long way to go.

This summary comparison indicates that in many aspects, hardware design and software design are based on dual concepts. Hence, being able to effectively transition from one world of design to the other is an important asset for a hardware/software codesigner. In this book, we will rely on this dualism, and attempt to combine the best concept of hardware design with the best concepts in software design. Our objective is not only to excel as a hardware designer or a software designer; our objective is to excel as a system designer.

## 1.6  Modeling Abstraction Level

The abstraction level of a model is the amount of detail that is available in a model. A lower abstraction level has more details, but constructing a design at lower abstraction level requires more effort.

This section defines the abstraction levels used in this book. We will differentiate the abstraction levels based on time-granularity. A smaller time-granularity typically implies that activities are expressed in a larger amount of (usually small) time steps. There are five abstraction levels commonly used by computer engineers for the design of electronic hardware-software systems. Starting at the lowest abstraction level, we enumerate the five levels. Figure 1.8 illustrates the hierarchy among these abstraction levels.

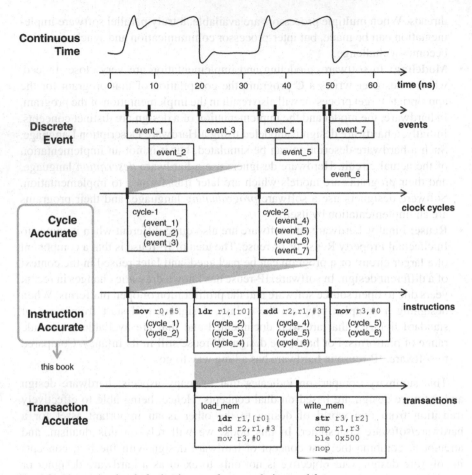

**Fig. 1.8** Abstraction levels for hardware-software codesign models

1. **Continuous time:** The lowest abstraction level describes operations as continuous actions. For example, electric networks can be described as systems of interacting differential equations. The voltages and currents in such electric networks can be found by solving these differential equations. The continuous-time model is a very detailed level, useful to analyze analog effects. However, this level of abstraction is not used to describe typical hardware-software systems.

2. **Discrete-event:** At the next abstraction level, activities are lumped together at discrete points in time called events. Those events can be irregularly spaced. For example, when the inputs of a digital combinatorial circuit change, the effect of those changes will ripple from input to output, changing the values at intermediate circuit nodes. Each change on a node can be thought of as an event: a (value, timestamp) tuple. Discrete-event simulation is commonly used to model digital hardware at low abstraction level. Discrete-event models avoid the complexity of continuous-time simulation, yet they capture relevant information

such as glitches and clock cycle edges. Discrete-event simulation is also used to model systems at high abstraction level, to simulate abstract events with irregular spacing in time. For example, discrete-event simulation can be used to model customer queues at a bank. In the context of hardware-software system design however, we will use discrete-event modeling to refer to digital hardware models at low abstraction level.

3. **Cycle-accurate:** Single-clock synchronous hardware circuits have the important property that all interesting things happen at regularly-spaced intervals, defined by the circuit clock period. This abstraction is important enough to merit its own abstraction level, and it is called cycle-accurate modeling. A cycle-accurate model does not capture propagation delays or glitches. All activities that fall in between two clock edges are concentrated at a single point in time. In a cycle-accurate model, activities happen either immediately (for combinatorial circuits for example), or else after an integral number of clock cycles (for sequential circuits). The cycle-accurate level is very important for hardware-software system modeling, and very often serves as the 'golden reference' for a hardware-software implementation. Cycle-accurate modeling will be used extensively throughout this book. Hardware models at cycle-accurate model are frequently called register-transfer level models, or RTL models.

4. **Instruction-accurate:**  For the simulation of complex systems, the cycle-accurate level may still be too slow. For example, your laptop's processor executes several billion clock cycles per second. Clearly, the simulation of even a single second of real time operation will take a significant amount of machine resources. In case when a microprocessor needs to be simulated, it is convenient to express the activities within the model in terms of one microprocessor instruction. Each instruction lumps several cycles of processing together. Instruction-accurate simulators are used extensively to verify complex software systems, such as complete operating systems. Instruction-accurate simulators keep track of an instruction count, but not of a cycle count. Thus, unless you map instructions back to clock cycles, this abstraction level may not reveal the real-time performance of a model.

5. **Transaction-accurate:**  For very complex systems, even instruction-accurate models may be too slow or require too much modeling effort. For these models, yet another abstraction level is introduced: the transaction-accurate level. In this type of model, the model is expressed in terms of the interactions between the components of a system. These interactions are called transactions. For example, one could model a system with a disk drive and a user application, and create a simulation that focuses on the commands exchanged between the disk drive and the user application. A transaction-accurate model allows considerable simplification of the internals of the disk drive and of the user application. Indeed, in between two transactions, millions of instructions can be lumped together and simulated as a single, atomic function call. Transaction-accurate models are important in the exploratory phases of a design. Transaction-accurate models enable a designer to define the overall characteristics of a design without going through the effort of developing a detailed model.

In summary, there are five abstraction levels that are commonly used for hardware-software modeling: transaction-accurate, instruction-accurate, cycle-accurate, event-driven and continuous-time. In this book, the emphasis is on cycle-accurate and instruction-accurate levels.

## 1.7  Concurrency and Parallelism

Concurrency and parallelism are terms that often occur in the context of hardware-software codesign. They have a different meaning. Concurrency is the ability to execute simultaneous operations because these operations are completely independent. Parallelism is the ability to execute simultaneous operations because the operations can run on different processors or circuit elements. Thus, concurrency relates to an application, while parallelism relates to the implementation of that application.

Hardware is always parallel. Software on the other hand can be sequential, concurrent or parallel. Sequential and concurrent software requires a single processor, parallel software requires multiple processors. The software running on your laptop (email, WWW, word processing, and so on) is concurrent. The software running on the 65536-processor IBM Blue Gene/L is parallel.

Making efficient use of parallelism (in the architecture) requires that you have an application which contains sufficient concurrency. There is a well-known law in supercomputing, called Amdahl's law, which states that the maximal speedup for any application that contains q % sequential code is $1/(q/100)$. For example, if your application is 33 % of its runtime executing like a sequential process, the maximal speedup is 3. It is easy to see why. Given sufficient hardware, the concurrent part of the application can complete arbitrarily fast, by implementing it on a parallel architecture. However, the sequential part cannot benefit from parallelism. If one third of the overall code is sequential, the speedup through a parallel implementaion cannot exceed three.

Surprisingly, even algorithms that seem sequential at first can be executed (and specified) in a parallel fashion. The following examples are discussed by Hillis and Steele. They describe the 'Connection Machine' (CM), a massively parallel processor. The CM contains a network of processors, each with their own local memory, and each processor in the network is connected to each other processor. The original CM machine contained 65536 processors, each of them with 4 Kbits of local memory. Interestingly, while the CM dates from the 1980s, multiprocessor architectures recently regained a lot of interest with the modern design community. Figure 1.9 illustrates an eight-node CM.

The question relevant to our discussion is: how hard is it to write programs for such a CM? Of course, you can write individual C programs for each node in the network, but that is not easy, nor is it very scalable. Remember that the original CM had 64 K nodes! Yet, as Hillis and Steele have shown, it is possible express algorithms in a concurrent fashion, such that they can map to a CM.

**Fig. 1.9** Eight node
connection machine

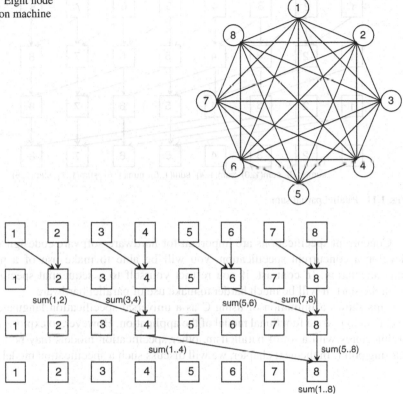

**Fig. 1.10** Parallel sum

Consider taking the sum of an array of numbers, illustrated in Fig. 1.10. To take the sum, we distribute the array over the CM processors so that each processor holds one number. We can now take the sum over the entire array in $log2(n)$ steps (n being the number of processors) as follows. We perform $2^i$ parallel additions per time step, for $i$ going from $log2(n-1)$ downto 0. For example, the sum of eight numbers can be computed in three time steps on a CM machine. In Fig. 1.10, time steps are taken vertically and each processor is drawn left to right. The communication activities between the processors are represented by means of arrows.

Compare the same algorithm running on a sequential processor. In that case, the array of numbers would be stored in a single memory and the processor needs to iterate over each element, requiring a minimum of eight time steps. You can also see that the parallel sum still wastes a lot of potential computing power. We have in total $3*8 = 24$ computation time-steps available, and we are only using seven of them. One extension of this algorithm is to evaluate all partial sums (i.e. the sum of the first two, three, four, etc numbers). A parallel algorithm that performs this in three time-steps, using 17 computation time-steps, is shown in Fig. 1.11.

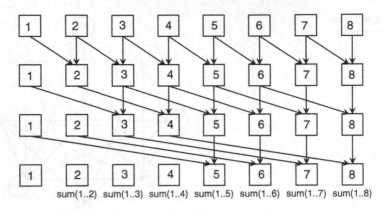

sum(1..2)  sum(1..3)  sum(1..4)  sum(1..5)  sum(1..6)  sum(1..7)  sum(1..8)

**Fig. 1.11** Parallel partial sum

Concurrent specifications are important for hardware-software codesign. If you develop a concurrent specification, you will be able to make use of a parallel implementation. In contrast, if you restrict yourself to a sequential specification from the start, it will be much harder to make use of parallel hardware.

This shows a limitation of using C as a universal specification language. C is excellent to make a functional model of the application. However, to explore system architectures with a lot of parallelism, other specification models may be a better starting point. In the next chapter, we will discuss such a specification model.

## 1.8   Summary

We have been able to benefit from an ever growing selection of programmable components, and design using these components has become a challenge by itself. Hardware/software codesign is the collection of techniques that deals with design using programmable components. We have defined hardware/software codesign as the partitioning and design of an application in terms of fixed and flexible parts. The flexible parts run as programs on those programmable components. Traditional microprocessors are only one of the many options, and we briefly described other components including FPGA's, DSP's and ASIP's. Platform selection is the job of determining which programmable component (or combination of components) is the best choice for a given application. Application mapping is the effort of transforming a specification into a program. Platform programming is the effort of converting an application program into low-level instructions for each programmable component. We also discussed the modeling abstraction levels for hardware and software, and we highlighted cycle-based synchronous RTL for hardware, and single-thread C for software as the golden level for this book. Finally, we also made careful distinction between a parallel implementation and a concurrent program.

## 1.9 Further Reading

Many authors have pointed out the advantages of dedicated hardware solutions when it comes to Energy Efficiency. A comprehensive coverage of the problem can be found in (Rabaey 2009).

Figure 1.4 is based on results published between 2003 and 2008 by various authors including (Good and Benaissa 2007; Bogdanov et al. 2007; Satoh and Morioka 2003; Leander et al. 2007; Kaps 2008; Meiser et al. 2007; Ganesan et al. 2003; Karlof et al. 2004).

Further discussion on the driving forces that require chips to become programmable is found in (Keutzer et al. 2000). A nice and accessible discussion of what that means for the hardware designer is given in (Vahid 2003).

Hardware-software codesign, as a research area, is at least two decades old. Some the early works are collected in (Micheli et al. 2001), and a retrospective of the main research problems is found in (Wolf 2003). Conferences such as the *International Conference on Hardware/Software Codesign and System Synthesis* (CODES+ISSS) cover the latest evolutions in the field.

Hardware/software codesign, as an educational topic, is still evolving. A key challenge seems to be to find a good common ground to jointly discuss hardware and software. Interesting ideas are found, for example in (Madsen et al. 2002) and (Vahid 2007b).

Despite its age, the paper on data-parallel algorithms by Hillis and Steel is still a great read (Hillis and Steele 1986).

## 1.10 Problems

**Problem 1.1.** Use the World Wide Web to find a data sheet for the following components. What class of components are they (RISC/FPGA/DSP)? How does one write software for each of them? What tools are used to write that software?

- TMS320DM6446
- EP4CE115
- SAM7S512
- ADSP-BF592
- XC5VFX100T

**Problem 1.2.** Develop a sorting algorithm for an 8-node Connection Machine, which can handle up to 16 numbers. Show that an N-node Connection Machine can complete the sorting task in a time proportional to N.

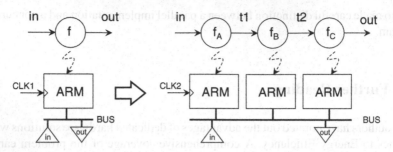

**Fig. 1.12** Multiprocessor system for problem 1.3

**Problem 1.3.** A single-input, single-output program running on an ARM processor needs to be rewritten such that it will run on three parallel ARM processors. As shown in Fig. 1.12, each ARM has its own, independent data- and instruction memory. For this particular program, it turns out that it can be easily rewritten as a sequence of three functions fA, fB and fC which are also single-input, single-output. Each of these three functions can be executed on a separate ARM processor, so that we get an arrangement as shown below. The sub-functions fA, fB, and fC contain 40, 20, and 40 % respectively of the instructions of the original program. You can ignore the time needed for communication of variables (out, in, t1, and t2 are integers).

(a) Assume that all ARMs have the same clock frequency (CLK1 = CLK2). Find the maximal speedup that the parallel system offers over the single-ARM system. For example, a speedup of 2 would mean that the parallel system could process two times as much input data as the single-ARM system in the same amount of time.

(b) For the parallel system of three ARM described above, we can reduce the power consumption by reducing their clock frequency CLK and their operating voltage V. Assume that both these quantities scale linearly (i.e. Reducing the Voltage V by half implies that the clock frequency must be reduced by half as well). We will scale down the voltage/clock of the parallel system such that the scaled-down parallel system has the same performance as the original, single-ARM sequential system. Find the ratio of the power consumption of the original sequential system to the power consumption of the scaled-down, parallel system (i.e. find the power-savings factor of the parallel system). You only need to consider dynamic power consumption. Recall that Dynamic Power Consumption is proportional to the square of the voltage and proportional to the clock frequency.

**Problem 1.4.** Describe a possible implementation for each of the following C statements in hardware. You can assume that all variables are integers, and that each of them is stored in a register.

(a) `a = a + 1;`

(b) if (a > 20) a = 20;
(c) while (a < 20) a = a + 1;

**Problem 1.5.** The function in Listing 1.6 implements a CORDIC algorithm. It evaluates the cosine of a number with integer arithmentic and using only additions, subtractions, comparisons. The angles[] variable is an array of constants. Answer each of the following questions. Motivate your answer.

- Do you think it is possible to compute this function in hardware within 1,000 clock cycles?
- Do you think it is possible to compute this function in hardware within 1,000 ms?
- Do you think it is possible to compute this function in hardware within one clock cycle?
- Do you think it is possible to compute this function in hardware within 1 ms?

**Listing 1.6** Listing for Problem 1.5.

```
1   int cordic_cos(int target) {
2       int X, Y, T, current;
3       unsigned step;
4       X        = AG_CONST;
5       Y        = 0;
6       current = 0;
7       for(step=0; step < 20; step++) {
8           if (target > current) {
9               T            = X - (Y >> step);
10              Y            = (X >> step) + Y;
11              X            = T;
12              current  += angles[step];
13          } else {
14              T            = X + (Y >> step);
15              Y            = -(X >> step) + Y;
16              X            = T;
17              current  -= angles[step];
18          }
19      }
20      return X;
21  }
```

**Problem 1.6.** Listing 1.7 shows a simplified version of the CORDIC algorithm in C. After compiling this code to an Intel architecture (x86), the assembly code from Listing 1.8 is generated. In this listing, arguments starting with the % sign are registers. Study the C and the assembly code, and answer the following questions.

- What register is used to store the variable current?
- What assembly instruction corresponds to the comparison of the variables target and current in the C code?
- What register is used to store the loop counter step?

**Listing 1.7** C Listing for Problem 1.6.

```
1   extern int angles[20];
2
```

```
3    int cordic(int target) {
4        int current;
5        unsigned step;
6        current = 0;
7        for(step=0; step < 20; step++) {
8            if (target > current) {
9                current   += angles[step];
10           } else {
11               current   -= angles[step];
12           }
13       }
14       return current;
15   }
```

**Listing 1.8**  Assembly Listing for Problem 1.6.

```
1    cordic:
2                pushl   %ebp
3                xorl    %edx, %edx
4                movl    %esp, %ebp
5                xorl    %eax, %eax
6                movl    8(%ebp), %ecx
7                jmp     .L4
8    .L9:
9                addl    angles(,%edx,4), %eax
10               addl    $1, %edx
11               cmpl    $20, %edx
12               je      .L8
13   .L4:
14               cmpl    %eax, %ecx
15               jg      .L9
16               subl    angles(,%edx,4), %eax
17               addl    $1, %edx
18               cmpl    $20, %edx
19               jne     .L4
20   .L8:
21               popl    %ebp
22               ret
```

# Chapter 2
# Data Flow Modeling and Transformation

## 2.1 Introducing Data Flow Graphs

By nature, hardware is parallel and software is sequential. As a result, software models (C programs) are not very well suited to capture hardware implementations, and vice versa, hardware models (RTL programs) are not a good abstraction to describe software. However, designers frequently encounter situations for which a given design may use either hardware or software as a target. Trying to do both (writing a full C program *as well as* a full hardware design) is not an option; it requires the designer to work twice as hard. An alternative is to use a high-level model, which enables the designer to express a design without committing to a hardware or a software implementation. Using a high-level model, the designer can gain further insight into the specification, and decide on the right path for implementation.

In the design of signal processing systems, the need for modeling is well known. Signal processing engineers describe complex systems, such as digital radios and radar processing units, using *block diagrams*. A block diagram is a high-level representation of the target system as a collection of smaller functions. A block diagram does not specify if a component should be implemented as hardware or software; it only expresses the operations performed on data signals. We are specifically interested in *digital* signal processing systems. Such systems represent signals as streams of discrete samples rather than continuous waveforms.

Figure 2.1a shows the block diagram for a simple digital signal processing system. It's a pulse-amplitude modulation (PAM) system, and it is used to transmit digital information over bandwidth-limited channels. A PAM signal is created from binary data in two steps. First, each word in the file needs to be mapped to PAM symbols, an alphabet of pulses of different heights. An entire file of words will thus be converted to a stream of PAM symbols or pulses. Next, the stream of pulses needs to be converted to a smooth shape using pulse-shaping. Pulse-shaping ensures that the bandwidth of the resulting signal does not exceed the PAM symbol rate. For example, if a window of 1,000 Hz transmission bandwidth is available, then we

P.R. Schaumont, *A Practical Introduction to Hardware/Software Codesign*,     31
DOI 10.1007/978-1-4614-3737-6_2, © Springer Science+Business Media New York 2013

a

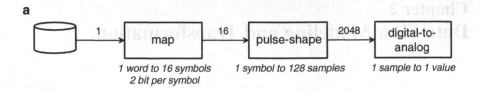

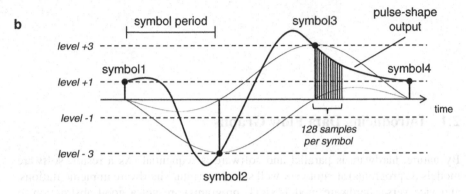

**Fig. 2.1** (**a**) Pulse-amplitude modulation system. (**b**) Operation of the pulse-shaping unit

can transmit 1,000 PAM symbols per second. In a digital signal processing system, a *smooth* curve is achieved by *oversampling*: calculating many closely-spaced discrete samples. The output of the pulse-shaping unit produces many samples for each input symbol pulse, but it is still a stream of discrete samples. The final module in the block diagram is the digital-to-analog converter, which will convert the stream of discrete samples into a continuous signal.

Figure 2.1a shows a PAM-4 system, which uses four different symbols. Since there are four different symbols, each PAM symbol holds 2 bits of source information. A 32-bit word from a data source is encoded with 16 PAM-4 symbols. The first block in the PAM transmission system makes the conversion of a single word to a sequence of 16 PAM-4 symbols. Figure 2.1b shows that each PAM-4 symbol is mapped to a pulse with four possible signal levels: $\{-3, -1, 1, 3\}$. Once the PAM-4 signals are available, they are shaped to a smooth curve using a pulse-shape filter. The input of this filter is a stream of symbol pulses, while the output is a stream of samples at a much higher rate. In this case, we generate 128 samples for each symbol.

Figure 2.1b illustrates the operation of the pulse-shape filter. The smooth curve at the output of the pulse-shape filter connects the top of each pulse. This is achieved by an interpolation technique, which extends the influence of a single symbol pulse over many symbol periods. The figure illustrates two such interpolation curves, one for `symbol2` and one for `symbol3`. The pulse-shape filter will produce 128 samples for each symbol entered into the pulse-shape filter.

**Listing 2.1** C example

```
1   extern int read_from_file();
2   extern int map_to_symbol(int, int);
3   extern int pulse_shape(int, int);
4   extern void send_to_da(int);
5
6   int main() {
7     int word, symbol, sample;
8     int i, j;
9     while (1) {
10      word = read_from_file();
11      for (i=0; i<16; i++) {
12        symbol = map_to_symbol(word, i);
13        for (j=0; j<128; j++)
14          sample = pulse_shape(symbol, j);
15        send_to_da(sample);
16      }
17    }
18  }
```

Now let's consider the construction of a simulation model for this system. We focus on capturing its functionality, and start with a C program as shown in Listing 2.1. We will ignore the implementation details of the function calls for the time being, and only focus on the overall structure of the program.

The program in Listing 2.1 is fine as a system simulation. However, as a model for the implementation, this C program is too strict, since it enforces *sequential* execution of all functions. If we observe Fig. 2.1a carefully, we can see that the block diagram does not require a sequential execution of the symbol mapping function and the pulse shaping function. The block diagram *only* specifies the flow of data in the system, but not the execution order of the functions. The distinction is subtle but important. For example, in Fig. 2.1a, it is possible that the map module and the pulse-shape module work in parallel, each on a different symbol. In Listing 2.1 on the other hand, the map_to_symbol() function and the pulse_shape() function will *always* execute sequentially. In hardware-software codesign, the implementation target could be either parallel or else sequential. The program in Listing 2.1 favors a sequential implementation, but it does not encourage a parallel implementation in the same manner as a block diagram.

This illustrates how the selection of a modeling technique can constrain the solutions that may be achieved starting from that model. In general, building a sequential implementation for a parallel model (such as a block diagram) is much easier than the opposite – building a parallel implementation from a sequential model. Therefore, we favor modeling styles that enable a designer to express parallel activities at the highest level of abstraction.

In this chapter we will discuss a modeling technique, called Data Flow, which achieves the objective of a parallel model. Data Flow models closely resemble

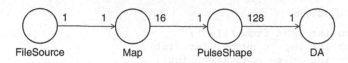

**Fig. 2.2** Data flow model for the pulse-amplitude modulation System

block diagrams. The PAM-4 system, as a Data Flow model, is shown in Fig. 2.2. In this case, the different functions of the system are mapped as individual entities or *actors* such as `FileSource`, `Map`, `PulseShape` and `DA`. These actors are linked through communication channels or *queues*. The inputs and outputs of each actor are marked with the relative rate of communications. For example, there are 16 samples produced by `Map` for each input sample. Each actor is an *independent unit*, continuously checking its input for the availability of data. As soon as data appears, each actor will calculate the corresponding output, passing the result to the next actor in the chain. In the remainder of this chapter, we will discuss the precise construction details of data flow diagrams. For now, we only point out the major differences of this modeling style compared to modeling in C.

- A strong point of Data Flow models, and the main reason why signal processing engineers love to use them, is that a Data Flow model is a *concurrent* model. Indeed, the actors in Fig. 2.2 operate and execute as individual concurrent entities. A concurrent model can be mapped to a parallel or a sequential implementation, and so they can target hardware as well as software.
- Data Flow models are distributed, and there is no need for a central controller or 'conductor' in the system to keep the individual system components in pace. In Fig. 2.2, there is no central controller that tells the actors when to operate; each actor can determine for itself when it's time to work.
- Data Flow models are modular. We can develop a design library of data flow components and then use that library in a plug-and-play fashion to construct data flow systems.
- Data Flow models can be analyzed. Certain properties, such as their ability to avoid deadlock, can be determined directly from the design. Deadlock is a condition sometimes experienced by real-time computing systems, in which the system becomes unresponsive. The ability to analyze the behavior of models at high abstraction level is an important advantage. C programs, for example, do not offer such convenience. In fact, a C designer typically determines the correctness of a C program by running it, rather than analyzing it!

Data Flow has been around for a surprisingly long time, yet it has been largely overshadowed by the stored-program (Von Neumann) computing model. Data Flow concepts have been explored since the early 1960s. By 1974, Jack Dennis had developed a language for modeling data flow, and described data flow using graphs, similar to our discussion in this chapter. In the 1970s and 1980s, an active research community was building not only data flow-inspired programming languages and

**Fig. 2.3** Data flow model of
an addition

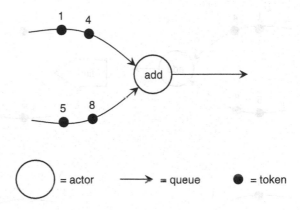

tools, but also computer architectures that implement data flow computing models. The work from Arvind was seminal in this area, resulting in several different computer architectures and tools (see Further Reading at the end of this chapter).

Today, data flow remains very popular to describe signal processing systems. For example, commercial tools such as Simulink® are based on the ideas of data flow. A interesting example of an academic environment is the Ptolemy project at UC Berkeley (http://ptolemy.eecs.berkeley.edu/ptolemyII/index.htm). The Ptolemy design environment can be used for many different types of system specification, including data flow. The examples on the website can be run inside of a web browser as Java applets.

In the following sections, we will consider the elements that make up a data flow model. We will next discuss a particular class of data flow models called Synchronous Data Flow Graphs (SDF). We will show how SDF graphs can be formally analyzed. Later, we will discuss transformations on SDF graphs, and show how transformations can lead to better, faster implementations.

## 2.1.1 Tokens, Actors, and Queues

Figure 2.3 shows the data flow model of a simple addition. This model contains the following elements.

- *Actors* contain the actual operations. Actors have a bounded behavior (meaning that they have a precise beginning and ending), and they iterate that behavior from start to completion. One such iteration is called an actor firing. In the example above, each actor firing would perform a single addition.
- *Tokens* carry information from one actor to the other. A token has a value, such as '1', '4', '5' and '8' in Fig. 2.3.

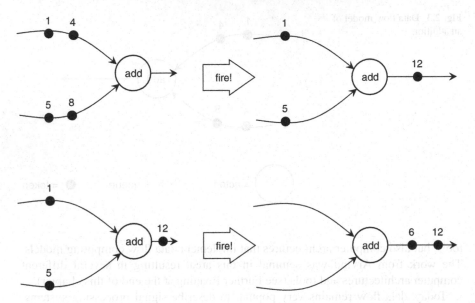

**Fig. 2.4** The result of two firings of the add actor, each resulting in a different marking

- *Queues* are unidirectional communication links that transport tokens from one
  actor to the other. Data Flow queues have an infinite amount of storage, so that
  tokens will never get lost in a queue. Data Flow queues are first-in first-out. In
  Fig. 2.4, there are two tokens in the upper queue, one with value '1' and one with
  value '4'. The '4' token was entered first into the queue, the '1' token was entered
  after that. When the 'add' actor will read a token from that queue, the actor will
  first read the token with value '4' and next the token with value '1'.

When a data flow model executes, the actors will read tokens from their input
queues, read the value of these tokens and calculate the corresponding output value,
and generate new tokens on their output queues. Each single execution of an actor
is called the firing of that actor. Data flow execution then is expressed as a sequence
of (possibly concurrent) actor firings.

Conceptually, data flow models are untimed. The firing of an actor happens
instantaneously, although any real implementation of an actor does require a finite
amount of time. *Untimed* does not mean zero time; it only means that time is
irrelevant for data flow models. Indeed, in data flow, the execution is guided by the
presence of data, not by a program counter or by a clock signal. An actor will never
fire if there's no input data, but instead it will wait until sufficient data is available
at its inputs.

A graph with tokens distributed over queues is called a *marking* of a data
flow model. When a data flow model executes, the entire graph goes through
a series of markings that drive data from the inputs of the data flow model to

**Fig. 2.5**  Data flow actor with
production/consumption rates

**Fig. 2.6**  Can this model do
any useful computations?

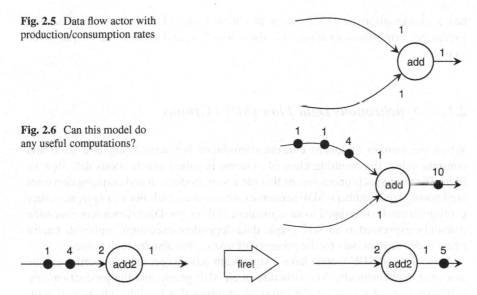

**Fig. 2.7**  Example of a multi-rate data flow model

the outputs. Each marking corresponds to a different state of the system, and the execution of a data flow model is defined by a sequence of markings. To an external observer, the marking (i.e., the distribution of the tokens on the queues) is the only observable state in the system. This is a crucial observation! It implies that an actor cannot use internal state variables that would affect the execution of the system, and thus the marking sequence. If we want to model system state that would affect the execution, we need to express it using tokens.

## 2.1.2   Firing Rates, Firing Rules, and Schedules

When should an actor fire? The *firing rule* describes the necessary and sufficient conditions for an actor to fire. Simple actors such as the add actor can fire when there is a single token on each of its queues. A firing rule thus involves testing the number of tokens on each of its input queues. The required number of tokens can be annotated to the actor input. Similarly, the amount of tokens that an actor produces per firing can be annotated to the actor output. These numbers are called the token consumption rate (at the actor inputs) and token production rate (at the actor outputs). The production/consumption rates of the add actor could be written such as shown in Figs. 2.5 or 2.6.

Data Flow actors may consume or produce more than one token per actor firing. Such models are called multirate data flow models. For example, the actor in Fig. 2.7

has a consumption rate of 2 and a production rate of 1. It will consume two tokens per firing from its input, add them together, and produce an output token as result.

## 2.1.3  Synchronous Data Flow (SDF) Graphs

When the number of tokens consumed/produced per actor firing is a fixed and constant value, the resulting class of systems is called synchronous data flow or SDF. The term synchronous means that the token production and consumption rates are known, fixed numbers. SDF semantics are not universal. For example, not every C program can be translated to an equivalent SDF graph. Data-dependent execution cannot be expressed as an SDF graph: data-dependent execution implies that actor firing is defined not only by the presence of tokens, but also by their value.

Nevertheless, SDF graphs have a significant advantage: their properties can be analyzed mathematically. The structure of an SDF graph, and the production/consumption rates of tokens on the actors, determines if a feasible schedule of actor firings is possible. We will demonstrate a technique that can analyze an SDF graph, and derive such a feasible schedule, if it exists.

## 2.1.4  SDF Graphs are Determinate

Assuming that each actor implements a deterministic function, then the entire SDF execution is determinate. This means that results, computed using an SDF graph, will always be the same. This property holds regardless of the firing order (or schedule) of the actors. Figure 2.8 illustrates this property. This graph contains actors with unit production/consumption rates. One actor adds tokens, the second actor increments the value of tokens. As we start firing actors, tokens are transported through the graph. After the first firing, an interesting situation occurs: both the add actor as well as the plus1 actor can fire. A first case, shown on the left of Fig. 2.8, assumes that the plus1 actor fires first. A second case, shown on the right of Fig. 2.8, assumes that the add actor fires first. However, regardless what path is taken, the graph marking eventually converges to the result shown at the bottom.

Why is this property so important? Assume that the add actor and the plus1 actor execute on two different processors, a slow one and a fast one. Depending upon which actor runs on the fast processor, the SDF execution will follow the left marking or else the right marking of the figure. Since SDF graphs are determinate, it doesn't matter which processor executes what actor: the results will be always the same. In other words, an SDF system will work as specified, regardless of the technology used to implement it. Of course, actors must be completely and correctly implemented, firing rule and all. Determinate behavior is vital in many embedded applications, especially in applications that involve risk.

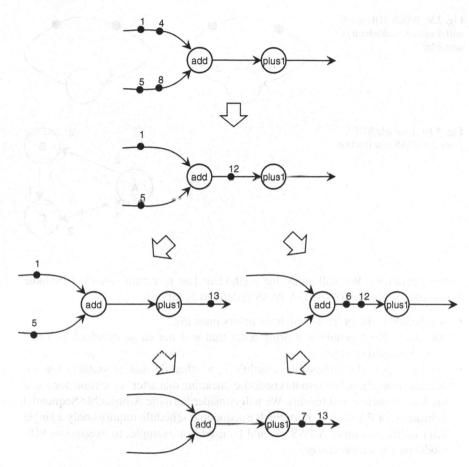

**Fig. 2.8** SDF graphs are determinate

## 2.2 Analyzing Synchronous Data Flow Graphs

An admissible schedule for an SDF graph is one that can run forever without causing deadlock and without overflowing any of the communication queues. The term *unbounded execution* is used to indicate that a model runs forever; the term *bounded buffer* is used to indicate that no communication queue needs infinite depth. A deadlock situation occurs when the SDF graph ends up in a marking in which it is no longer possible to fire any actor.

Figure 2.9 shows two SDF graphs where these two problems are apparent. Which graph will deadlock, and which graph will result in an infinite amount of tokens?

Given an arbitrary SDF graph, it is possible to test if it is free from deadlock, and if it only needs bounded storage under unbounded execution. The nice thing about this test is that we don't need to use any simulation; the test can be done using basic

**Fig. 2.9** Which SDF graph
will deadlock, and which is
unstable?

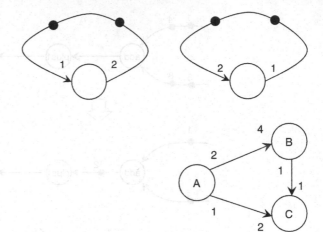

**Fig. 2.10** Example SDF
graph for PASS construction

matrix operations. We will study the method of Lee to create so-called Periodic
Admissible Schedules (PASS). A PASS is defined as follows.

- A schedule is the order in which the actors must fire.
- An admissible schedule is a firing order that will not cause deadlock and that
  yields bounded storage.
- Finally, a periodic admissible schedule is a schedule that is suitable for unbounded execution, because it is periodic (meaning that after some time, the same
  marking sequence will repeat). We will consider Periodic Admissible Sequential
  Schedules, or PASSs for short. Such a sequential schedule requires only a single
  actor to fire at a time. A PASS would be used, for example, to execute an SDF
  model on a microprocessor.

### 2.2.1 Deriving Periodic Admissible Sequential Schedules

We can create a PASS for an SDF graph (and test if one exists) with the following
four steps.

1. Create the topology matrix G of the SDF graph;
2. Verify the rank of the matrix to be one less than the number of nodes in the graph;
3. Determine a firing vector;
4. Try firing each actor in a round robin fashion, until it reaches the firing count as
   specified in the firing vector.

   We will demonstrate each of these steps using the example of the three-node SDF
graph shown in Fig. 2.10.

**Step 1.** Create a topology matrix for this graph. This topology matrix has as many rows as graph edges (FIFO queues) and as many columns as graph nodes. The entry $(i, j)$ of this matrix will be positive if the node j produces tokens into graph edge i. The entry $(i, j)$ will be negative if the node $j$ consumes tokens from graph edge $i$. For the above graph, we thus can create the following topology matrix. Note that $G$ does not have to be square – it depends on the amount of queues and actors in the system.

$$G = \begin{bmatrix} 2 & -4 & 0 \\ 1 & 0 & -2 \\ 0 & 1 & -1 \end{bmatrix} \begin{matrix} \leftarrow edge(A,B) \\ \leftarrow edge(A,C) \\ \leftarrow edge(B,C) \end{matrix} \qquad (2.1)$$

**Step 2.** The condition for a PASS to exist is that the rank of $G$ has to be one less than the number of nodes in the graph. The proof of this theorem is beyond the scope of this book, but can be consulted in (Lee and Messerschmitt 1987). The rank of a matrix is the number of independent equations in $G$. It can be verified that there are only two independent equations in $G$. For example, multiply the first column with $-2$ and the second column with $-1$, and add those two together to find the third column. Since there are three nodes in the graph and the rank of $G$ is 2, a PASS is possible.

Step 2 verifies that tokens cannot accumulate on any of the edges of the graph. We can find the resulting number of tokens by choosing a firing vector and making a matrix multiplication. For example, assume that $A$ fires two times, and $B$ and $C$ each fire zero times. This yields the following firing vector:

$$q = \begin{bmatrix} 2 \\ 0 \\ 0 \end{bmatrix} \qquad (2.2)$$

The residual tokens left on the edges after these firings are two tokens on $edge(A,B)$ and a token on $edge(A,C)$:

$$b = Gq = \begin{bmatrix} 2 & -4 & 0 \\ 1 & 0 & -2 \\ 0 & 1 & -1 \end{bmatrix} \begin{bmatrix} 2 \\ 0 \\ 0 \end{bmatrix} = \begin{bmatrix} 2 \\ 1 \\ 0 \end{bmatrix} \qquad (2.3)$$

**Step 3.** Determine a periodic firing vector. The firing vector indicated above is not a good choice to obtain a PASS: each time this firing vector executes, it adds three tokens to the system. Instead, we are interested in firing vectors that leave no additional tokens on the queues. In other words, the result must equal the zero-vector.

$$Gq_{PASS} = 0 \qquad (2.4)$$

**Fig. 2.11** A deadlocked
graph

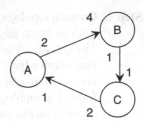

Since the rank of $G$ is less than the number of nodes, this system has an infinite number of solutions. Intuitively, this is what we should expect. Assume a firing vector $(a,b,c)$ would be a solution that can yield a PASS, then also $(2a,2b,2c)$ will be a solution, and so is $(3a,3b,3c)$, and so on. You just need to find the simplest one. One possible solution that yields a PASS is to fire A twice, and B and C each once:

$$q_{PASS} = \begin{bmatrix} 2 \\ 1 \\ 1 \end{bmatrix} \qquad (2.5)$$

The existence of a PASS firing vector does not guarantee that a PASS will also exist. For example, just by changing the direction of the $(A,C)$ edge, you would still find the same $q_{PASS}$, but the resulting graph is deadlocked since all nodes are waiting for each other. Therefore, there is still a fourth step: construction of a valid PASS.

**Step 4.**  Construct a PASS. We now try to fire each node up to the number of times specified in $q_{PASS}$. Each node which has the adequate number of tokens on its input queues will fire when tried. If we find that we can fire no more nodes, and the firing count of each node is less than the number specified in $q_{PASS}$, the resulting graph is deadlocked.

We apply this on the original graph and using the firing vector $(A = 2, B = 1, C = 1)$. First we try to fire $A$, which leaves two tokens on $(A,B)$ and one on $(A,C)$. Next, we try to fire $B$ – which has insufficient tokens to fire. We also try to fire $C$ but again have insufficient tokens. This completes our first round through – $A$ has fired already one time. In the second round, we can fire $A$ again (since it has fired less than two times), followed by $B$ and $C$. At the end of the second round, all nodes have reached the firing count specified in the PASS firing vector, and the algorithm completes. The PASS we are looking for is $(A,A,B,C)$.

The same algorithm, when applied to the deadlocked graph in Fig. 2.11, will immediately abort after the first iteration, because no node was able to fire.

Note that the determinate property of SDF graphs implies that we can try to fire actors in any order of our choosing. So, instead of trying the order $(A,B,C)$ we can also try $(B,C,A)$. In some SDF graphs (but not in the one discussed above), this may lead to additional PASS solutions.

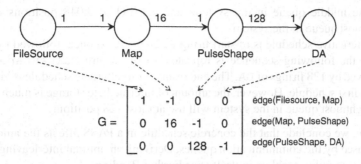

**Fig. 2.12** Topology matrix for the PAM-4 system

## 2.2.2 Example: Deriving a PASS for the PAM-4 System

At the start of this chapter, we discussed the design of a digital pulse-amplitude modulation system for the generation of PAM4 signals. The system, shown in Fig. 2.12, is modeled with four data flow actors: a word source, a symbol mapper, a pulse shaper, and a digital-to-analog converter. The system is a multi-rate data flow system. For every byte modulated, $16.128 = 2048$ output samples are generated.

Our objective is to derive a PASS for this data flow system. The first step is to derive the topology matrix $G$. The matrix has three rows and four columns, corresponding to the three queues and four actors in the system. The second step in deriving the PASS is to verify that the rank of this topology matrix equals the number of data flow actors minus one. It's easy to demonstrate that $G$ indeed consists of three independent equations: no row can be created as a linear combination of two others. Hence, we confirm that the condition for a PASS to exist is fulfilled. Third, we have to derive a feasible firing vector for this system. This firing vector, $q_{PASS}$, needs to yield a zero-vector when multiplied with the topology matrix. The solution for $q_{PASS}$ is to fire the `Filesource` and `Map` actors one time, the `PulseShape` actor 16 times, and the `DA` actor 2,048 times.

$$G.q_{PASS} = \begin{bmatrix} +1 & -1 & 0 & 0 \\ 0 & 16 & -1 & 0 \\ 0 & 0 & 128 & -1 \end{bmatrix} \begin{bmatrix} 1 \\ 1 \\ 16 \\ 2{,}048 \end{bmatrix} = 0 \qquad (2.6)$$

The final step is to derive a concrete schedule with the derived firing rates. We discuss two alternative solutions.

- By inspection of the graph in Fig. 2.12, we conclude that firing the actors from left to right according to their $q_{PASS}$ firing rate will result in a feasible solution that ensures sufficient tokens in each queue. Thus, we start by firing `FileSource` once, followed by `Map`, followed by 16 firings of `PulseShape`, and finally 2,048 firings of `DA`. This particular schedule will require a FIFO of 16 positions for the middle queue in the system, and a FIFO of 2,048 positions for the rightmost queue in the system.

- An alternative schedule is to start firing FileSource once, followed by Map. Next, the following sequence is repeated 16 times: fire PulseShape once, followed by 128 firings of DA. The end result of this alternate schedule is identical to the first schedule. However, the amount of intermediate storage is much lower: the rightmost queue in the system will use at most 128 positions.

Hence, we conclude that the concrete schedule in a PASS affects the amount of storage used by the communication queues. Deriving an optimal interleaving of the firings is a complex problem in itself (See Further Reading).

This completes our discussion of PASS. SDF has very powerful properties, which enable a designer to predict critical system behavior such as determinism, deadlock, and storage requirements. Yet, SDF is not a universal specification mechanism; it is not a good replacement for any type of application. The next part will further elaborate on the difficulty of implementing control-oriented systems using data flow modeling.

## 2.3   Control Flow Modeling and the Limitations of Data Flow Models

SDF systems are distributed, data-driven systems. They execute whenever there is data to process, and remain idle when there is nothing to do. However, SDF seems to have trouble to model control-related aspects. Control appears in many different forms in system design, for example:

- **Stopping and restarting**. An SDF model never terminates; it just keeps running. Stopping and re-starting is a control-flow property that cannot be addressed well with SDF graphs.
- **Mode-switching**. When a cell-phone switches from one standard to the other, the processing (which may be modeled as an SDF graph) needs to be reconfigured. However, the topology of an SDF graph is fixed and cannot be modified at runtime.
- **Exceptions**. When catastrophic events happen, processing may suddenly need to be altered. SDF cannot model exceptions that affect the entire graph topology. For example, once a token enters a queue, the only way of removing it is to read the token out of the queue. It is not possible to suddenly flush the queue on a global, exceptional condition.
- **Run-time Conditions**. A simple if-then-else statement (choice between two activities depending on an external condition) is troublesome for SDF. An SDF node cannot simply 'disappear' or become inactive – it is always there. Moreover, we cannot generate conditional tokens, as this would violate SDF rules which require fixed production/consumption rates. Thus, SDF cannot model conditional execution such as required for if-then-else statements.

**Fig. 2.13** Emulating
if-then-else conditions in SDF

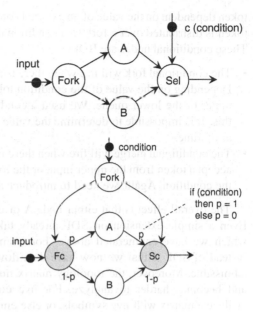

**Fig. 2.14** Implementing
if-then-else using Boolean
Data Flow

There are two solutions to the problem of control flow modeling in SDF. The first one is to emulate control flow using SDF, at the cost of some modeling overhead. The second one is to extend the semantics of SDF. We give a short example of each strategy.

## 2.3.1 Emulating Control Flow with SDF Semantics

Figure 2.13 shows an example of an if-then-else statement, SDF-style. Each of the actors in the above graph are SDF actors. The last one is a selector-actor, which will transmit either the A or B input to the output depending on the value of the input condition. Note that when Sel fires, it will consume a token from each input, so both A and B have to run for each input token. This is thus not really an if-then-else in the same sense as in C programming. The approach taken by this graph is to implement both the if-leg and the else-leg and afterwards transmit only the required result. This approach may work when there is sufficient parallelism available. For example, in hardware design, the equivalent of the Sel node would be a multiplexer.

## 2.3.2 Extending SDF Semantics

Researchers have also proposed extensions on SDF models. One of these extensions was proposed by Joseph Buck, and is called BDF (Boolean Data Flow) (Lee and Seshia 2011). The idea of BDF is to make the production and consumption-rate of a token dependent on the value of an external control token. In Fig. 2.14, the condition token is distributed over a fork to a conditional fork and a conditional merge node. These conditional nodes are BDF.

- The conditional fork will fire when there is an input token and a condition token. Depending on the value of the condition token, it will produce an output on the upper or the lower queue. We used a conditional production rate p to indicate this. It is impossible to determine the value of p upfront – this can only be done at runtime.
- The conditional merge will fire when there is a condition token. If there is, it will accept a token from the upper input or the lower input, depending on the value of the condition. Again, we need to introduce a conditional consumption rate.

The overall effect is that either node A or else node B will fire, but never both. Even a simple extension on SDF already takes jeopardizes the basic properties which we have enumerated above. For example, a consequence of using BDF instead of SDF is that we now have data flow graphs that are only conditionally admissible. Moreover, the topology matrix now will include symbolic values (p), and becomes harder to analyze. For five conditions, we would have to either analyze a matrix with five symbols, or else enumerate all possible condition values and analyze 32 different matrices (each of which can have a different series of markings). In other words, while BDF can help solving some of practical cases of control, it quickly becomes impractical for analysis.

Besides BDF, researchers have also proposed other flavors of control-oriented data flow models, such as Dynamic Data Flow (DDF) which allows variable production and consumption rates, and Cyclo-Static Data Flow (CSDF) which allows a fixed, iterative variation on production and consumption rates. All of these extensions break down the elegance of SDF graphs to some extent. SDF remains a very popular technique for Digital Signal Processing applications. But the use of BDF, DDF and the like has been limited.

## 2.4  Adding Time and Resources

So far, we have treated data flow graphs as untimed: the analysis of data flow graphs was based only on their marking (distribution of tokens), and not on the time needed to complete a computation. However, we can also use the data flow model to do performance analysis. By introducing a minimal resource model (actor execution time and bounded FIFO queues), we can analyze the system performance of a data flow graph. Furthermore, we can analyze the effect of performance-enhancing transformations on the data flow graph.

**Fig. 2.15** Enhancing the
SDF model with resources:
execution time for actors, and
delays for FIFO queues

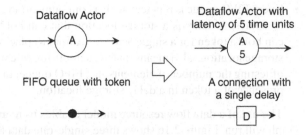

Dataflow Actor

Dataflow Actor with
latency of 5 time units

FIFO queue with token

A connection with
a single delay

## 2.4.1 Real-Time Constraints and Input/Output Sample Rate

A data flow graph is a model for a repeating activity. For example, the PAM-4
modulation system described in this chapter's introduction transforms an infinite
stream of input samples (words) into an infinite stream of output samples. The model
shows how one single sample is processed, and the streaming character is implicit.

The input sample rate is the time period between two adjacent input-samples
from the stream. The sample rate typically depends on the application. CD audio
samples, for example, are generated at 44,100 samples per second. The input sample
rate thus sets a design constraint for the performance of the data flow system: it
specifies how quickly the data flow graph must be computed in order to achieve
real-time performance. A similar argument can be made for the output sample rate.
In either case, when there's a sample-rate involved with the input or the output of a
data flow graph, there is also a real-time constraint on the computation speed for the
data flow graph.

We define the input *throughput* as the amount of input samples per second.
Similarly, we define the output throughput as the amount of output samples per
second. The *latency* is the time required to process a single token from input to
output. Throughput and latency are two important system constraints.

## 2.4.2 Data Flow Resource Model

In this section, we are interested in performance analysis of data flow graphs. This
requires the introduction of time and resources. Figure 2.15 summarizes the two
enhancements needed.

- Every actor is decorated with an execution latency. This is the time needed by
  the actor to complete a computation. We assume that the actor's internal program
  requires all inputs to be available at the start of the execution. Similarly, we
  assume that the actor's internal program produces all outputs simultaneously
  after the actor latency. Latency is expressed in time units, and depending on the
  implementation target, a suitable unit can be chosen – clock cycles, nanoseconds,
  and so on.

- Every FIFO queue is replaced with a communication channel with a fixed number of delays. A delay is a storage location that can hold one token. A single delay can hold a token for a single actor execution. Replacing a FIFO queue with delay storage locations also means that the actor firing rule needs to be changed. Instead of testing the number of elements in a FIFO queue, the actor will now test for the presence of a token in a delay storage location.

The use of a data flow resource model enables us to analyze how fast a data flow graph will run. Figure 2.16 shows three single-rate data flow graphs, made with two actors $A$ and $B$. Actor $A$ needs five units of latency, while actor $B$ requires three units of latency. This data flow graph also has an input and an output connection, through which the system can accept a stream of input samples, and deliver a stream of output samples. For our analysis, we do not define an input or output sample rate. Instead, we are interested to find out how fast these data flow graphs can run.

The easiest way to analyze this graph is to evaluate the latency of samples as they are processed through the data flow graph. Eventually, this analysis yields the time instants when the graph reads from the system input, or writes to the system output.

In the graphs of Fig. 2.16a, b, there is a single delay element in the loop. Data input/output is defined by the combined execution time of actor $A$ and actor $B$. The time stamps for data production for the upper graph and the middle graph are different because of the position of the delay element: for the middle graph, actor $B$ can start at system initialization time, since a token is available from the delay element.

In the graph of Fig. 2.16c, there are two delay elements in the loop. This enables actor $A$ and actor $B$ to operate in parallel. The performance of the overall system is defined by the slowest actor $A$. Even though actor $B$ completes in three time units, it needs to wait for the next available input until actor $A$ has updated the delay element at its output.

Hence, we conclude that the upper two graphs have a throughput of 1 sample per 8 time units, and that the lower graph has a throughput of 1 sample per 5 time units.

### 2.4.3  Limits on Throughput

The example in the previous section illustrates that the number of delays, and their distribution over the graph, affects the throughput of the data flow system. A second factor that affects the latency are the feedback links or loops in the data flow graph. Together, loops and delays determine an upper bound on the computation speed of a data flow graph.

We define two quantities to help us analyze the throughput limits of a data flow system: the *loop bound* and the *iteration bound*. The loop bound is the round-trip delay in a given loop of a data flow graph, divided by the number of delays in that loop. The iteration bound is the highest loop bound for any loop in a given data flow graph. The iteration bound sets an upper limit for the computational throughput of a data flow graph.

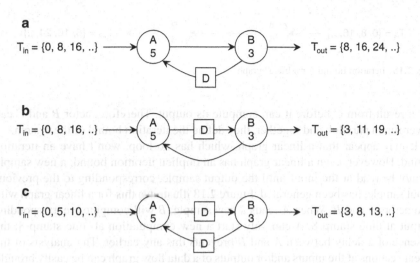

**Fig. 2.16** Three data flow graphs

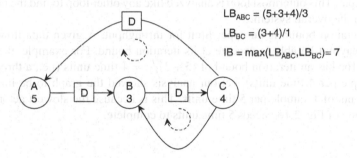

**Fig. 2.17** Calculating loop bound and iteration bound

These concepts are explained with an example as shown in Fig. 2.17. This graph has three actors $A$, $B$, and $C$, each with different execution times. The graph has two loops: $BC$ and $ABC$. The loop bounds are:

$$LB_{BC} = \frac{3+4}{1} = 7$$

$$LB_{ABC} = \frac{5+3+4}{3} = 4$$

The iteration bound of the system in Fig. 2.17 is the maximum of these two, or seven. This iteration bound implies that the implementation of this data flow graph will need at least 7 time units to process every iteration. The iteration bound thus sets an upper limit on throughput. If we inspect this graph closely, we conclude that loop $BC$ is indeed the bottleneck of the system. Actors $A$ and $C$ have delays at their inputs, so that they can always execute in parallel. Actor $B$ however, needs to wait

$T_{in} = \{0, 8, 16, ..\}$ → (A 5) → (B 3) → $T_{out} = \{8, 16, 24, ..\}$

**Fig. 2.18** Iteration bound for a linear graph

for a result from $C$ before it can compute its output. Therefore, actor $B$ and $C$ can never run in parallel, and together, they define the iteration bound of the system.

It may appear that a linear graph, which has no loop, won't have an iteration bound. However, even a linear graph has an implicit iteration bound: a new sample cannot be read at the input until the output sample, corresponding to the previous input sample, has been generated. Figure 2.18 illustrates this for a linear graph with two actors $A$ and $B$. When $A$ reads a new sample, $B$ will compute a corresponding output at time stamp 8. $A$ can only start a new computation at time stamp 8; the absence of a delay between $A$ and $B$ prevents this any earlier. The analysis of the linear sections at the inputs and/or outputs of a data flow graph can be easily brought into account by assuming an implicit feedback from each output of a data flow graph to each input. This outermost loop is analyzed like any other loop to find the iteration bound for the overall system.

The iteration bound is an upper-limit for throughput. A given data flow graph may or may not be able to execute at its iteration bound. For example, the graph in Fig. 2.16c has an iteration bound of $(5+3)/2 = 4$ time units (i.e., a throughput of 1 sample per 4 time units), yet our analysis showed the graph to be limited at a throughput of 1 sample per 5 time units. This is because the slowest actor in the critical loop of Fig. 2.16c needs 5 time units to complete.

## 2.5   Transformations

Using performance analysis on data flow graphs, we can now evaluate suitable transformations to improve the performance of slow data flow graphs. We are interested in transformations that maintain the functionality of a data flow graph, but that increase the throughput and/or decrease the latency. This section will present several transformations which are frequently used to enhance system performance. Transformations don't affect the steady-state behavior of a data flow graph but, as we will illustrate, they may introduce transient effects, typically at startup. We cover the following transformations.

- **Multi-rate Expansion** is used to convert a multi-rate synchronous data flow graph to a single-rate synchronous data flow graph. This transformation is helpful because the other transformations assume single-rate SDF systems.
- **Retiming** considers the redistribution of delay elements in a data flow graph, in order to optimize the throughput of the graph. Retiming does not change the latency or the transient behavior of a data flow graph.

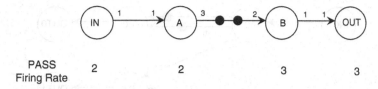

PASS Firing Rate: 2      2      3      3

**Fig. 2.19** Multi-rate data flow-graph

- **Pipeling** introduces additional delay elements in a data flow graph, with the intent of optimizing the iteration bound of the graph. Pipelining changes the throughput, and the transient behavior of a data flow graph.
- **Unfolding** increases the computational parallelism in a data flow graph by duplicating actors. Unfolding does not change the transient behavior of a data flow graph, but may modify the throughput.

### 2.5.1 Multirate Expansion

It is possible to transform a multi-rate SDF graph systematically to a single-rate SDF graph. The following steps to convert a multi-rate graph to a single-rate graph.

1. Determine the PASS firing rates of each actor
2. Duplicate each actor the number of times indicated by its firing rate. For example, given an actor $A$ with a firing rate of 2, we create $A0$ and $A1$. These actors are two identical copies of the same generic actor $A$.
3. Convert each multi-rate actor input/output to multiple single-rate input/outputs. For example, if an actor input has a consumption rate of 3, we replace it with three single-rate inputs.
4. Re-introduce the queues in the data flow system to connect all actors. Since we are building a PASS system, the total number of actor inputs will be equal to the total number of actor outputs.
5. Re-introduce the initial tokens in the system, distributing them sequentially over the single-rate queues.

Consider the example of a multirate SDF graph in Fig. 2.19. Actor $A$ produces three tokens per firing, actor $B$ consumes two tokens per firing. The resulting firing rates are 2 and 3, respectively.

After completing steps 1–5 discussed above, we obtain the SDF graph shown in Fig. 2.20. The actors have duplicated according to their firing rates, and all multi-rate ports were converted to single-rate ports. The initial tokens are redistributed over the queues connecting instances of $A$ and $B$. The distribution of tokens follows the sequence of queues between A,B (ie. follows the order a, b, etc.).

Multi-rate expansion is a convenient technique to generate a specification in which every actor needs to run at the same speed. For example, in a hardware

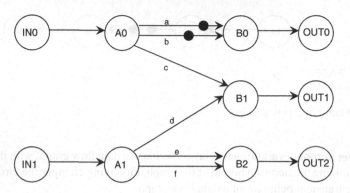

**Fig. 2.20** Multi-rate SDF graph expanded to single-rate

implementation of data flow graphs, multi-rate expansion will enable all actors to run from the same clock signal.

## 2.5.2  Retiming

Retiming is a transformation on data flow graphs which doesn't change the total number of delays between input and output of a data flow graph. Instead, retiming is the redistribution the delays in the data flow graph. This way, the immediate dependency between actors can be broken, allowing them to operate in parallel. A retimed graph may have an increased system throughput. The retiming transformation is easy to understand. The transformation is obtained by evaluating the performance of successive markings of the data flow graph, and then selecting the one with the best performance.

Figure 2.21 illustrates retiming using an example. The top data flow graph, Fig. 2.21a, illustrates the initial system. This graph has an iteration bound of 8. However, the actual data output period of Fig. 2.21a is 16 time units, because actors A, B, and C need to execute as a sequence. If we imagine actor A to fire once, then it will consume the tokens (delays) at its inputs, and produce an output token. The resulting graph is shown in Fig. 2.21b. This time, the data output period has reduced to 11 time units. The reason is that actor A and the chain of actors B and C, can each operate in parallel. The graph of Fig. 2.21b is functionally identical to the graph of Fig. 2.21a: it will produce the same identical stream of output samples when given the same stream of input samples. Finally, Fig. 2.21c shows the result of moving the delay across actor B, to obtain yet another equivalent marking. This implementation is faster than the previous one; as a matter of fact, this implementation achieves the iteration bound of 8 time units per sample. No faster implementation exists for the given graph and the given set of actors.

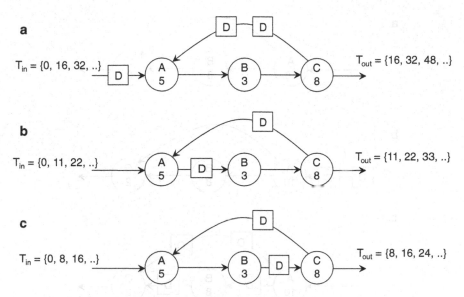

**Fig. 2.21** Retiming: (**a**) Original Graph. (**b**) Graph after first re-timing transformation. (**c**) Graph after second re-timing transformation

Shifting the delay on the edge *BC* further would result in a delay on the outputs of actor *C*: one on the output queue, and one in the feedback loop. This final transformation illustrates an important property of retiming: it's not possible to increase the number of delays in a loop by means of retiming.

### 2.5.3 Pipelining

Pipelining increases the throughput of a data flow graph at the cost of increased latency. Pipelining can be easily understood as a combination of retiming and adding delays. Figure 2.22 demonstrates pipelining on an example. The orginal graph in Fig. 2.22a is extended with two pipeline delays in Fig. 2.22b. Adding delay stages at the input increases the latency of the graph. Before the delay stages, the system latency was 20 time units. After adding the delay stages, the system latency increases to 60 time units (3 samples with a latency of 20 time units each). The system throughput is 1 sample per 20 time units. We can now increase the system throughput by retiming the pipelined graph, so that we obtain Fig. 2.22c. The throughput of this graph is now 1 sample per 10 time units, and the latency is 30 time units (3 times 10 time units). This analysis points out an important property of pipelining: the slowest pipeline stage determines the throughput of the overall system.

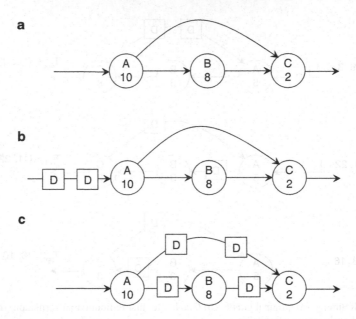

**Fig. 2.22** Pipelining: (**a**) Original graph. (**b**) Graph after adding two pipeline stages. (**c**) Graph after retiming the pipeline stages

### 2.5.4  Unfolding

The final transformation we discuss is unfolding. The idea of unfolding is the parallel implementation of multiple instances of a given data flow graph. For example, assume a data flow graph $G$ which processes a stream of samples. The two-unfolded graph $G2$ consists of two instances of $G$; this graph $G2$ processes two samples at a time.

The rules of unfolding are very similar to the rules of multi-rate expansion. Each actor $A$ of the unfolded system is replicated the number of times needed for the unfolding. Next, the interconnections are made while respecting the sample sequence of the original system. Finally, the delays are redistributed over the interconnections.

The unfolding process is formalized as follows.

- Assume a graph $G$ with an actor $A$ and an edge $AB$ carrying $n$ delays.
- The $v$-unfolding of the graph $G$ will replicate the actor $A$ $v$ times, namely $A_0$, $A_1$, .., $A_{v-1}$. The interconnection $AB$ is replicated $v$ times as well, $AB_0$, $AB_1$, .., $AB_{v-1}$.
- Edge $AB_i$ connects $A_i$ with $B_k$, for which $i : 0..v-1$ and $k = (i+n)\%v$.
- Edge $AB_i$ carries $\lfloor (i+n)/v \rfloor$ delays. If $n < v$, then there will be $v - n$ edges without a delay.

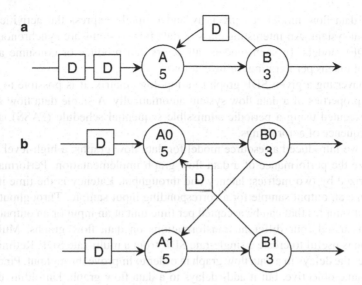

**Fig. 2.23** Unfolding: (**a**) Original graph. (**b**) Graph after two-unfolding

Figure 2.23 illustrates the unfolding process with an example graph, unfolded two times. You can notice that the unfolded graph has two inputs and two outputs, and hence is able to accept twice as much data per iteration as the original data flow graph. On the other hand, unfolding the graph seems to slow it down. The critical loop now includes $A_0$, $B_0$, $A_1$ and $B_1$, but there is still only a single delay element in the overall loop. Hence, the iteration bound of a $v$-unfolded graph has increased $v$ times.

Unfolding of data-flow graphs is used to process data streams with very high sample rates. In this case, the high-speed stream is expanded into $v$ parallel streams. Stream $i$ carries sample $s_i, s_{i+v}, s_{i+2v}, ..$ from the original stream. For example, in Fig. 2.23, the even samples would be processed by $A_0$ while the odd samples would be processed by $A_1$. However, because unfolding decreases the iteration bound, the overall computation speed of the system may be affected.

This completes our discussion on data flow graph transformations. Pipelining, retiming and unfolding are important performance-enhancing manipulations on data flow graphs, and they have a significant impact on the quality of the final implementation.

## 2.6   Data Flow Modeling Summary

Data flow models express concurrent systems in such as way that the models can map into hardware as well as in software. Data flow models consist of actors which communicate by means of tokens which flow over queues from one actor to the other. A data flow model can precisely and formally express the activities of a concurrent system. An interesting class of data flow systems are synchronous data flow (SDF) models. In such models, all actors can produce or consume a fixed amount of tokens per iteration (or invocation).

By converting a given SDF graph to a topology matrix, it is possible to derive stability properties of a data flow system automatically. A stable data flow system can be executed using a periodic admissible sequential schedule (PASS), a fixed period sequence of actor firings.

Next, we introduced a resource model for data flow graphs: a high-level model to analyze the performance of a data flow graph implementation. Performance is characterized by two metrics: latency and throughput. Latency is the time it takes to compute an output sample for a corresponding input sample. Throughput is the amount of samples that can be accepted per time unit at an input or an output.

We discussed four different transformations on data flow graphs. Multi-rate expansion is useful to create a single-rate SDF from a multi-rate SDF. Retiming redistributes the delays in a data flow graph in order to improve throughput. Pipelining has the same objective, but it adds delays to a data flow graph. Unfolding creates parallel instances from a data flow graph, in order to process multiple samples from a data stream in parallel.

Data flow modeling remains an important and easy-to-understand design and modeling technique. They are very popular in signal-processing application, or any application where infinite streams of signal samples can be captured as token streams. Data flow modeling is highly relevant to hardware-software codesign because of the clear and clean manner in which it captures system specifications.

## 2.7   Further Reading

Data flow analysis and implementation has been well researched over the past few decades, and data flow enjoys a rich body of literature.

In the early 1970s, data flow has been considered as a replacement for traditional instruction-fetch machines. Actual data flow computers were build that operate very much according to the SDF principles discussed here. Those early years of data flow have been documented very well at a retrospective conference called *Data flow to Synthesis Retrospective*. The conference honored Arvind, one of data flows' pioneers, and the online proceedings include a talk by Jack Dennis (Dennis 2007).

In the 1980s, data flow garnered attention because of its ability to describe signal processing problems well. For example, Lee and Messerschmit described SDF scheduling mechamisms (Lee and Messerschmitt 1987). Parhi and Messerschmit discussed retiming, pipelining and unfolding transformations of SDF graphs (Parhi and Messerschmitt 1989). Lee as well as Parhi have each authored an excellent textbook that includes data flow modeling and implementation as part of the material, see (Lee and Seshia 2011) and (Parhi 1999). The work from Lee eventually gave rise to the Ptolemy environment (Eker et al. 2003). Despite these successes,

**Fig. 2.24** SDF graph
for Problem 2.1

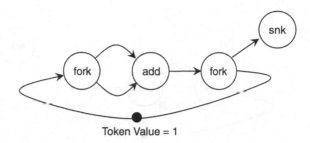

Token Value = 1

data flow never became truly dominant compared to existing control oriented paradigms. This is regrettable. a well-known, but difficult to solve, design problem that affects hardware and software designers alike is how to build parallel versions of solutions originally conceived as sequential (C) programs.

## 2.8 Problems

**Problem 2.1.** Consider the single-rate SDF graph in Fig. 2.24. The graph contains three types of actors. The fork actor reads one token and produces two copies of the input token, one on each output. The add actor adds up two tokens, producing a single token that holds the sum of the input tokens. The snk actor is a token-sink which records the sequence of tokens appearing at its input. A single initial token, with value 1, is placed in this graph. Find the value of tokens that is produced into the snk actor. Find a short-hand notation for this sequence of numbers.

**Problem 2.2.** The Fibonacci Number series F is defined by $F(0)=0$, $F(1)=1$, $F(i)=F(i-1)+F(i-2)$ when i is greater then 1. By changing the marking of the SDF graph in Fig. 2.27, it is possible to generate the Fibonacci series into the snk actor. Find the location and the initial value of the tokens in the modified graph.

**Problem 2.3.** Consider the SDF graph in Fig. 2.25. Transform that graph such that it will produce the same sequence of tokens as tuples instead of as a sequence of singletons. To implement this, replace the snk actor with snk2, an actor which requires two tokens on two different inputs in order to fire. Next make additional transformations to the graph and its marking so that it will produce this double-rate sequence into snk2.

**Problem 2.4.** Data Flow actors cannot contain state variables. Yet, we can 'simulate' state variables with tokens. Using only an adder actor, show how you can implement an accumulator that will obtain the sum of an infinite series of input tokens.

**Problem 2.5.** For the SDF graph of Fig. 2.26, find a condition between x and y for a PASS to exist.

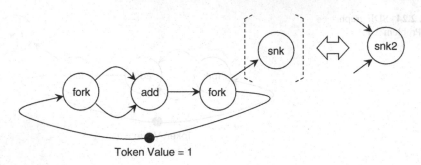

Token Value = 1

**Fig. 2.25** SDF graph for Problem 2.3

**Fig. 2.26** SDF graph
for Problem 2.5

**Fig. 2.27** SDF graph
for Problem 2.6

**Fig. 2.28** SDF graph
for Problem 2.7

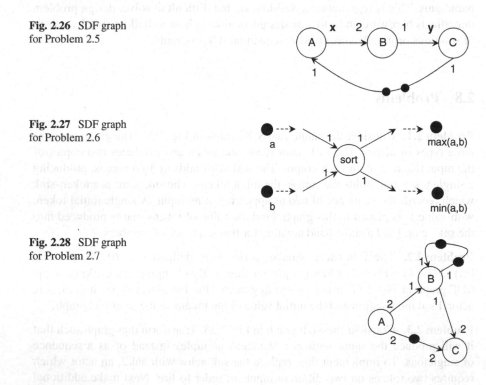

**Problem 2.6.** Given the two-input sorting actor shown in Fig. 2.27. Using this actor, create a SDF graph of a sorting network with four inputs and four outputs.

**Problem 2.7.** Draw the multi-rate expansion for the multirate SDF given in Fig. 2.28. Don't forget to redistribute the initial tokens on the multirate-expanded result.

**Problem 2.8.** The data flow diagram in Fig. 2.29 demonstrates *reconvergent* edges: edges which go around one actor. Reconvergent edges tend to make analysis of a data flow graph a bit harder because they may imply that multiple critical loops may be laying on top of one another. This problem explores this effect.

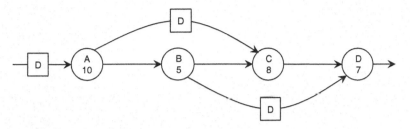

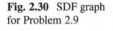

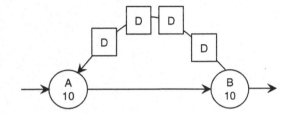

**Fig. 2.29** SDF graph for Problem 2.8

**Fig. 2.30** SDF graph
for Problem 2.9

(a) Determine the loop bounds of all loops in the graph of Fig. 2.29. Take the
     input/output constraints into account by assuming an implicit loop from output
     to input.
(b) Using the results of the previous part, determine the iteration bound for this
     graph.
(c) Find the effective throughput of this graph, based on the distribution of delays
     as shown in Fig. 2.29.
(d) Does the system as shown in Fig. 2.29 achieve the iteration bound? If not, apply
     the retiming transformation and improve the effective throughput so that you
     get as close as possible to the iteration bound.

**Problem 2.9.** Unfold the graph in Fig. 2.30 three times. Determine the iteration
bound before and after the unfolding operation.

# Chapter 3
# Data Flow Implementation in Software and Hardware

## 3.1 Software Implementation of Data Flow

The design space to map data flow in software is surprisingly broad. Nevertheless, a dataflow implementation will always start from the same semantics model with dataflow actors and dataflow queues.

### 3.1.1 Converting Queues and Actors into Software

Let's first recall the essential features of SDF graphs. SDF graphs represent concurrent systems, and they use actors which communicate over FIFO queues. Actor firing only depends on the availability of data (tokens) in the FIFO queues; the firing conditions are captured in the firing rule for that actor. The amount of tokens produced/consumed per firing at the output/input of an actor is specified by the production rate/consumption rate for that output/input. When implementing an SDF graph in software, we have to map *all* elements of the SDF graph in software: actors, queues, and firing rules. Under some conditions, the software implementation may be optimized. For example, when a fixed execution order of actors can be found (a PASS, as discussed in Chap. 2), it may be possible to skip testing of the firing rules. However, the principle remains that the implementation needs to follow the rules of dataflow semantics: optimizations (such as skipping the testing of firing rules) need to be motivated from analysis of the data flow graph.

Figure 3.1 demonstrates several different approaches to map dataflow into software. We'll distinguish the mapping of data flow to a multi-processor system from the mapping of data flow to a single processor system. When a data flow graph requires implementation in a multi-processor system, a designer will need to partition the actors in the graph over the processors. This partitioning is driven by several criteria. Typically, designers ensure that the computational load is balanced over the processors, and that the inter-processor communication is minimized.

P.R. Schaumont, *A Practical Introduction to Hardware/Software Codesign*,
DOI 10.1007/978-1-4614-3737-6_3, © Springer Science+Business Media New York 2013

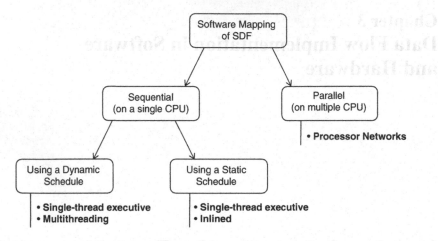

**Fig. 3.1** Overview of possible approaches to map dataflow into software

However, the focus of this section will be on the single-processor implementation of a data flow graph. The key objective of a single-processor implementation of a data flow system is the efficient implementation of a sequential schedule. There are two methods to implement such a sequential schedule.

- We can use a *dynamic schedule*, which evaluates the execution order of actors during the execution of the SDF graph. Thus, at runtime, the software will evaluate the actors' firing rule and decide if the actor body should execute or not. A dynamic schedule can be implemented using a single-thread executive or else using a multi-thread executive.
- We can also use a *static schedule*, which means that the execution order of the actors is determined at design-time. A static schedule can be implemented using a single-threaded executive. However, because the static schedule fixes the execution order of the actors, there is an additional important optimization opportunity: we can treat the firing of multiple actors as a single firing. Eventually, this implies that we can inline the entire dataflow graph in a single function.

We'll start with the easy stuff: we show how to implement FIFO queues and dataflow actors in C. After that, we will be ready to implement static and dynamic schedules in software.

### 3.1.1.1 FIFO Queues

An SDF system requires, in principle, infinitely large FIFO queues. Such queues cannot be implemented; in practice, an implementation has to have bounded storage. If we know a PASS, we can derive a static schedule and determine the maximum number of tokens on each queue, and then appropriately choose the size for each queue.

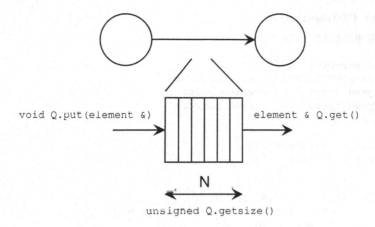

**Fig. 3.2** A software queue

Figure 3.2 shows a software interface to a queue object. The software interface has two parameters and three methods.

- The number of elements N that can be stored by the queue (parameter).
- The data type element of a queue elements (parameter).
- A method to put elements into the queue.
- A method to get elements from the queue.
- A method to test the number of elements in the queue.

The storage organization can be done with a standard data structure such as a circular queue. A circular queue is a data structure consisting of an array of memory locations, a write-pointer and a read-pointer. Figure 3.3 illustrates the operation of a two-element circular queue. Such a queue uses an array of three locations. The read- and write-pointers map relative queue addresses to array addresses using modulo addressing. The head of the queue is at Rptr. Element I of the queue is at $(Rptr + I)$ mod 3. The tail of the queue is at $(Wptr - 1)$ mod 3.

Listing 3.1 shows the definition of a FIFO object in C. This example uses static allocation of the array, which implies that the maximum length of each queue is fixed before the simulation starts. For large systems, with many different queues, this strategy may be too pessimistic and too greedy on system resources. Another approach may be to use dynamically expanding queues. In that case, the FIFO object starts with a small amount of storage. Each time the FIFO queue would overflow, the amount of allocated storage is doubled. This technique is useful when a static schedule for the data flow system is unknown, or cannot be derived Problem 3.1.

### 3.1.1.2 Actors

A data flow actor can be implemented as a C function, with some additional support to interface with the FIFO queues. Designers will often differentiate between the

**Listing 3.1** FIFO object in C

```c
#define MAXFIFO 1024

typedef struct fifo {
  int data[MAXFIFO]; // token storage
  unsigned wptr;      // write pointer
  unsigned rptr;      // read pointer
} fifo_t;

void init_fifo(fifo_t *F) {
  F->wptr = F->rptr = 0;
}

void put_fifo(fifo_t *F, int d) {
  if (((F->wptr + 1) % MAXFIFO) != F->rptr) {
    F->data[F->wptr] = d;
    F->wptr = (F->wptr + 1) % MAXFIFO;
    assert(fifo_size(F) <= 10);
  }
}

int get_fifo(fifo_t *F) {
  int r;
  if (F->rptr != F->wptr) {
    r = F->data[F->rptr];
    F->rptr = (F->rptr + 1) % MAXFIFO;
    return r;
  }
  return -1;
}

unsigned fifo_size(fifo_t *F) {
  if (F->wptr >= F->rptr)
    return F->wptr - F->rptr;
  else
    return MAXFIFO - (F->rptr - F->wptr) + 1;
}

int main() {
  fifo_t F1;
  init_fifo(&F1); // resets wptr, rptr;
  put_fifo(&F1, 5); // enter 5
  put_fifo(&F1, 6); // enter 6
  printf("%d_%d\n", fifo_size(&F1), get_fifo(&F1));
  // prints: 2 5
  printf("%d\n", fifo_size(&F1)); // prints: 1
}
```

internal activities of an actor, and the input-output behavior. The behavior corresponding to actor firing can be implemented as a simple C function. The firing-rule logic evaluates the firing condition, and calls the actor body when the condition is true.

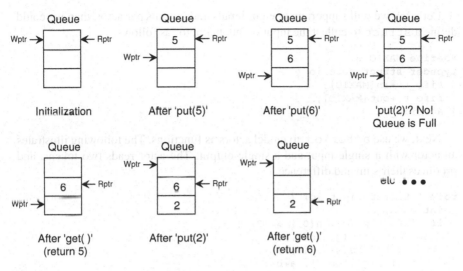

Fig. 3.3 Operation of the circular queue

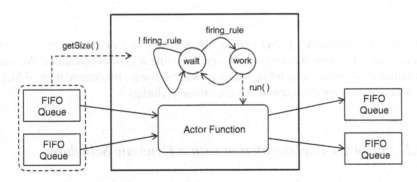

Fig. 3.4 Software implementation of the datafow actor

Figure 3.4 shows that the firing rule logic is implemented as a small, local controller inside of the actor. The local controller goes through two states. In the wait state the actor remains idle, but it tests the firing rule upon each invocation of the actor (an invocation is equivalent to calling the C function that implements the dataflow actor). When the firing rule evaluates true, the actor proceeds to the work state. In this state, the actor invocation will read tokens from the input queue(s), extract their values, and feed these to the actor body. Next, the resulting output values are written into the output queues, and the actor returns to the wait state. When implementing the actor, we must take care to implement the firing as specified. SDF actor firing implies that the actor has to read all input queues according to the specified consumption rates, and it has to write all output queues according to the specified production rates.

Let's say we will support up to eight inputs and outputs per actor, then we could define a struct to collect the input/output per actor as follows.

```
#define MAXIO 8
typedef struct actorio {
  fifo_t *in[MAXIO];
  fifo_t *out[MAXIO];
} actorio_t;
```

Next, we use actorio_t to model actors as functions. The following illustrates an actor with a single input and a single output. The actor reads two tokens, and produces their sum and difference.

```
void fft2(actorio_t *g) {
  int a, b;
  if (fifo_size(g->in[0]) >= 2) {
    a = get_fifo(g->in[0]);
    b = get_fifo(g->in[0]);
    put_fifo(g->out[0], a+b);
    put_fifo(g->out[0], a-b);
  }
}
```

Finally, the actorio_t and queue objects can be instantiated in the main program, and the actor functions can be called using a system scheduler. We will first introduce dynamic scheduling techniques for software implementation of SDF, and next demonstrate an example that uses these techniques.

## 3.1.2  Software Implementation with a Dynamic Scheduler

A software implementation of SDF is obtained by combining several different actor descriptions, by interconnecting those actors using FIFO queues, and by executing the actors through a system schedule. In a *dynamic* system schedule, the firing rules of the actors will be tested at runtime; the system scheduling code consists of the firing rules, as well as the order in which the firing rules are tested.

Following the FIFO and actor modeling in C, as discussed in Sect. 3.1.1, we can implement a system schedule as a function that instantiates all actors and queues, and next calls the actors in a round-robing fashion.

```
void main() {
  fifo_t q1, q2;
  actorio_t fft2_io  = {{&q1}, {&q2}};
  ...
  init_fifo(&q1);
  init_fifo(&q2);
  ...
```

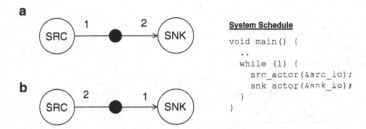

**Fig. 3.5** (a) A graph which will simulate under a single rate system schedule, (b) a graph which will cause extra tokens under a single rate schedule

```
while (1) {
  fft2_actor(&fft2_io);
  // .. call other actors
  }
}
```

The interesting question, of course, is: what is the most appropriate invocation order of the actors in the system schedule? First, note that it is impossible to invoke the actors in the 'wrong' order, because each of them has a firing rule that prevents them from firing when there is no data available. Consider the example in Fig. 3.5a. Even though snk will be called as often as src, the firing rule of snk will only allow that actor to run when there is sufficient data available. This means that the snk actor will only fire every other time the main function invokes it.

While a dynamic scheduling loop can prevent actors from firing prematurely, it is still possible that some actors fire too often, resulting in the number of tokens on the interconnection queues continuously growing. This happens, for example, in Fig. 3.5b. In this case, the src actor will produce two tokens each time the main function invokes it, but the snk actor will only read one of these tokens per invocation. This means that, sooner or later, the queue between src and snk in Fig. 3.5b can overflow.

The problem of the system schedule in Fig. 3.5b is that the firing rate provided by the system schedule is different from the required firing rate for a PASS. Indeed, the PASS for this system would be (src, snk, snk). However, the dynamic system schedule, given by the code on the right of Fig. 3.5, cannot make the firing rate of SNK higher than that of SRC. This problem can be addressed in several ways.

- **Solution 1:** We could adjust the system schedule to reflect the firing rate predicted by the PASS. In this case, the code for the system scheduler becomes:

```
void main() {
  ..
  while (1) {
    src_actor(&src_io);
    snk_actor(&snk_io);
```

**a**                                                          **b**

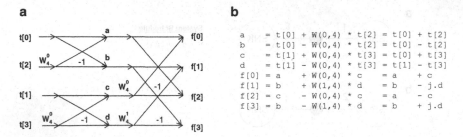

```
a    = t[0] + W(0,4) * t[2] = t[0] + t[2]
b    = t[0] - W(0,4) * t[2] = t[0] - t[2]
c    = t[1] + W(0,4) * t[3] = t[0] + t[3]
d    = t[1] - W(0,4) * t[3] = t[1] - t[3]
f[0] = a    + W(0,4) * c    = a    + c
f[1] = b    + W(1,4) * d    = b    - j.d
f[2] = c    - W(0,4) * c    = a    - c
f[3] = b    - W(1,4) * d    = b    + j.d
```

**Fig. 3.6** (a) Flow diagram for a four-point Fast Fourier Transform (b) Equivalent set of operations

```
        snk_actor(&snk_io);
    }
}
```

This solution is not very elegant, because it destroys the idea of a dynamic scheduler that automatically converges to the PASS firing rate. It also makes the `main` loop dependent on the topology of the SDF graph.

- **Solution 2:** We could adjust the code for the `snk` actor to continue execution as long as there are tokens present. Thus, the code for the `snk` actor becomes:

```
void snk_actor(actorio_t *g) {
    int r1, r2;
    while ((fifo_size(g->in[0]) > 0)) {
        r1 = get_fifo(g->in[0]);
        ... // do processing
    }
}
```

This is a better solution than the previous one, because it keeps the advantages of a dynamic system schedule.

### 3.1.3 Example: Four-Point Fast Fourier Transform as an SDF System

Figure 3.6a shows a four-point Fast Fourier Transform (FFT). It takes an array of 4 signal samples t[0] through t[3] and converts it to a spectral representation f[0] through f[3]. FFT's are extensively used in signal processing to do spectral analysis, filtering, and more. The references include a few pointers to detailed descriptions of the FFT algorithm.

In this example, we are interested in the mapping of a four-point FFT into a data flow graph. An FFT is made up out of 'butterfly' operations. An equivalent set of operations corresponding to the graph from Fig. 3.6a is shown in Fig. 3.6b. The twiddle factor W(k,N), or $W_N^k$, is a complex number defined as $e^{-j2\pi k/N}$. Obviously, W(0,4)=1 and W(1,4)=-j. The FFT thus produces complex numbers at

**Fig. 3.7** Synchronous
dataflow diagram for a
four-point Fast Fourier
Transform

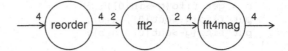

the output. However, if the input values v[0] through v[3] are real values, then
the output values V[1] and V[3] will be complex conjugate: there is thus some
redundancy in the output.

Figure 3.7 shows a data flow model for the same flow diagram. It consists of
three data flow actors: reorder, fft2, and fft4mag.

- reorder reads four tokens and reshuffles them according to the requirements
  of an FFT. In Fig. 3.6a, you can observe that the elements of the input array
  are not processed in linear order: t[0] and t[2] are processed by the first
  butterfly, while t[1] and t[3] are processed by the second butterfly. The
  reorder actor thus converts the sequence t[0], t[1], t[2], t[3]
  into the sequence t[0], t[2], t[1], t[3].
- fft2 calculates the butterflies for the left half of Fig. 3.6a. This actor reads two
  tokens, computes the butterfly, and produces two tokens.
- fft4mag calculates the butterflies of the right half of Fig. 3.6a. This actor reads four tokens, computes two butterflies, and produces four tokens.
  The fft4mag actor computes the magnitude vector real(V[0]*V[0]),
  real(V[1]*V[1]), real(V[2]*V[2]), real(V[3]*V[3]).

To implement the FFT as a data flow system, we would first need to compute a
valid schedule for the actors. It's easy to see that a stable firing vector for this set
of actors is $[q_{reorder}, q_{fft2}, q_{fft4mag}] = [1, 2, 1]$. Listing 3.2 shows a description for
the reorder and fft4mag actors, as well as a main program to implement this
schedule.

**Listing 3.2** 4-point FFT as an SDF system

```
void reorder(actorio_t *g) {
  int v0, v1, v2, v3;
  while (fifo_size(g->in[0]) >= 4) {
    v0 = get_fifo(g->in[0]);
    v1 = get_fifo(g->in[0]);
    v2 = get_fifo(g->in[0]);
    v3 = get_fifo(g->in[0]);
    put_fifo(g->out[0], v0);
    put_fifo(g->out[0], v2);
    put_fifo(g->out[0], v1);
    put_fifo(g->out[0], v3);
  }
}

void fft2(actorio_t *g) {
  int a, b;
  while (fifo_size(g->in[0]) >= 2) {
    a = get_fifo(g->in[0]);
```

```
      b = get_fifo(g->in[0]);
      put_fifo(g->out[0], a+b);
      put_fifo(g->out[0], a-b);
   }
}

void fft4mag(actorio_t *g) {
   int a, b, c, d;
   while (fifo_size(g->in[0]) >= 4) {
      a = get_fifo(g->in[0]);
      b = get_fifo(g->in[0]);
      c = get_fifo(g->in[0]);
      d = get_fifo(g->in[0]);
      put_fifo(g->out[0], (a+c)*(a+c));
      put_fifo(g->out[0], b*b - d*d);
      put_fifo(g->out[0], (a-c)*(a-c));
      put_fifo(g->out[0], b*b - d*d);
   }
}

int main() {
   fifo_t q1, q2, q3, q4;
   actorio_t reorder_io = {{&q1}, {&q2}};
   actorio_t fft2_io    = {{&q2}, {&q3}};
   actorio_t fft4_io    = {{&q3}, {&q4}};

   init_fifo(&q1);
   init_fifo(&q2);
   init_fifo(&q3);
   init_fifo(&q4);

   // test vector fft([1 1 1 1])
   put_fifo(&q1, 1);
   put_fifo(&q1, 1);
   put_fifo(&q1, 1);
   put_fifo(&q1, 1);

   // test vector fft([1 1 1 0])
   put_fifo(&q1, 1);
   put_fifo(&q1, 1);
   put_fifo(&q1, 1);
   put_fifo(&q1, 0);

   while (1) {
      reorder(&reorder_io);
      fft2    (&fft2_io);
      fft4mag(&fft4_io);
   }

   return 0;
}
```

The use of the `actorio_t` in the main program simplifies the interconnection of FIFO queues to actors. In this case, connections are made per actor, making it easy to follow which actor reads and writes which queue.

The actor descriptions use `while` loops, as discussed earlier, to ensure that the dynamic scheduler can achieve the PASS firing rate for each actor. In addition, thanks to the individual firing rules within the actors, and the deterministic property of SDF, we can write this scheduling loop in any order. For example, the following loop would yield the same results in the output queue q4.

```
int main() {
    ...
    while (1) {
        fft2     (&fft2_io);
        reorder(&reorder_io);
        fft4mag(&fft4_io);
    }
}
```

### 3.1.3.1 Multi-thread Dynamic Schedules

The actor functions, as described above, are captured as plain C functions. As a result, actors can maintain state using local variables in between invocations. We have to use global variables, or static variables. Another approach to create dynamic schedules is to use multi-threaded programming. We will discuss a solution based on multi-threaded programming, in which each actor executes in a separate thread.

A multithreaded C program is a program that has two concurrent threads of execution. For example, in a program with two functions, one thread could be executing the first function, while the other thread could be executing the second function. Since there is only a single processor to execute this program, we need to switch the processor back and forth between the two threads of control. This is done with a thread scheduler. Similar to a scheduler used for scheduling actors, a thread scheduler will switch the processor between threads.

We will illustrate the use of *cooperative* multithreading. In this model, the threads of control indicate at which point they release control back to the scheduler. The scheduler then decides which thread can run next.

Figure 3.8 shows an example with two threads. Initially, the user has provided the starting point of each thread using `create()`. Assume that the upper thread (`thread1`) is running and arrives at a `yield( )` point. This is a point where the thread returns control to the scheduler. The scheduler maintains a list of threads under its control, and therefore knows that the lower thread (`thread2`) is ready to run. So it allows `thread2` to run until that thread, too, comes at a yield point. Now the scheduler sees that each thread had a chance to run, so it goes back to the first thread. The first thread then will continue just after the yield point.

**Fig. 3.8** Example of cooperative multi-threading

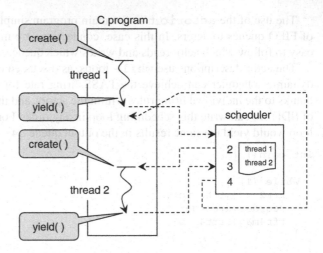

Consequently, two functions are enough to implement a threaded system: `create( )` and `yield( )`. The scheduler can apply different strategies to select each thread, but the simplest one is to let each thread run in turn – this is called a 'round-robin' scheduling strategy. We will use a cooperative multithreading library called Quickthreads. The Quickthreads API (Application Programmers' Interface) consists of four function calls.

- `stp_init( )` initializes the theading system
- `stp_create(stp_userf_t *F, void *G)` creates a thread that will start execution with user function F. The function will be called with a single argument G. The thread will terminate when that function completes, or when the thread aborts.
- `stp_yield( )` releases control over the thread to the scheduler.
- `stp_abort( )` terminates a thread, so that it will be no more scheduled.

Here is a small program that uses the QuickThreads library.

**Listing 3.3** Example of QuickThreads

```
#include "../qt/stp.h"
#include <stdio.h>

void hello(void *null) {
  int n = 3;
  while (n-- > 0) {
    printf("hello\n");
    stp_yield();
  }
}

void world(void *null) {
  int n = 5;
  while (n-- > 0) {
    printf("world\n");
```

```
    stp_yield();
  }
}

int main(int argc, char **argv) {
  stp_init();
  stp_create(hello, 0);
  stp_create(world, 0);
  stp_start();
  return 0;
}
```

This program creates two threads (lines 21 and 22), one which starts at function hello, and another which starts at function world. Function hello (lines 3–9) is a loop that will print "hello" three times, and yield after each iteration. After the third time, the function will return, which also terminates the thread. Function world (lines 11–17) is a loop that will print "world" five times, and yield at end of each iteration. When all threads are finished, the main function will terminate. We compile and run the program as follows. The references include a link to the source code of QuickThreads.

```
>gcc -c ex1.c -o ex1 ../qt/libstp.a ../qt/libqt.a
./ex1
hello
world
hello
world
hello
world
world
world
```

The printing of hello and world are interleaved for the first three iterations, and then the world thread runs through completion.

We can now use this multi-threading system to create a multi-thread version of the SDF scheduler. Here is the example of a fft2 actor, implemented using the cooperative threading model.

```
void fft2(actorio_t *g) {
  int a, b;
  while (1) {
    while (fifo_size(g->in[0]) >= 2) {
      a = get_fifo(g->in[0]);
      b = get_fifo(g->in[0]);
      put_fifo(g->out[0], a+b);
      put_fifo(g->out[0], a-b);
    }
    stp_yield();
  }
}
```

The system scheduler now creates threads rather than directly invoking actors:

```
int main() {
```

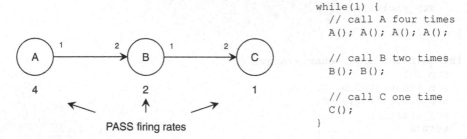

```
while(1) {
    // call A four times
    A(); A(); A(); A();

    // call B two times
    B(); B();

    // call C one time
    C();
}
```

**Fig. 3.9** System schedule for a multirate SDF graph

```
fifo_t q1, q2;
actorio_t fft2_io = {{&q1}, {&q1}};
...
stp_create(fft2, &fft2_io); // create thread
...
stp_start(); // run the schedule
}
```

The execution rate of the actor code must be equal to the PASS firing rate in order to avoid unbounded growth of tokens in the system. A typical cooperative multi-threading system uses round-robin scheduling.

### 3.1.4   Sequential Targets with Static Schedule

When we have completed the PASS analysis for an SDF graph, we know at least one solution for a feasible sequential schedule. We can use this to optimize the implementation in several ways.

- First, we are able to write an exact schedule, such that every actor fires upon every invocation. Hence, we are able to remove the firing rules of the actors. This will yield a small performance advantage. Of course, we can no longer use such actors with dynamic schedulers.
- Next, we can also investigate the optimal interleaving of actors such that the storage requirements for the queues are reduced.
- Finally, we can create a fully inlined version of the SDF graph, by exploiting our knowledge on the static, periodic behavior of the system as much as possible. We will see that this not only allows us to get rid of the queues, but also allows us to create a fully inlined version of the entire SDF system.

Consider the example in Fig. 3.9. From this SDF topology, we know that the relative firing rates of A, B, and C must be 4, 2, and 1 to yield a PASS. The right side of the figure shows an example implementation of this PASS. The A, B, C actors are called in accordance to their PASS rate. For this particular interleaving, it is easy to see that in a steady state condition, the queue AB will carry a maximum of four tokens, while the queue BC will contain a maximum of two tokens. This is not the most optimal interleaving. By calling the actors in the sequence (A, A, B, A, A, B, C), the maximum amount of tokens on any queue is reduced to two. Finding an optimal interleaving in an SDF graph is an optimization problem. While an in-depth discussion of this problem is beyond the scope of this book, remember that actor interleaving will affect the storage requirements for the implementation.

Implementing a truly *static* schedule means that we will no longer test firing rules when calling actors. In fact, when we call an actor, we will have to guarantee that the required input tokens are available. In a system with a static schedule, all SDF-related operations get a fixed execution order: the actor firings, and the sequences of put and get operations on the FIFO queues. This provides the opportunity to optimize the resulting SDF system.

We will discuss optimization of single-thread SDF systems with a static schedule using an example we discussed before – the four-point Fast Fourier Transform.

The system uses three actors – reorder, fft2 and fft4rev – which have PASS firing rates of 1, 2 and 1 respectively. A feasible static and cyclic schedule could be, for example: [reorder, fft2, fft2, fft4rev].

Next, let's optimize the system description from Listing 3.2 as follows.

1. Because the firing order of actors can be completely fixed, the access order on queues can be completely fixed as well. This latter fact will allow the queues themselves to be optimized out and replaced with fixed variables. Indeed, assume for example that we have determined that the access sequence on a particular FIFO queue will always be as follows:

```
loop {
   ...
   q1.put (value1);
   q1.put (value2);
   ...
   .. = q1.get ();
   .. = q1.get ();
}
```

   In this case, only two positions of FIFO q1 are occupied at a time. Hence, FIFO q1 can be replaced by two single variables.

```
loop {
   ...
   r1 = value1;
   r2 = value2;
   ...
   .. = r1;
   .. = r2;
}
```

2. As a second optimization, we can inline actor code inside of the main program and the main scheduling loop. In combination with the above optimization, this will allow to remove the firing rules and to collapse an entire dataflow graph

**Listing 3.4** Inlined data flow system for the four-point FFT

```
void dfsystem(int     in0,     in1,     in2,     in3,
                      *out0,   *out1,   *out2,   *out3) {
  int reorder_out0, reorder_out1, reorder_out2, reorder_out3;
  int fft2_0_out0,  fft2_0_out1,  fft2_0_out2,  fft2_0_out3;
  int fft2_1_out0,  fft2_1_out1,  fft2_1_out2,  fft2_1_out3;
  int fft4mag_out0, fft4mag_out1, fft4mag_out2, fft4mag_out3;

  reorder_out0 = in0;
  reorder_out1 = in2;
  reorder_out2 = in1;
  reorder_out3 = in3;

  fft2_0_out0  = reorder_out0 + reorder_out1;
  fft2_0_out1  = reorder_out0 - reorder_out1;

  fft2_1_out0  = reorder_out2 + reorder_out3;
  fft2_1_out1  = reorder_out2 - reorder_out3;

  fft4mag_out0 = (fft2_0_out0 + fft2_1_out0)*
                 (fft2_0_out0 + fft2_1_out0);
  fft4mag_out1 =  fft2_0_out1*fft2_0_out1 -
                  fft2_1_out1*fft2_1_out1;
  ftt4mag_out2 = (fft2_0_out0 - fft2_1_out0)*
                 (fft2_0_out0 - fft2_1_out0);
  fft4mag_out3 =  fft2_0_out1*fft2_0_out1 -
                  fft2_1_out1*fft2_1_out1;
}
```

in a single function. In case an actor would have a PASS firing rate above one, multiple instances of the actor body are needed. If needed, the multi-rate expansion technique discussed in Sect. 2.5.1 can be used to identify the correct single-rate system topology.

When we apply these optimizations to the four-point FFT example, each queue can be replaced by a series variables. The optimized system is shown in Listing 3.4. We can expect the runtime of this system to decrease significantly: there are no firing rules, no FIFO manipulations and no function boundaries. This is possible because we have determined a valid PASS for the initial data flow system, and because we have chosen a fixed schedule to implement that PASS.

## 3.2  Hardware Implementation of Data Flow

In this section, we are particularly interested in simple, optimized implementations. The use of hardware FIFO's will be discussed later.

### 3.2.1 Single-Rate SDF Graphs into Hardware

The simplest case is the mapping from a single-rate SDF graph to hardware. In a single rate schedule, the relative firing rate of all actors is equal to 1. This means that all actors will execute at the same rate. If the actors are implemented in hardware, all of them will be working at the same clock frequency.

We will be mapping such a single-rate graph into hardware using the following three rules.

1. All actors are implemented as combinational circuits.
2. All communication queues are implemented as wires (without storage).
3. Each initial token on a communication queue is replaced by a register.

The following observations can be made for circuits developed in this fashion. A combinational circuit is a circuit which can finish a computation within a single clock cycle (assuming no internal feedback loops exist). By implementing an actor as a combinational circuit, we ensure that it can do a complete firing within a single clock cycle. Now imagine two combinational circuits, implemented back-to-back. Such a combination will still operate as a single combinational circuit, and thus it will still complete computation within a single clock cycle. When two data flow actors, implemented as combinational circuits, are placed back-to-back, they will finish within a single clock cycle. Hence, a complete iteration of the schedule still takes only a single clock cycle. If we perform a scheduling analysis and identify a PASS, then the hardware circuit, created by the mapping rules described above, will be a valid implementation.

There is, however, a catch. The speed of computation of a combinational circuit is finite. Therefore, the speed of the overall hardware implementation created from a dataflow graph is limited as well. When actors need to compute sequentially, their computation times will add up. Hence, when actors are implemented as combinational circuits, their combined delay needs to remain below the clock period used for the hardware system. There is therefore a limit on the maximum clock frequency used for the system. Note, however, that a precise analysis of this problem is more complex than simply making the sum of computation times of all actors in the PASS.

We can define the *critical path* in a dataflow graph in terms of the resource model introduced in Sect. 2.4.2. First, define the latency of a path as the sum of actor latencies included in the path. Next, call a combinational path, a path which does not contain initial tokens on any of the communication queues included in the path. Finally, call the critical path of a dataflow graph the longest combinational path in the graph.

We next illustrate the hardware implementation of single-rate SDF graphs with an example: an SDF system for Euclid's Greatest Common Divisor algorithm. The SDF in Fig. 3.10 evaluates the greatest common divisor of two numbers $a$ and $b$. It uses two actors: sort and diff.

The sort actor reads two numbers, sorts them and copies them to the output. The diff actor subtracts the smallest number from the largest one, as long as they

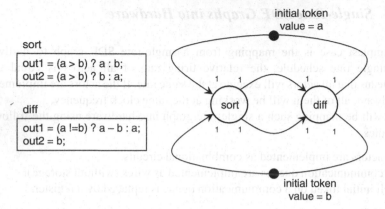

**Fig. 3.10** Euclid's greatest common divisor as an SDF graph

are different. If this system runs for a while, the value of the tokens moving around converge to the greatest common divisor of the two numbers $a$ and $b$. For example, assume

$$(a_0, b_0) = (16, 12) \qquad (3.1)$$

then we see the following sequence of token values.

$$(a_1, b_1) = (4, 12), (a_2, b_2) = (8, 4), (a_3, b_3) = (4, 4), \ldots \qquad (3.2)$$

$a_i$ and $b_i$ are the token values upon iteration $i$ of the PASS. Since this sequence converges to the tuple $(4, 4)$, the greatest common divisor of 12 and 16 is 4.

We now demonstrate a PASS for this system. The topology matrix $G$ for this graph is shown below. The columns, left to right, correspond to each node from the SDF graph, left to right.

$$G = \begin{bmatrix} +1 & -1 \\ +1 & -1 \\ -1 & +1 \\ -1 & +1 \end{bmatrix} \begin{matrix} \leftarrow edge(sort, diff) \\ \leftarrow edge(sort, diff) \\ \leftarrow edge(diff, sort) \\ \leftarrow edge(diff, sort) \end{matrix} \qquad (3.3)$$

The rank of this matrix is one, since the columns complement each other. There are two actors in the graph, so we conclude that the condition for PASS (i.e. rank$(G)$ = nodes $-$ 1) is fulfilled. A valid firing vector for this system is one in which each actor fires exactly once per iteration.

$$q_{PASS} = \begin{bmatrix} 1 \\ 1 \end{bmatrix} \qquad (3.4)$$

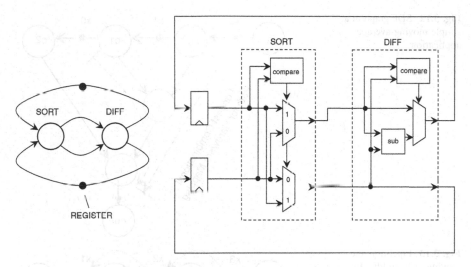

**Fig. 3.11** Hardware implementation of euclid's algorithm

Based on this analysis, we can now proceed with the hardware implementation of the Euclid design. As discussed earlier, we use the following transformation.

1. Map each communication queue to a wire.
2. Map each queue containing a token to a register. The initial value of the register must equal the initial value of the token.
3. Map each actor to a combinational circuit, which completes a firing within a clock cycle. Both the sort and diff actors require no more than a comparator module, a few multiplexers and a subtractor.

Figure 3.11 illustrates how this works out for the Euclid example. In every single clock cycle, the sort actor and the diff actor are computed. The speed of computation of the overall circuit is determined by the combined computation speed of sort and diff. Indeed, the critical path of this graph is a path starting at sort and ending at diff. Assume that sort requires 40 ns of time to compute, and diff requires 60 ns of time, then the critical path of this system is 100 ns. Therefore, the maximum clock frequency of this design is 10 MHz.

### 3.2.2 *Pipelining*

It should be no surprise that some of the transformations, discussed earlier in Sect. 2.5, can also be used to enhance the throughput of hardware implementations. Pipelining is a very good example.

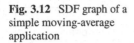

**Fig. 3.12** SDF graph of a
simple moving-average
application

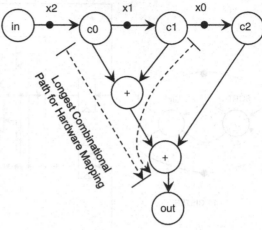

**Fig. 3.13** Pipelining the
moving-average filter by
inserting additional tokens (1)

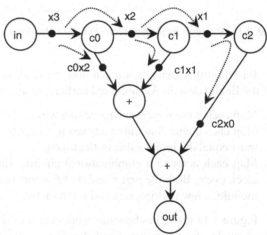

Figure 3.12 shows a data flow specification of a digital filter. It evaluates a
weighted sum of samples of an input stream, with the sum defined as $out = x0.c2 + x1.c1 + x2.x0$.

It can be seen from this graph that the critical path is equal to a constant
multiplication (with $c0$ or $c1$) and two additions. We would like to 'push down'
initial tokens into the adder tree. With the rules of data flow execution, this is
easy. Consider a few subsequent markings of the graph. Assume the `in` actor fires
additional tokens, and the $c0$, $c1$, $c2$ and `add` actors fire as well so that additional
tokens start to appear on queues that have no such tokens. For example, assume that
the `in` actor produces a single additional token $x3$. Then the resulting graph looks
as in Fig. 3.13.

In this graph, the critical path is reduced to only two additions. By letting the
`in` actor produce another token, we will be able to reduce the critical path to a
single addition, as shown in Fig. 3.14. The resulting pipelined SDF graph thus can
be implemented as shown in Fig. 3.15.

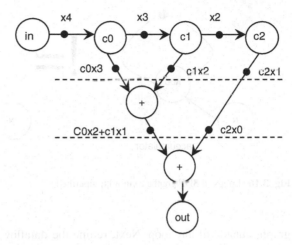

**Fig. 3.14** Pipelining the moving-average filter by inserting additional tokens (2)

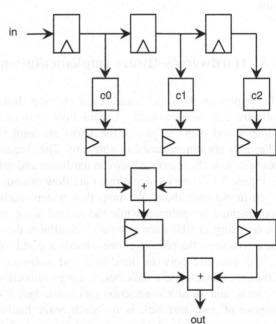

**Fig. 3.15** Hardware implementation of the moving-average filter

Remember that pipelining requires you to introduce additional tokens. This may change the behavior of the dataflow graph. The change in behavior is obvious in the case of feedback loops, such as shown in the accumulator circuit in Fig. 3.16. Using a single token in the feedback loop of an add actor will accumulate all input samples. Using two tokens in the feedback loop will accumulate the odd samples and even samples separately. To avoid introducing accidental tokens in a loop, you can also perform pipelining as follows: introduce initial tokens at the input or output of the

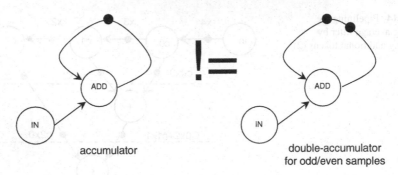

**Fig. 3.16** Loops in SDF graphs cannot be pipelined

graph, outside of any loop. Next, retime the dataflow graph to reduce the critical path.

## 3.3   Hardware/Software Implementation of Data Flow

The implementation techniques used to map data flow graphs in hardware or software can be combined. A data flow system with multiple actors can be implemented such that part of the actors are implemented in hardware, while the other half are implemented in software. This section illustrates, by means of an example, how the interface between hardware and software can be handled.

Figure 3.17 shows a single rate data flow system with two actors and an initial token in between them. We map this system such that the first actor, `ctr`, is implemented in hardware, while the second actor, `snk`, is implement in software. We are using an 8051 microcontroller. Similar to the example of Sect. 1.1.3, we will use microcontroller ports to connect hardware and software.

The interface between hardware and software physically consists of three different connections: a data bus, a `req` connection (request) from hardware to software, and an `ack` connection (acknowledge) from software to hardware. The purpose of `req` and `ack` is to synchronize hardware and software when they communicate a token. A communication queue in data flow also needs storage; this storage is implemented on the 8051 processor as a FIFO queue.

Listing 3.5 shows a GEZEL system description of the data flow design of Fig. 3.17. The hardware actor is included on lines 1–24; the rest of the listing includes an 8051 processor, and communication ports to connect the hardware actor

**Fig. 3.17** Hybrid
hardware/software
implementation of a dataflow
graph

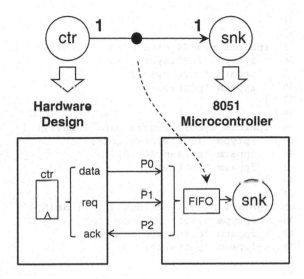

**Listing 3.5** GEZEL hardware description of data flow example of Fig. 3.17

```
1   dp send_token(out dout    : ns(8);
2                 out req     : ns(1);
3                 in  ack     : ns(1)) {
4     reg ctr    : ns(8);
5     reg rack   : ns(1);
6     reg rreq   : ns(1);
7     always {
8       rack = ack;
9       rreq = rack ? 0 : 1;
10      ctr  = (rack & rreq) ? ctr + 1 : ctr;
11      dout = ctr;
12      req  = rreq;
13    }
14    sfg transfer {
15      $display($cycle, " token ", ctr);
16    }
17    sfg idle {}
18  }
19  fsm ctl_send_token(send_token) {
20    initial s0;
21    state s1;
22    @s0 if (rreq & rack) then (transfer) -> s0;
23                         else (idle)     -> s0;
24  }
25
26  ipblock my8051 {
27    iptype "i8051system";
28    ipparm "exec=df.ihx";
29    ipparm "verbose=1";
30    ipparm "period=1";
```

```
31  }
32
33  ipblock my8051_data(in data : ns(8)) {
34    iptype "i8051systemsink";
35    ipparm "core=my8051";
36    ipparm "port=P0";
37  }
38
39  ipblock my8051_req(in data : ns(8)) {
40    iptype "i8051systemsink";
41    ipparm "core=my8051";
42    ipparm "port=P1";
43  }
44
45  ipblock my8051_ack(out data : ns(8)) {
46    iptype "i8051systemsource";
47    ipparm "core=my8051";
48    ipparm "port=P2";
49  }
50
51  dp sys {
52    sig data, req, ack : ns(8);
53    use my8051;
54    use my8051_data(data);
55    use my8051_req (req);
56    use my8051_ack (ack);
57    use send_token (data, req, ack);
58  }
59
60  system S {
61    sys;
62  }
```

**Listing 3.6**  Software description of data flow example of Fig. 3.17

```
 1  #include <8051.h>
 2  #include "fifo.c"
 3
 4  void collect(fifo_t *F) {
 5    if (P1) {              // if hardware has data
 6      put_fifo(F, P0); //   then accept it
 7      P2 = 1;          //      indicate data was taken
 8      while (P1 == 1); //     wait until the hardware
                                    acknowledges
 9      P2 = 0;          //      and reset
10    }
11  }
12
13  unsigned acc;
14  void snk(fifo_t *F) {
15    if (fifo_size(F) >= 1)
16      acc += get_fifo(F);
17  }
18
```

```
19   void main() {
20     fifo_t F1;
21
22     init_fifo(&F1);
23     put_fifo(&F1, 0); // initial token
24     acc = 0;
25
26     while (1) {
27       collect(&F1);
28       snk(&F1);
29     }
30   }
```

to the 8051. The hardware actor uses a so-called handshake protocol to ensure synchronization with the software. Synchronization protocols will be discussed in detail in Chap. 11.

Listing 3.6 shows the 8051 software to interface the hardware actor. The schedule in the main function invokes two functions, collect and snk. The first, collect, implements the synchronization with hardware. Every token received from the hardware is entered into a FIFO queue. The other function, snk, is a standard data flow actor.

Simulation of this design proceeds in the same way as the example in Sect. 1.1.3. First, compile the software, and next, run the cosimulation. The first 50,000 cycles of the simulation look as follows.

```
> sdcc dfsys.c
> /opt/gezel/bin/gplatform -c 50000 dfsys.fdl
i8051system: loading executable [df.ihx]
0x00     0x01     0x01     0xFF
17498 token 0/1
0x01     0x00     0x00     0xFF
0x01     0x01     0x01     0xFF
26150 token 1/2
0x02     0x00     0x00     0xFF
0x02     0x01     0x01     0xFF
34802 token 2/3
0x03     0x00     0x00     0xFF
0x03     0x01     0x01     0xFF
43454 token 3/4
0x04     0x00     0x00     0xFF
Total Cycles: 50000
```

It takes several thousand cycles to transfer a single token; this overhead can be completely attributed to software on the 8051. At 12 cycles per instruction, each token requires 712 instruction of the 8051. However, after applying software optimization techniques such as inlining, the performance of the system will improve dramatically; See Problem 3.8.

## 3.4  Summary

We discussed three possible target implementation for data flow systems: software, hardware, and combined hardware/software. The design space is broad and offers many alternatives.

For a sequential software implementation, we can use either threads or else static scheduling of C functions to capture the concurrent behavior of a data flow system. Optimization techniques, such as inlining of static data flow, yield compact and efficient implementations of software data flow systems.

For hardware implementation, a simple one-to-one conversion technique translates single-rate SDF graphs into hardware. The data flow transformation techniques we discussed earlier, including pipelining and retiming, can be applied to optimize the performance of graphs with a long critical path.

Finally, we discussed how both of these techniques in hardware and software can be combined into a hybrid system. In this case, we need to build synchronization between the hardware part and the software part of the data flow system.

Data flow modeling and – implementation is an important system design technique, and its ideas are present in many aspects of hardware/software design.

## 3.5  Further Reading

As much as you should learn the concepts, elegance, and advantages of data flow modeling, you should also understand their limitations: when to use them, and when not to use them.

An excellent place to start studying on dataflow implementations would be to look at the Ptolemy environment (Eker et al. 2003), and look at how it gets around the limitations of data flow with additional models of computation.

Dataflow excels in the description of streaming processing, and therefore it remains very popular for signal processing applications. In particular, the recent trend towards multi-processors has spurred a new interest in streaming applications. System specification is done in a dataflow-variant or language, and an automatic design environment maps this to a multiprocessor target. Some of the recent work in this area includes StreamIt (which maps to an IBM Cell Processor) (Thies 2008) and Brook (which maps to a Graphics Processor) (Stanford Graphics Lab 2003). Mature versions of data flow modeling system, with extensions for control modeling and event modeling, are also available in commercial environments such as Matlab's Simulink and National Instruments' Labview.

The Quickthreads threading system discussed in this chapter was developed by D. Keppel (Keppel 1994). I've used it for its simplicity and elegance, not for its completeness! If you're interested in threads, you'll have to consult a book on parallel and/or concurrent programming, for example (Butenhof 1997).

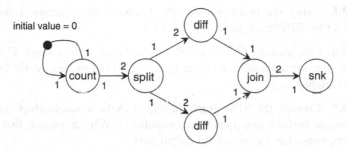

initial value = 0

**Fig. 3.18** SDF graph for Problem 3.4

## 3.6 Problems

**Problem 3.1.** Add the following improvement to the FIFO queue design of Listing 3.1. First, make the initial FIFO size very small, for example two elements. Next, implement dynamic FIFO sizing: when the FIFO overflows upon put_fifo, double the allocated memory for the FIFO. Use dynamic memory allocation, such as malloc in C. Implement changes to put_fifo and get_fifo as needed to keep track of a variable data flow queue depth.

**Problem 3.2.** Using an accumulator actor, as derived in Problem 2.4, implement the following C program as a SDF graph. The graph has a single input token, in, and produces a single output token out, corresponding to the return value of the function.

```
int graph(int in) {
  int i, j, k = 0;
  for (i=0; i<10; i++)
    for (j=0; j<10; j++)
      k = k + j * (i + in);
  return k;
}
```

**Problem 3.3.** Assume a C function with only expressions on scalar variables (no pointers) and for-loops. Show that such a C function can be translated to a SDF graph if and only if the loop-bound expressions are manifest, that is, they only depend on compile-time constant values and loop counters.

**Problem 3.4.** Using a PASS analysis, find a stable firing rate for each actor of the SDF graph in Fig. 3.18. The six actors in this graph have the following functionality. count increments a token at the input, and produces a copy of the incremented value on each of its outputs. split reads two tokens at the input and distributes these tokens over each output. diff reads two tokens at the input and produces the difference of these tokens (first minus last) at the output. join reads a token on each input and produces a merged stream. join is the complement of split. snk prints the input token.

**Problem 3.5.** Using the quickthreads API defined earlier, create a data flow simulation for the SDF graph shown in Fig. 3.18.

**Problem 3.6.** Optimize the SDF graph shown in Fig. 3.18 to a single C function by implementing the schedule at compile-time, and by optimizing the FIFO queues into single variables.

**Problem 3.7.** Convert the SDF graph in Fig. 3.18 to a single-clock hardware implementation. Perform first a multi-rate expansion. You can assume that the snk actor is implemented as a system-level output port.

**Problem 3.8.** Optimize the C program in Listing 3.6 with inlining. Implement and run the cosimulation before and after the optimization, and evaluate the gain in cycle count.

# Chapter 4
# Analysis of Control Flow and Data Flow

## 4.1 Data and Control Edges of a C Program

In the previous chapter, we discussed the data flow model of computation. Fundamental to this model is the decomposition of a system into individual nodes (*actors*), which communicate through unidirectional, point-to-point channels (*queues*). The resulting system model is represented as a graph. The data flow model of computation describes concurrent computations. We discussed techniques to create a hardware or a software implementation starting from the same data flow model.

Our objective in this chapter is to think of a C program in a similar target-independent fashion. For a software designer, a C program is software, and sequential. For a hardware-software codesigner however, a C program may be hardware or software, depending on the requirements and needs of the application. Obviously one cannot make a direct conversion of C into hardware – a major roadblock is that hardware is parallel by nature, while C is sequential.

However, there may be a different way of looking at a C program. We can think of a C program as a high-level description of the *behavior* of an implementation, without specifically pinning down the exact implementation details. Thus, we seek to understand the structure of the C program in terms of the individual operations it contains, and in terms of the relations between those operations.

We define two types of relationships between the operations of a C program: data edges and control edges. At first glance, data edges and control edges are quite similar.

> A **data edge** is a relation between two operations, such that data which is produced by one operation is consumed by the other.
>
> A **control edge** is a relation between two operations, such that one operation has to execute after the other.

P.R. Schaumont, *A Practical Introduction to Hardware/Software Codesign*,
DOI 10.1007/978-1-4614-3737-6_4, © Springer Science+Business Media New York 2013

This looks similar, but it's not identical. Let's try to identify data edges and control edges in the following C function, which finds the maximum of two variables.

```c
int max(int a, b) {
  int r;
  if (a > b)
    r = a;
  else
    r = b;
  return r;
}
```

This function contains two assignment statements and an if-then-else branch. For the purpose of this analysis, we will equate statements in C with 'operations'. In addition, we define the entry points and exit points of the function as two additional operations. Therefore, the max function contains five operations:

```c
int max(int a, b) {      // operation 1 - enter the function
  int r;
  if (a > b)             // operation 2 - if-then-else
    r = a;               // operation 3
  else
    r = b;               // operation 4
  return r;              // operation 5 - return max
}
```

To find the control edges in this function, we need to find the sequence of operations in this function. There may be more than one possible sequence, when the program contains if-then-else statements and loops. The control edges should capture all possible paths. In the example, operation 2 will always execute after operation 1. Therefore, there is a control edge from operations 1 to 2. An if-then-else statement introduces two control edges, one for each of the possible outcomes of the if-then-else test. If a > b is true, then operation 3 will follow operation 2, otherwise operation 4 will follow operation 2. There is a control edge from operation 2 to each of operations 3 and 4. Finally, operation 5 will follow either operation 3 or 4. There is a control edge from each of operations 3 and 4 to operation 5. Summing up, finding control edges corresponds to finding the possible execution paths in the C program, and linking up the operations in these execution paths with edges.

To find the data edges in this function, we examine the data production/consumption patterns of each operation.

```c
int max(int a, b) {      // operation 1 - produce a, b
  int r;
  if (a > b)             // operation 2 - consume a, b
    r = a;               // operation 3 - consume a and (a>b),
                         //               produce r
  else
    r = b;               // operation 4 - consume b and (a>b),
```

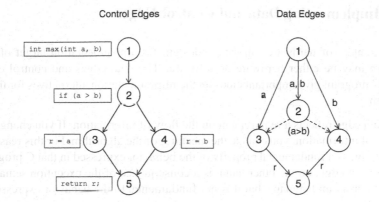

**Fig. 4.1** Control edges and data edges of a simple C program

```
                    //                    produce r
   return r;        // operation 5 - consume r
}
```

The data edges are defined between operations of corresponding production/consumption. For example, operation 1 defines the value of a and b. Several operations will make use of those values. The value of a is used by operations 2 and 3. Therefore there is a data edge from operations 1 to 2, as well as a data edge from operations 1 to 3. The same goes for the value of b, which is produced in operation 1 and consumed in operations 2 and 4. There is a data edge for b from operations 1 to 2, as well as from operations 1 to 4.

Control statements in C may produce data edges as well. In this case, the if-then-else statement evaluates a flag, and the *value* of that flag is needed before subsequent operations can execute. For example, operation 3 will only execute when the conditional expression (a>b) is true. We can think of a boolean flag carrying the value of (a>b) from operations 2 to 3. Similarly, operation 4 will only execute when the conditional expression (a>b) is false. There is a boolean flag carrying the value of (a>b) from operations 2 to 4.

The data edges and control edges of the operations from the max function can be arranged in a graph, where each operation represents a node. The result is shown in Fig. 4.1, and it represents the control flow graph (CFG) and the data flow graph (DFG) for the program. Control edges express a general relation between two nodes (operations), while data edges express a relation between two nodes for a specific variable. Therefore, data edges are labeled with that variable.

We will now explore the properties of control edges and data edges more carefully, and evaluate how the CFG and DFG can be created systematically for a more complex C program.

## 4.2   Implementing Data and Control Edges

In the context of hardware-software codesign, the implementation target of a C program may be either hardware or software. The data edges and control edges of the C program give important clues on the implementation alternatives for that C program.

- A data edge reflects a requirement on the flow of information. If you change the flow of information, you change the meaning of the algorithm. For this reason, a data edge is a fundamental property of the behavior expressed in that C program.
- A control edge, on the other hand, is a consequence of the execution semantics of the program language, but it is not fundamental to the behavior expressed in that C program.

In hardware-software codesign, we are looking to design the architecture that fits best to a given algorithm. Even though we may start from a C program, the target of this program may not be a processor. It may be a processor with a coprocessor, or a full hardware implementation. One question then inevitably arises: what are the important parts of a C program that will be present in any implementation of that program? The answer to this question is given by the control edges and data edges of the program, and it is summarized as follows.

> A **data edge** must always be implemented regardless of the underlying architecture.
> A **control edge** may be removed if the underlying architecture can handle the resulting concurrency.

In other words, control edges can be removed, by building sufficient parallelism into the underlying architecture. For example, modern microprocessors are able to run multiple instructions in parallel, even when they would belong to two different sequential C statements. These microprocessors are able to analyze and modify the control edges of the flow of instructions at runtime. They do this such that the data edges within that instruction flow are never broken.

Here is another example. The following function adds up three numbers using multiple operations.

```c
int sum(int a, b, c) {   // operation 1
  int v1;
  v1 = a + b;            // operation 2
  v2 = v1 + c;           // operation 3
  return v2;             // operation 4
}
```

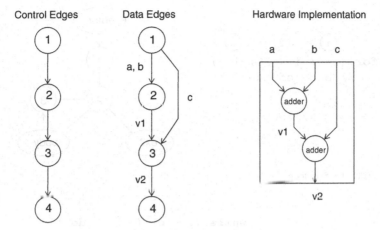

**Fig. 4.2** Hardware implementation of a chained addition

It is straightforward to draw a fully parallel hardware implementation of this function. This implementation is shown, together with the data flow graph and control flow graph of the function, in Fig. 4.2. The similarity between the set of *data* edges and the interconnection pattern of the hardware is obvious. The control edges, however, carry no meaning for the hardware implementation, since hardware is parallel. The structure shown on the right of Fig. 4.2 will complete the addition in a single clock cycle.

The next section will introduce a systematic method to derive the control flow graph and the data flow graph of C programs.

## 4.3   Construction of the Control Flow Graph

A C program can be systematically converted into an intermediate representation called a Control Flow Graph (CFG). A CFG is a graph that contains all the control edges of a program. Each node in the graph represents a single operation (or C statement). Each edge of the graph indicates a control edge, i.e. an execution order for the two operations connected by that edge.

Since C executes sequentially, this conversion is straightforward. However, some cases require further attention. Control statements (such as loops) may require multiple operations. In addition, when decision-making is involved, multiple control edges may originate from a single operation.

Consider the for loop in C, as illustrated next.

```
for (i=0; i < 20; i++) {
    // body of the loop
}
```

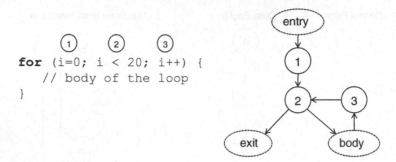

**Fig. 4.3** CFG of a `for` loop

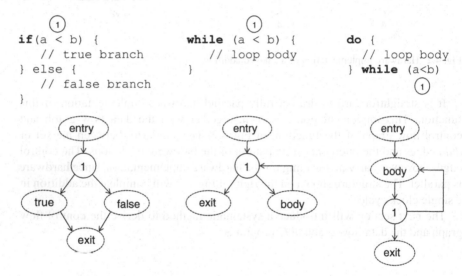

**Fig. 4.4** CFG of `if-then-else`, `while-do`, `do-while`

This statement includes four distinct parts: the loop initialization, the loop condition, the loop-counter increment operation, and the body of the loop. The `for` loop thus contributes three operations to the CFG, as shown in Fig. 4.3. The dashed nodes in this figure (`entry`, `exit`, `body`) represent other parts of the C program, each of which is a complete single-entry, single-exit CFG.

The `do-while` loop and the `while-do` loop are similar iterative structures. Figure 4.4 illustrates a template for each of them, as well as for the `if-then-else` statement.

As an example, let's create the CFG of the following C function. This function calculates the Greatest Common Divisor (GCD) using Euclid's algorithm.

```
int gcd(int a, int b) {
  while (a != b) {
    if (a > b)
      a = a - b;
```

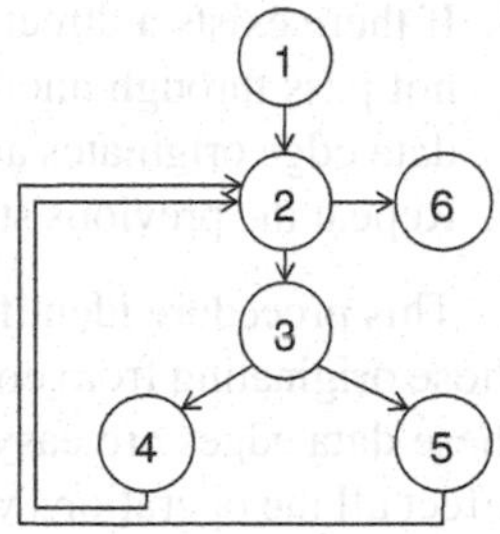

**Fig. 4.5** CFG of the CGD program

```
1:  int gcd(int a, int b) {
2:      while (a != b) {
3:          if (a > b)
4:              a = a - b;
            else
5:              b = b - a;
        }
6:      return a;
    }
```

```
    else
        b = b - a;
    }
    return a;
}
```

To construct the CFG of this program, we convert each statement to one or more operations in the CFG, and then connect the operations using control edges. The result of this conversion is shown in Fig. 4.5.

In a CFG it is useful to define a *control path*, a path between two nodes in the CFG. For example, each non-terminating iteration of the while loop of the C program will follow either the path 2-3-4-2 or else 2-3-5-2. Control paths will be important in the construction of the data flow graph (DFG), which is discussed next.

## 4.4 Construction of the Data Flow Graph

A C program can be systematically converted into a data structure called a Data Flow Graph (DFG). A DFG is a graph that reflects all the data edges of a program. Each node in the graph represents a single operation (or C statement). Each edge of the graph indicates a data edge, i.e. a production/consumption relationship between two operations in the program.

Obviously, the CFG and the DFG will contain the same set of nodes. Only the edges will be different. Since a variable in C can be written-to/read-from an arbitrary number of times, it can be difficult to find matching read-write pairs in the program. The easiest way to construct a DFG is to first construct the CFG, and then use the CFG in combination with the C program to derive the DFG. The trick is to trace control paths, and at the same time identify corresponding read- and write operations of variables.

Let's assume that we're analyzing programs without array expressions and pointers; we will extend our conclusions later to those other cases as well. The procedure to recover the data edges related to assignment statements is as follows.

1. In the CFG, select a node where a variable is used as an operand in an expression. Mark that node as a read-node.
2. Find the CFG nodes that assign that variable. Mark those nodes as write-nodes.

3. If there exists a direct control path from a write-node into a read-node that does not pass through another write-node, then you have identified a data edge. The data edge originates at the write-node and ends at the read-node.
4. Repeat the previous steps for all variable-read operations in every node.

This procedure identifies all data edges related to assignment statements, but not those originating from conditional expressions in control flow statements. However, these data edges are easy to find: they originate from the condition evaluation and affect all the operations whose execution depends on that condition.

Let's derive the data flow graph of the GCD program given in Fig. 4.5. According to the procedure, we pick a node where a variable is read. For example, node 5 in the CFG reads variables a and b.

```
    b = b - a;
```

First, concentrate on the b operand. We need to find all nodes that write into variable b. In the CFG, we can trace precedessor nodes for this node until we hit one that writes into variable b. The predecessors of node 5 include: node 3, node 2, node 1, node 4, and node 5. Both node 1 and 5 write into b. In addition, there is a path from node 1 to 5 (e.g. 1-2-3-5), and there is also a path from node 5 to 5 (e.g. 5-2-3-5). In each of these paths, no other nodes write into b apart from the final node 5. Thus, there is a data edge for variable b from node 1 to 5 and from node 5 to 5. Starting from the same read-node 5, we can also find all predecessors that define the value of operand a. In this case, we find that nodes 1 and 4 write into variable a, and that there is a direct path from node 1 to 5, as well as from node 4 to 5. Therefore, there is a data edge for variable a from node 1 to 5, and from node 4 to 5.

To complete the set of data edges into node 5, we also need to identify all conditional expressions that affect the outcome of node 5. Considering the control statements in this function, we see that node 5 depends on the condition evaluated in node 3 (a > b) as well as the condition evaluated in node 2 (a != b). There is a data edge from each of nodes 2 and 3 to node 5, carrying the outcome of this condition. The collection of all data edges into node 5 can now be annotated into the DFG, resulting in the partial DFG of Fig. 4.6.

We can repeat this procedure for each other node of the graph in order to construct the complete DFG. The result of this analysis is shown in Fig. 4.7. This graph does not contain the data edges originating from conditional expressions.

How do we draw a DFG of a program with pointers and arrays? There are possible several approaches, and they depend on the objectives of the analysis and the level of detail desired.

First, observe that an indexed variable is not really different from a scalar variable as long as we can exactly determine the value of the index during the data-flow analysis. Similarly, data edges resulting from pointers are easy to find if we can exactly determine the value of the pointer. However, in practice, this may be difficult or impossible. An indexed variable may have a complex index expression that depends on multiple loop counters, or the index expression may contain a variable which is unknown at compile time.

**Fig. 4.6** Incoming data edges for node 5 in the CGD program

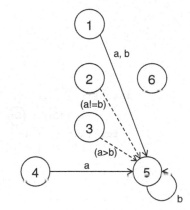

**Fig. 4.7** Data edges for all nodes in the GCD program, apart from edges carrying condition variables

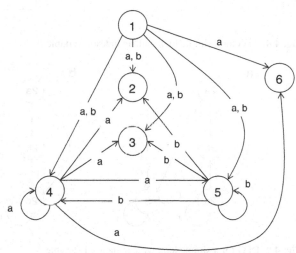

We may be able to relax the analysis requirements, and simplify the data-flow analysis. In many applications, the upper bound and lower bound of an index expression is known. In that case, we may consider any *write* operation into the range of indices as a single write, and any *read* operation into the range of indices as a single read. For cases when an entire range of indices would map into a single memory (a single register file, or a single-port RAM memory), this type of data-flow analysis may be adequate.

We illustrate this approach using the following example. The CFG of the following loop is shown in Fig. 4.8.

```
int L[3] = {10, 20, 30};
for (int i=1; i<3; i++)
    L[i] = L[i] + L[i-1];
```

To create a DFG for this program, proceed as before. For each node that reads from a variable, find the nodes that write into that variable over a direct path in the

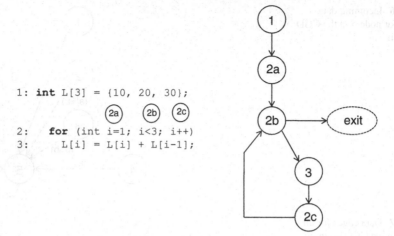

```
1: int L[3] = {10, 20, 30};

            2a    2b    2c

2:    for (int i=1; i<3; i++)
3:      L[i] = L[i] + L[i-1];
```

**Fig. 4.8** CFG for a simple loop with an indexed variable

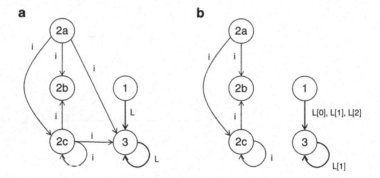

**Fig. 4.9** DFG for a simple loop with an indexed variable

CFG. As discussed above, we can handle the analysis of the indexed variable L in two different ways. In the first approach, we look upon L as a single monolithic variable, such that a read from any location from L is treated as part of the same data edge. In the second approach, we distinguish individual locations of L, such that each location of L may contribute to a different data edge. The first approach is illustrated in Fig. 4.9a, while the second approach is illustrated in Fig. 4.9b.

When the individual locations of L cannot be distinguished with a data edge, additional information is needed to extract the entry of interest. For this reason, node 3 in Fig. 4.9a has an additional data edge to provide the loop counter i. Thus, in Fig. 4.9a, reading entry L[i] means: read all the entries of L and then select one using i. In Fig. 4.9b, reading entry L[i] means three different read operations, one for each value of i.

Index analysis on arbitrary C programs quickly becomes very hard to solve. Yet, hardware-software codesigners often only have a C program to start their design with. Insight into the data-flow of a complex C program is essential for a successful hardware-software codesign.

This concludes an introduction to control-flow and data-flow analysis of C programs. The next section shows an application for the techniques we've covered so far. By deriving the CFG and the DFG, we can translate simple C programs systematically into hardware. This technique is by no means a universal mechanism to translate C; its' purpose is to clarify the meaning of control edges and data edges.

## 4.5 Application: Translating C to Hardware

A nice application of analysis of data-flow and control-flow in a C program is the systematic translation of C into hardware. This problem is very complex, when our objective is to solve the translation of general C. Therefore, we will focus on a simplified version of this problem. We show how to translate one particular flavor of C into one particular flavor of hardware circuit. We are making the following two assumptions.

- We will translate only scalar C code (no pointers and no arrays).
- We implement each C statement in a single clock cycle.

### 4.5.1 Designing the Datapath

Starting from the C program, we first create the CFG and the DFG. Next, we use the data edges and control edges to implement the hardware. The data edges will help us to define the datapath components, and their connectivity. The control edges will help us to define the control signals used by the datapath. With the CFG and DFG available, the following rules will define the implementation of the hardware datapath.

1. Each variable in the C program is translated into a register with a multiplexer in front of it. The multiplexer is needed when multiple sources may update the register. By default, the register will update itself. The selection signals of the multiplexer will be driven by the controller.
2. For each C expression embedded in a node of the CFG, create an equivalent combinational circuit to implement that expression. For example, if a node in the CFG corresponds to the C statement a = b - a, then the C expression embedded in that statement is b - a. The combinational circuit required to implement this expression is a subtractor. Conditional expressions generate datapath elements, too. The outputs of these expressions become the flags used by the hardware controller of this datapath.
3. The datapath circuit and the register variables are connected based on the data edges of the DFG. Each assignment operation connects a combinational

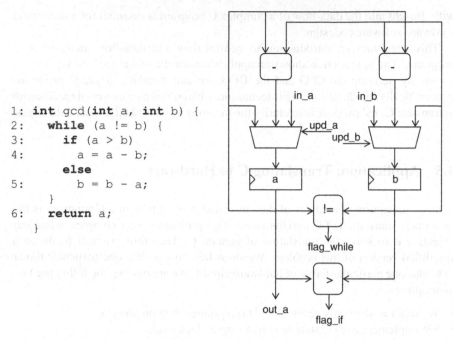

```
1:  int gcd(int a, int b) {
2:      while (a != b) {
3:          if (a > b)
4:              a = a - b;
            else
5:              b = b - a;
        }
6:      return a;
    }
```

**Fig. 4.10**  Hardware implementation of GCD datapath

circuit with a register. Each data edge connects a register with the input of a combinational circuit. Finally, we also connect the system-inputs and system-outputs to inputs of datapath circuits and register outputs respectively.

The GCD program can now be converted into a hardware implementation as follows. We need two registers, for each of the variables a and b. The conditional expressions for the if and while statement need an equality-comparator and a bigger-then comparator. The subtractions b-a and a-b are implemented using a subtractor. The connectivity of the components is defined by the data edges of the DFG.

The resulting datapath has two data inputs (in_a and in_b), and one data output (out_a). The circuit requires two control variables (upd_a and upd_b) to operate, and it produces two flags (flag_while and flag_if). The control variables and the flags are the ouputs and inputs, respectively, of the controller of this datapath, see Fig. 4.10.

## 4.5.2  Designing the Controller

How do we create the controller for this datapath such that it implements the GCD algorithm? This control information is present in the C program, and is captured

**Fig. 4.11** Control
specification for the GCD
datapath

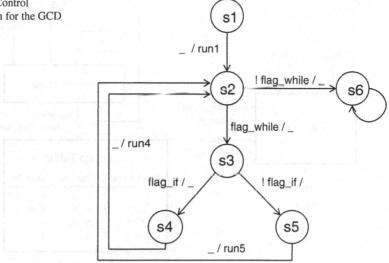

in the CFG. In fact, we can translate the CFG almost directly into hardware, by considering it to be a finite state machine (FSM) specification.

A finite-state machine (FSM) specification for the GCD algorithm is shown in Fig. 4.11. The correspondence with the CFG is obvious. Each of the transitions in this FSM takes one clock cycle to complete. The activities of the FSM are expressed as condition/command tuples. For example, _/run1 means that during this clock cycle, the condition flags are don't-care, while the command for the datapath is the symbol run1. Similarly, flag_while/_ means that this transition is conditional on flag_while being true, and that the command for the dapath is a hold operation. A *hold operation* is one which does not change the state of the datapath, including registers. The command set for this FSM includes (_, run1, run4, run5). Each of these symbols represents the execution of a particular node of the CFG. The datapath control signals can be created by additional decoding of these command signals. In this case of the GCD, the datapath control signals consist of the selection signals of the datapath multiplexers.

A possible implementation of the GCD controller is shown in Fig. 4.12. Each clock cycle, the controller generates a new command based on the current state and the value of flag_while and flag_if. The commands run1, run4 and run5 are decoded into upd_a and upd_b. The table in Fig. 4.12 indicates how each command maps into these control signals. The resulting combination of datapath and finite state machine, as illustrated in Fig. 4.12 is called a Finite State Machine with Datapath (FSMD). This concept is central to custom hardware design, and we will discuss design and modeling of FSMD in detail in Chap. 5.

The operation of this hardware circuit is illustrated with an example in Table 4.1. Each row of the table corresponds to one clock cycle. It takes eight clock cycles to evaluate the greatest common divisor of 6 and 4.

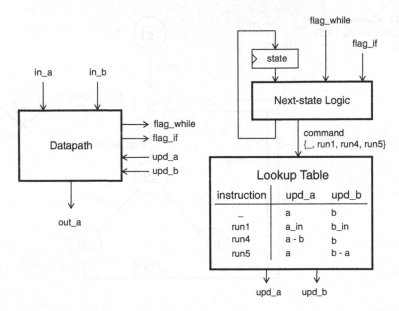

**Fig. 4.12** Controller implementation for the GCD datapath

**Table 4.1** Operation of the hardware to evaluate GCD(4,6)

| Cycle | a | b | State | flag_if | flag_while | Next state | upd_a | upd_b |
|---|---|---|---|---|---|---|---|---|
| 1 | _ | _ | s1 | _ | _ | s2 | in_a | in_b |
| 2 | 6 | 4 | s2 | 1 | 1 | s3 | a | b |
| 3 | 6 | 4 | s3 | 1 | 1 | s4 | a | b |
| 4 | 6 | 4 | s4 | 1 | 1 | s2 | a-b | b |
| 5 | 2 | 4 | s2 | 0 | 1 | s3 | a | b |
| 6 | 2 | 4 | s3 | 0 | 1 | s5 | a | b |
| 7 | 2 | 4 | s5 | 0 | 1 | s2 | a | b-a |
| 8 | 2 | 2 | s2 | 0 | 0 | s6 | a | b |
| 9 | 2 | 2 | s6 | _ | _ | s6 | a | b |

In conclusion, the DFG and CFG of a C program can be used to create and implement a hardware circuit. Of course, there are many sub-optimal elements left. First, we did not address the use of arrays and pointers. Second, the resulting implementation in hardware is not very impressive: the resulting parallelism is limited to a single C statement per clock cycle, and operations cannot be shared over operator implementations. For instance, two substractors are implemented in hardware in Fig. 4.10, but only one is used at any particular clock cycle.

## 4.6  Single-Assignment Programs

Converting C programs into hardware at one cycle per C-statement is not very impressive, in particular because most microprocessors can already do that. Can we do any better than this? That is, can we create a translation strategy that will allow the execution of multiple C statements per clock cycle? To answer this question, we need to understand the limitations of our current translation strategy, as described in Sect. 4.5.

In that approach, each C statement takes a single clock cycle to execute because each variable is mapped into a register. A data value takes a full clock cycle to propagate from the input of a register to the output. Therefore, each variable assignment takes a full clock cycle to take effect; it takes a full clock cycle before the value of a variable assignment is available at the corresponding register output. Thus, the mapping of each variable into a register, which by itself seems a sensible decision, also introduces a performance bottleneck. If we want to run at a faster pace then one statement per clock cycle, we will need to revise this variable-to-register mapping strategy.

The above observation triggers another question, namely: why did we map each variable into a register in the first place? The reason is, of course, to make the evaluation of expressions, and the design of control, easy. By mapping each variable into a register, it is as if we're concentrating all data-flow edges related to a given variable to go through a single, global storage location, so we always know where to find the value of a given variable. This strategy, however, hurts performance.

We can do better than that with the following technique. A C program can be translated into a *single-assignment program*. The key property of a single-assignment program is exactly what its' name refers says: each variable in that program is assigned only a single time within a single lexical instance of the program. Let's illustrate the conversion with a simple example. Assume that we have a C snippet that looks as follows.

```
a = a + 1;
a = a * 3;
a = a - 2;
```

This section of code contains three assignments on a. Using our previous strategy, we would need three clock cycles to execute this fragment. Instead, we can rewrite this program so that each variable is assigned only once. This requires the introduction of additional variables.

```
a2 = a1 + 1;
a3 = a2 * 3;
a4 = a3 - 2;
```

The difference with the previous program is that each assignment is matched by a different read operation. In other words, the single-assignment form of the program visualizes the data edges of the program in the source code: assigning a given variable indicates the start of a data edge, while reading the same variable indicates the end of the data edge.

After a program is in single-assignment form, the register-assignment strategy can be improved. For instance, in the previous example, the cycle count may be reduced by mapping a2 and a3 to a wire, while keeping a4 in a register. This would group the three C statements in a single clock cycle.

In converting C programs to single-assignment form, all assignments to a variable must be taken into account. In particular, when variables are assigned under different control conditions, or in different levels of a loop nesting structure, the single-assignment form may become ambiguous. Consider the following example: a loop which makes the sum of the numbers 1–5.

```
a = 0;
for (i = 1; i < 6; i++)
  a = a + i;
```

In the single-assignment form, the assignments to a can be made unique, but it remains unclear what version of a should be read inside of the loop.

```
a1 = 0;
for (i = 1; i < 6; i++)
  a2 = a + i;   // which version of a to read?
```

The answer is that both a1 and a2 are valid solutions for this program: it depends on the iteration within the loop. When we first enter the loop, we would write:

```
a2 = a1 + 1;
```

After the first loop iteration, we would write instead:

```
a2 = a2 + i;   // when i > 1
```

To resolve this ambiguity, single-assignment programs use a merge function, an operation that can merge multiple data edges into one. We can introduce a new variable a3 to hold the result of the merge function, and now formulate the program into single-assignment form as follows.

```
a1 = 0;
for (i = 1; i < 6; i++) {
  a3 = merge(a2, a1);
  a2 = a3 + i;
```

The hardware equivalent of the merge function would be a multiplexer, under control of an appropriate selection signal. In this case, (i==0) would be an appropriate selection signal. The above translation rules can be used to more complicated programs as well. For example, the GCD program can be converted as follows.

```
int gcd(int a, int b) {
  while (a != b) {
    if (a > b)
      a = a - b;
    else
      b = b - a;
  }
  return a;
}
```

**Fig. 4.13** GCD datapath
from single-assignment code

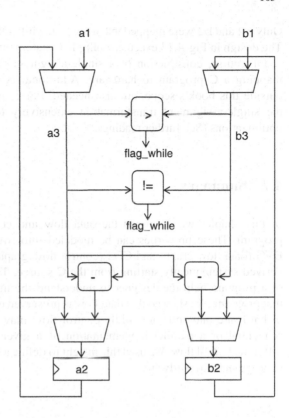

The equivalent single-assignment form of the GCD is shown below. The conditional expression in the `while` statement uses variables from either the function input or else the body of the loop. For this reason, the conditional expression uses the `merge` function as an operand.

```
int gcd(int a1, int b1) {
  while (merge(a1, a2) != merge(b1, b2)) {
    a3 = merge(a1, a2);
    b3 = merge(b1, b2);
    if (a3 > b3)
      a2 = a3 - b3;
    else
      b2 = b3 - a3;
  }
  return a2;
}
```

A single assignment program such as this one is valuable because it visualizes the data edges in the source code of the program, making the connection with hardware more obvious. Furthermore, the `merge` functions can be mapped into multiplexers in hardware. A possible datapath corresponding to this single-assignment version of GCD is shown in Fig. 4.13. This datapath looks very much like the previous design (Fig. 4.10), but this design was derived from a C program with four assignments.

Only a2 and b2 were mapped into a register, while other variables are simply wires. The design in Fig. 4.13 executes multiple C statements per clock cycle.

Of course, construction of a single-assignment program is only one step in mapping a C program to hardware. Additional compiler techniques, which are outside this book's scope, are also needed. For example, software compilers use the single-assignment transformation extensively to implement advanced code optimizations (See further reading).

## 4.7  Summary

In this chapter we discussed the data flow and control flow properties of a C program. These properties can be modeled into two graph structures, called the DFG (data flow graph) and CFG (control flow graph). The DFG and CFG can be derived systematically starting from the C source. The data flow and control flow of a program help the designer to understand the implementation alternatives for that program. We showed that data flow is preserved over different implementations in hardware and software, while control flow may change drastically. Indeed, a sequential or a parallel implementation of a given algorithm may have a very different control flow. We used this insight to define a simple mechanism to translate C programs into hardware.

## 4.8  Further Reading

The material discussed in this chapter can typically be found, in expanded from, in a textbook on compiler construction such as Muchnick (1997) or Appel (1997). In particular, these books provide details on the analysis of the control flow graph and data flow graph.

High-level synthesis is a research area that investigates the automated mapping of programs written in C and other high-level languages into lower-level architectures. In contrast to compilers, which target a processor with a fixed architecture, high-level synthesis does support some freedom in the target architecture. High-level synthesis has advantages and limitations; proponents and opponents. Refer to Gupta et al. (2004) to see what can be done; read Edwards (2006) as a reminder of the pitfalls.

During our discussion on the mapping of C programs into hardware, we did explicitly rule out pointers and arrays. In high-level synthesis, the design problems related to implementation of memory elements are collected under the term *memory management*. This includes for example the systematic mapping of array variables into memory elements, and the efficient conversion of indexed expressions into memory addresses. Refer to Panda et al. (2001) for an introduction to memory management issues.

The original work on Static Single Assignment (SSA) was by Cytron et al. (1991). A discussion on how the SSA form can assist in the translation of C software into hardware may be found in Kastner et al. (2003).

## 4.9  Problems

**Problem 4.1.** Do the following for the C program in Listing 4.1.

**Listing 4.1**  Program for Problem 4.1

```
int addall(int a, int b, int c, int d) {
  a = a + b;
  a = a + c;
  a = a + d;
  return a;
}
```

(a) Derive and draw the CFG and the DFG.
(b) The *length* of a path in a graph is defined as the number of edges in that path. Find the longest path in the DFG.
(c) Rewrite the program in Listing 4.1 so that the maximal path length in the DFG decreases. Assume that you can do only a single arithmetic operation per C statement. Draw the resulting DFG.

**Problem 4.2.** Draw the CFG and the DFG of the program in Listing 4.2. Include all control dependencies in the CFG. Include the data dependencies for the variables a and b in the DFG.

**Listing 4.2**  Program for Problem 4.2

```
int count(int a, int b) {
  while (a < b)
    a = a * 2;
  return a + b;
}
```

**Problem 4.3.** Design a datapath in hardware for the program shown in Listing 4.3. Allocate registers and operators. Indicate control inputs required by the data-path, and condition flags generated by the datapath.

**Listing 4.3**  Program for Problem 4.3

```
unsigned char mysqrt(unsigned int N) {
  unsigned int x,j;
  x = 0;
  for(j= 1<<7; j != 0; j>>=1) {
    x = x + j;
    if( x*x > N)
```

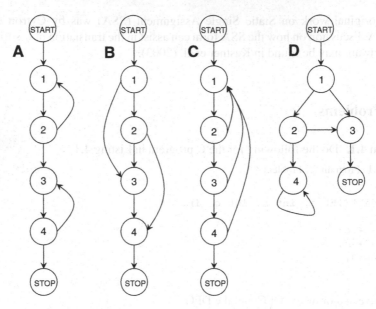

**Fig. 4.14** Four CFG for Problem 4.4

```
     x = x - j;
}
  return x;
}
```

**Problem 4.4.** A well-structured C program is a program that only contains the following control statements: if-then-else, while, do-while, and for. Consider the four CFG in Fig. 4.14. Which of the these CFG does correspond to a well-structured C program? Note that a single node in the CFG may contain more than a single statement, but it will never contain more than a single decision point.

**Problem 4.5.** Draw the DFG for the program in Listing 4.4. Assume all elements of the array a [ ] to be stored in a single resource.

**Listing 4.4** Program for Problem 4.5

```
int a[] = {1, 2, 3, 4, 5, 6, 7, 8, 9, 10};

int findmax() {
  int max, i;

  max = a[0];
  for (i=1; i<10; i++)
    if (max < a[i])
      max = a[i];

  return max;
}
```

**Fig. 4.15** A single-port
read-only memory, used to
solve Problem 4.6

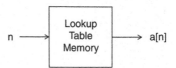

**Problem 4.6.** Design a hardware implementation (datapath and controller) for the program in Listing 4.4. Assuming that the elements of array a [ ] are all stored in a memory with a single read port. Figure 4.15 illustrates such a memory. The time to lookup an element is very short; thus, you can think of this memory as a combinational element.

**Problem 4.7.** Convert the program in Listing 4.3 to single-assignment form.

**Problem 4.8.** This problem requires access to a GNU Compiler (gcc) version 4.0 or above. Start by writing up the Listing of Problem 4.3 in a file can call the file mysqrt.c.

(a) Compile this function using the following command line.

```
gcc -c -fdump-tree-cfg mysqrt.c
```

The compiler generates an object file mysqrt.o, as well as a file with debug information. Under gcc 4.0.2, the name of that file is mysqrt.c.t13.cfg.

Open mysqrt.c.t13.cfg in a text editor. This is a textual representation of a CFG as produced by the GCC compiler. Compare this CFG to one you would draw by hand. In particular, comment on the following two observations: (1) Nodes in a CFG can be grouped together when they all belong to a single path of the CFG with a single exit point. (2) goto and if-then-else are adequate to capture all control statements in C (such as for, while, and so on).

(b) Compile this function using the following command line. O2 turns on the compiler optimizer, so that GCC will try to produce better code.

```
gcc -c -O2 -fdump-tree-ssa mysqrt.c
```

The compiler generates an object file, and a file with debug information. Under gcc 4.0.2, the name of that file is mysqrt.c.t16.ssa.

Open mysqrt.c.t16.ssa in a text editor. This is a textual representation of the SSA as produced by the GCC compiler. Find the merge functions in this file, and compare the number and location of these functions in the CFG. Did you find the same number of merge functions in Problem 4.7? Do they have the same location?

# Part II
# The Design Space of Custom Architectures

This second part of this book describes various hardware architectures, each with a varying degree of flexibility and sophistication. The objective of this part is to build insight in the nature of design problems that come with a particular architecture, and what it means to *customize* the architecture. Starting from very simple cycle-based hardware models, we gradually add control structures to increase their flexibility. We will cover FSMD (Finite State Machine with Datapath), microprogrammed architectures, general-purpose embedded cores, and finally system-on-chip architectures. We will demonstrate that each of these machines makes a trade-off between flexibility and performance. More flexible solutions, of course, will require more programming.

# Chapter 5
# Finite State Machine with Datapath

## 5.1 Cycle-Based Bit-Parallel Hardware

In this chapter, we develop a model to systematically describe custom hardware consisting of a controller and a datapath. Together with the model, we will also learn to capture hardware designs into a language, called GEZEL. This section, and the next one, describes how to create datapath modules. Further sections will explain the control model and the integration of control and datapath into an FSMD.

We will create cycle-based hardware models. In such models, the behavior of a circuit is expressed in steps of a single clock cycle. This abstraction level is very common in digital design, and it is captured with the term *synchronous* design. We will design circuits with a single, global clock signal.

### 5.1.1 Wires and Registers

We start with the variables that are used to describe synchronous digital hardware. The following example shows a 3-bit counter.

Listing 5.1 A three-bit counter module

```
1  reg a : ns(3); // a three-bit unsigned register
2  always {
3    a = a + 1;   // each clock cycle, increment a
4  }
```

This fragment represents a 3-bit register, called a. Each clock cycle, the value of a is incremented by one. Since a is a 3-bit value, the register will count from 0 to 7, and then wrap around. The initial value of a is not specified by this code. In GEZEL, all registers are initialized to zero.

Let's look closer at the expression 'a = a + 1' and describe precisely what it means. The right-hand side of this expression, a+1, reads the value of the register

P.R. Schaumont, *A Practical Introduction to Hardware/Software Codesign*,
DOI 10.1007/978-1-4614-3737-6_5, © Springer Science+Business Media New York 2013

**Fig. 5.1** Equivalent
hardware for the 3-bit counter

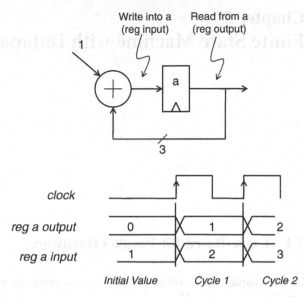

a, and adds one to that value. The left-hand side of this expression assigns that value
to a, and thus writes into the register. Figure 5.1 gives an equivalent circuit diagram
for 'a = a + 1', and it illustrates a key feature of a register: the input and the
output of a register can have a different value. The input and the output are each
connected to a different bus of wires. The timing diagram in Fig. 5.1 illustrates the
circuit operation. Before the very first clock edge, the output of a is initialized to
its initial value of 0. At the very first clock edge, the register will be incremented,
which means that the value at the input of a is copied to the output of a.

Going back to Listing 5.1, we see that the always statement describes the
activities in the model as a result of *updating* the registers. As shown in the diagram,
this happens on the upgoing clock edge. This is the only time-related aspect of the
model. The time required to execute a+1 is unspecified, and the expression will be
evaluated as soon as the output of the a register changes. Note also that Listing 5.1
does not contain an explicit clock signal: the simulation is completely defined using
the semantics of a register variable.

Besides a register variable, there is another variable type, called a *signal*. A signal
has the same meaning as a wire. A multi-bit signal corresponds to a bundle of wires.
Listing 5.2 illustrates how a signal is created and used. A signal instantly takes up
the value of the expression assigned to it. Thus, the value of b in Listing 5.2 will
instantly reflect the value of a+1. The circuit diagram corresponding to this program
looks identical to the diagram shown in Fig. 5.1. However, in this case, we have a
specific name for the value at the *input* of the register a, namely the signal b.

**Listing 5.2** Another 3-bit counter module

```
1  reg a : ns(3); // a three-bit unsigned register
2  sig b : ns(3); // a three-bit undersigned signal
```

```
3  always {
4    b = a + 1;
5    a = b;
6  }
```

A signal has no memory. When a signal value is used on the right-hand side of an expression, it will return the value assigned to the signal during that clock cycle. This has a particular effect on the program shown in Listing 5.2: the lexical order of expressions has no meaning. Only the data flow between reading/writing registers and signals is important. For example, the program in Listing 5.3 has exactly the same behavior as the program in Listing 5.2.

**Listing 5.3** Yet another 3-bit counter module

```
1  reg a : ns(3); // a three-bit unsigned register
2  sig b : ns(3); // a three-bit unsigned signal
3  always {
4    a = b;
5    b = a + 1;
6  }
```

One can think of the difference between registers and signals also as follows. When a register is used as an operand in an expression, it will return the value assigned to that register during the *previous* clock cycle. When a signal is used as an operand in an expression, it will return the value assigned to that signal during the *current* clock cycle. Thus, registers implement communication across clock cycles, while signals implement communication within a single clock cycle.

Because a signal has no memory, it cannot have an initial value. Therefore, the value of a signal remains undefined when it is not assigned during a clock cycle. It is illegal to use an undefined signal as an operand in an expression, and the GEZEL simulator will flag this as a runtime error. Another case which is unsupported is the use of signals in a circular definition, such as for example shown in Listing 5.4. It is impossible to determine a stable value for a or b during any clock cycle. This type of code will be rejected as well by the GEZEL simulator with a runtime error message. In Sect. 5.6, we define the rules for a properly formed FSMD more precisely.

**Listing 5.4** A broken 3-bit counter module

```
1  sig a : ns(3);
2  sig b : ns(3);
3  always {
4    a = b;
5    b = a + 1;  // this is not a valid GEZEL program!
6  }
```

## 5.1.2   Precision and Sign

In contrast to C, hardware registers and signals can have an arbitrary wordlength, from a single bit up to any value. It is also possible to mix multiple wordlengths in an expression. In addition, registers and signals can be unsigned or signed.

The wordlength and the sign of a register or signal are specified at the creation of that register or signal. The following example creates a 4 bit unsigned register a and a 3 bit signed signal b. The representation of b follows two's complement format: the weight of the most significant bit of b is negative.

```
reg a : ns(4);    // unsigned 4-bit value
sig b : tc(3);    // signed 3-bit value
```

In an expression, registers and signals of different wordlenghts can be combined. The rules that govern the precision of the resulting expression are as follows.

*   The evaluation of an expression does not loose precision. All operands will automatically adapt their precision to a compatible wordlength.
*   Assigning the result of an expression, or casting an expression type, will adjust the precision of the result.

Listing 5.5   Adding up 4 and 2 bit

```
1   reg a : ns(4);    // a four-bit unsigned number
2   sig b : ns(2);    // a two-bit unsigned number
3   always {
4      b = 3;          // assign 0b(011) to b
5      a = b + 9;      // add 0b(11) and 0b(1010)
6   }
```

As an example, the code shown in Listing 5.5 will store the value 12 in register a. Walking step by step through this code, the precision of each expression is evaluated as follows. First, the constant 3 needs to be assigned to b. A constant is always represented with sufficient bits to capture it as a two's complement number. In this case, you can express the constant 3 as a 3-bit two's complement number with the bit pattern 011. When assigning this 3-bit value to b, the lower 2 bits will be copied, and the bitpattern in b becomes 11. On line 5 of the code, we add the constant 9 to b. The bitpattern corresponding to the decimal constant 9 is 1001. To add the bitpattern 11 and 1001 as unsigned numbers, we extend 11 to 0011 and perform the addition to find 1100, or 12 in decimal. Finally, the bitpattern 1100 is assigned to a, which can accommodate all bits of the result.

When the length of an operand is extended, the rules of sign extension will apply. The additional bits beyond the position of the most significant bit are copies of the sign bit, in the case of two's complement numbers, or zeroes, in the case of unsigned numbers. In Listing 5.6, a is a 6-bit unsigned number, and b is a 2-bit *signed* number. After assigning the constant 3 to b, the value of b will be -1, and the bit pattern of b equals 11. The result of subtracting b and 3 is -4 in decimal,

**Table 5.1** Operations in GEZEL and equivalent hardware implementation. const is a constant number

| Operation | Operator | Implementation | Precedence |
|---|---|---|---|
| Addition | + | Adder | 4 |
| Subtraction | - | Subtractor | 4 |
| Unary minus | - | Subtractor | 7 |
| Multiplication | * | Multiplier | 5 |
| Right-shift | >> (variable) | Variable-shifter | 0 |
| Left-shift | << (variable) | Variable-shifter | 0 |
| Constant right-shift | >> const | Wiring | 4 |
| Constant left-shift | << const | Wiring | 4 |
| Lookup table | A(n) | Random logic | 10 |
| And | & | And-gate | 2 |
| Or | | | Or-gate | 2 |
| Xor | ^ | Xor-gate | 3 |
| Not | ~ | Not-gate | 8 |
| Smaller-then | < | Subtractor | 3 |
| Bigger-then | > | Subtractor | 3 |
| Smaller-equal-then | <= | Subtractor | 3 |
| Bigger-equal-then | >= | Subtractor | 3 |
| Equal-to | == | Comparator | 3 |
| Not-equal-to | != | Comparator | 3 |
| Bit selection | [const] | Wiring | 9 |
| Bit-vector selection | [const:const] | Wiring | 9 |
| Bit concatenation | # | Wiring | 4 |
| Type cast | (type) | Wiring | 6 |
| Precedence ordering | ( ) | | 11 |
| Selection | ? : | Multiplexer | 1 |

which is 100 as a bitpattern (with the msb counting as a sign bit). Finally, assigning -4 to a 6-bit number will result in the bitpattern 111100 to be stored in a. Since a is an unsigned number, the final result is the decimal number 60.

**Listing 5.6** Subtracting 2 and 4 bit

```
1  reg a : ns(6); // a six-bit unsigned number
2  sig b : tc(2); // a two-bit signed number
3  always {
4    b = 3;        // assign 0b(011) to b
5    a = b - 3;    // subtract 0b(11) and 0b(011)
6  }
```

The effect of an assignment can also be obtained immediately by means of a cast operation, expressed by writing the desired type between brackets in front of an expression. For example, (tc(1)) 1 has the value -1, while (ns(3)) 15 has the value 8.

## 5.1.3   Hardware Mapping of Expressions

For each expression involving signals and registers of a specified sign and precision, there is an equivalent hardware circuit. This circuit is easy to derive, once we know how each operator is mapped into hardware. We will discuss a list of common operations, and indicate how they map into hardware logic. Table 5.1 presents a summary.

**Arithmetic Operations.** Addition (+), subtraction (-), multiplication (*) are commonly used in datapath hardware design. The division (/) is not supported in GEZEL (as in most other synthesis-oriented hardware modeling languages). The modulo operation (%) is supported, but it is not synthesizable for arbitrary inputs (since that would require a division). Left-shift (<<) and right-shift (>>) are used to implement multiplication/division with powers of two. Constant-shifts are particularly advantageous for hardware implementation, since they translate to simple hardware wiring.

**Bitwise Operations.** All of the bitwise operations, including AND (&), OR (|), XOR (^) and NOT (~) have a direct equivalent to logic gates. Bitwise operations are defined as bit-by-bit operations. The same precision rules as for all other operators apply: when the operands of a bitwise operation are of unequal length, they will be extended until they match. For example, if w is a word and u is a bit, then the expression

```
w & (tc(1)) u
```

will AND each bit of w with the bit in u.

**Comparison Operations.** All of the comparison operations return a single unsigned bit (ns(1)). These operations use a subtractor to compare two numbers, and then use the sign/overflow flags of the result to evaluate the result of the comparison. Exact comparison (== or !=) can be done by matching the bitpattern of each operand. In contrast to arithmetic operations, the comparison operations are implemented differently for signed and unsigned numbers. Indeed, the bit pattern 111 is smaller than the pattern 001 for signed numbers, but the same pattern 111 is bigger than the pattern 001 for unsigned numbers.

**Bitvector Operations.** Single bits, or a vector of several bits, can be extracted out of a word using the bit-selection operator.

```
reg a : ns(5);
reg b : ns(1);
reg c : ns(2);
always {
  b = a[3];   // if a = 10111, then b = 0
  c = a[4:2]; // if a = 10111, then a[4:2] = 101, so c = 01
}
```

The type of a bit-selection operation is unsigned, and just wide enough to hold all the bits. The bits in a bit vector are counted from right to left, with bit 0 holding

the least significant bit. The opposite operation of bit-selection is bit-concatenation
( #), which sticks bits together in a larger word.

```
reg a : ns(5);
reg b : ns(1);
reg c : ns(2);
always {
  a = c # b; // if b = 0, c = 11, then a = 00110
}
```

**Selection.** The ternary operator a ? b : c is the equivalent notation for a
multiplexer. The result of the ternary operation will be b or c depending on the
value of a. The wordlength of the result will be long enough to accommodate the
largest word of either input of the multiplexer.

**Indexed storage.** There is no array construction in GEZEL. However, it is possible
to capture lookup tables. Lookup tables can be implemented in hardware with ROMs
or with random logic.

```
lookup T : ns(12) = {0x223, 0x112, 0x990};
reg a : ns(12);
always {
  a = T(2);   // a = 0x990
}
```

**Organization and Precedence.** Finally, brackets may be used to group expressions
and change the evaluation order. The default evaluation order is determined by the
precedence of each operator. The precedence is shown as a number in Table 5.1,
where a higher number corresponds to a higher precedence, meaning that operator
will be evaluated before others.

Each expression created using registers, signals and the operations of Table 5.1,
corresponds to a hardware datapath. A few examples are shown next. The first one,
in Listing 5.7, shows Euclid's Greatest Common Divisor algorithm. Two registers m
and n are compared, and each clock cycle, the smallest one is subtracted from the
largest one. Note that Listing 5.7 does not show how m and n are initialized.

**Listing 5.7** Datapath to evaluate greatest common divisor

```
1  reg m,n : ns(16);
2  always {
3    m = (m > n) ? (m - n) : m;
4    n = (n > n) ? (n - m) : m;
5  }
```

Describing datapaths with expressions results in compact hardware descriptions.
An excellent example are shift registers. Figure 5.2 illustrates a Linear Feedback
Shift Register, which is a shift register with a feedback loop created by XORing
bits within the shift register. The feedback pattern is specified by a polynomial, and

**Fig. 5.2** Linear feedback
shift register for
$p(x) = x^4 + x^3 + 1$

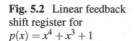
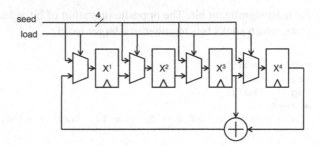

the polynomial used for Fig. 5.2 is $p(x) = x^4 + x^3 + 1$. LFSRs are used for pseudo-random sequence generation. If a so-called maximum-length polynomial is chosen, the resulting sequence of pseudorandom bits has a period of $2^n - 1$, where $n$ is the number of bits in the shift register. Thus, an LFSR is able to create a long non-repeating sequence of pseudorandom bits with a minimal amount of hardware. The shift register used to implement the LFSR must always contain at least one non-zero bit. It is easy to see in Fig. 5.2 that an all-zero pattern in the shift register will only reproduce itself. Therefore, an LFSR must be initialized with a non-zero seed value. The seed value is programmed using a multiplexer in front of each register.

Although the structure of Fig. 5.2 is complex to draw, it remains very compact when written using word-level expressions. This is shown in Listing 5.8. Line 6 of the code represents the shift-and-feedback operation. Line 7 of the code represents the loading of the seed value into the LFSR register.

**Listing 5.8** Linear feedback shift register

```
1   reg shft      : ns(4);
2   sig shft_new  : ns(4);
3   sig load      : ns(1);
4   sig seed      : ns(4);
5   always {
6       shft_new  = (shft << 1) | (shft[2] ^ shft[3]);
7       shft      = load ? seed : shft_new;
8   }
```

In summary, using two variable types (signals and registers), it is possible to describe synchronous hardware by means of expressions on those signals and registers. Remember that the order in which expressions are written is irrelevant: they will all execute within a single clock cycle. In the next section, we will group expressions into modules, and define input/output ports on those modules.

## 5.2  Hardware Modules

A hardware module defines a level of hierarchy for a hardware netlist. In order to communicate across levels of hierarchy, hardware modules define ports. Figure 5.3

**Fig. 5.3** Three-bit counter module

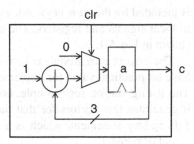

**Fig. 5.4** Hardware equivalent of Listing 5.10

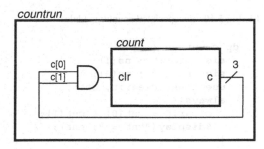

shows the 3-bit counter, encapsulated as a module. There is a single input port, clr, which synchronously clears the register. There is also a 3-bit output port c that holds the current count value. The equivalent description in GEZEL language of this structure is shown in Listing 5.9. The always block is included in a dp (datapath), which defines a list of in and out ports. There can be as many input and output ports as needed, and they can be created in any order. Registers and signals are local to a single module and invisible outside of the module boundary. Input ports and output ports are equivalent to wires, and therefore behave identical to signals. Input ports and output ports are subject to similar requirements as signals: it is not allowed to assign an output port more than once during a clock cycle, and each output must be assigned at least once during each clock cycle. We will further investigate this while discussing the formal properties of the FSMD model in Sect. 5.6.

**Listing 5.9** Three-bit counter module with reset

```
1  dp count(in   clr : ns(1);
2            out c   : ns(3)) {
3    reg a : ns(3);
4    always {
5      a = clr ? 0 : a + 1;
6      c = a;
7    }
8  }
```

After hardware is encapsulated inside of a module, the module itself can be used as a component in another hardware design. This principle is called *structural hierarchy*. As an example, Listing 5.10 shows how the 3-bit counter is included in a testbench structure that clears the counter as soon as it reaches three. The module is included by the use keyword, which also shows how ports should be connected to local signals and registers. The equivalent hardware structure of Listing 5.10 is shown in Fig. 5.4.

The countrun module in Listing 5.10 has no inputs nor outputs. Such modules have no practical value for implementation, but they may be useful for simulation. The listing shows, for example, how countrun encapsulates count, and how it generates test vectors for that module. The listing also illustrates the use of a $display statement, which is a simulation *directive* that prints the value of a signal or register.

**Listing 5.10**  Encapsulated counter module

```
1  dp countrun {
2     sig clearit : ns(1);
3     sig cnt     : ns(3);
4     use count(clearit, cnt);
5     always {
6        clearit = cnt[0] & cnt[1];
7        $display("cnt = ", cnt);
8     }
9  }
```

Once a module has been included inside of another one by means of the use statement, it cannot be included again: each module can be used only once. However, it is easy to create a duplicate of an existing module by means of a *cloning* statement. Listing 5.11 shows how to create three 3-bit counters, count0, count1 and count2.

**Listing 5.11**  Cloning of modules

```
1   dp count0(in  clr : ns(1);
2              out c   : ns(3)) {
3      reg a : ns(3);
4      always {
5         a = clr ? 0 : a + 1;
6         c = a;
7      }
8   }
9   dp count1 : count0
10  dp count2 : count0
```

## 5.3  Finite State Machines

We will next describe a mechanism to control hardware circuits. As discussed before, the expressions that are part of an always block are evaluated at each clock cycle, and it is not possible to conditionally evaluate an expression. Even the selection operator (c ? expr1 : expr2) will evaluate the true-part as well

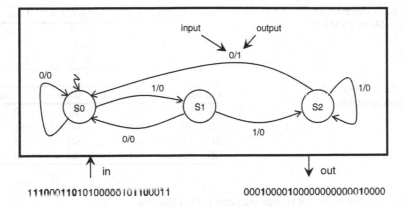

in

1110001101010000010110011

out

00010000100000000000010000

**Fig. 5.5** Mealy FSM of a recognizer for the pattern '110'

as the false-part regardless of the condition value c. If `expr1` and `expr2` would contain an expensive operator, then we would need two copies of that operator to implement `c ? expr1 : expr2`.

A control model, on the other hand, allows us to indicate what expressions should execute during each clock cycle. Very simple control models will select a sequence of expressions to execute, over multiple clock cycles. More complex control models will also allow decision making. Advanced control models also consider exceptions, recursion, out-of-order execution, and more. In this section, we describe a common control model for hardware description, called Finite State Machine (FSM). An FSM can be used to described sequencing and decision making. In the next section, we will combine the FSM with expressions in a datapath.

An FSM is a sequential digital machine which is characterized by

- A set of states;
- A set of inputs and outputs;
- A state transition function;
- An output function

An FSM has a current state, equal to an element from the set of states. Each clock cycle, the state transition function selects the next value for the current state, and the output function selects the value on the output of the FSM. The state transition function and the output function are commonly described in terms of a graph. In that case, the set of states becomes the set of nodes of the graph, and the state transitions become edges in the graph.

The operation of an FSM is best understood by means of an example. Suppose we need to observe a (possibly infinite) sequence of bits, one at a time. We need to determine at what point the sequence contains the pattern '110'. This problem is well suited for an FSM design. We can distinguish three relevant states for the FSM, by realizing that sequential observation of the input bits transforms this pattern recognition problem into an incremental process.

**Table 5.2** Conversion of
Mealy to Moore FSM for
Fig. 5.5

| Current-state | Input = 0 | Input = 1 |
|---|---|---|
| S0 | S0, 0 | S1, 0 |
| S1 | S0, 0 | S2, 0 |
| S2 | S0, 1 | S2, 0 |

**Table 5.3** Resulting Moore
state transition table

| Current-state | Input = 0 | Input = 1 | Output |
|---|---|---|---|
| SA = S0, 0 | SA | SC | 0 |
| SB = S0, 1 | SA | SC | 1 |
| SC = S1, 0 | SA | SD | 0 |
| SD = S2, 0 | SB | SD | 0 |

1. State S0: We have not recognized any useful pattern.
2. State S1: We have recognized the pattern '1'.
3. State S2: We have recognized the pattern '11'.

When we consider each state and each possible input bit, we can derive all state transitions, and thus derive the state transition function. The output function can be implemented by defining a successful recognition as an input bit of '0' when the FSM is in state S2. This leads to the state transition graph (or state transition diagram) shown in Fig. 5.5. The notation used for Fig. 5.5 is that of a *Mealy* FSM. The output of a Mealy FSM is defined by the present state, as well as the input.

There is a different formulation of a FSM known as a *Moore* state machine. The output of a Moore state machine is only dependent on the current state, and not on the current input. Both forms, Mealy and Moore, are equivalent formulations of FSM. A Mealy machine can be converted into an equivalent Moore machine using a simple conversion procedure.

To convert the Mealy form into a Moore form, make a state transition table for the Mealy FSM. This table contains the next-state of each state transition combined with the relevant output. For example, in state S1 of Fig. 5.5, the input '1' leads to state S2 with output 0. Annotate this in the Table as: S2, 0. The entire diagram can be translated this way, leading to Table 5.2.

Using the conversion table, an equivalent Moore FSM can be constructed as follows. There is one Moore state for each unique *(next-state, output)* pattern. From Table 5.2, we can find four Moore states: (S0, 0), (S0, 1), (S1, 0) and (S2, 0). To find the Moore FSM state transitions, we replicate the corresponding Mealy transitions for each Moore state. There may be multiple Moore transitions for a single Mealy transition. For example, Mealy state S0 is replicated into Moore states (S0, 0) and (S0, 1). Thus, each of the state transitions out of S0 will be replicated two times. The resulting Moore state transition table is shown in Table 5.3, while the resulting Moore FSM graph is drawn in Fig. 5.6. A small ambiguity was removed from Fig. 5.6 by making state SA = (S0, 0) the initial state of the Moore machine. That is, the Mealy machine does not specify the initial value of the output. Since the Moore machine ties the output to the current state, an initial output value must be assumed as well.

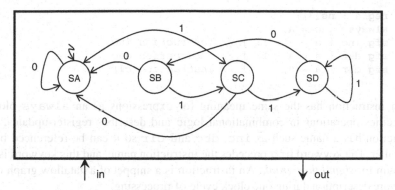

**Fig. 5.6** Moore FSM of a recognizer for the pattern '110'

In summary, a Finite State Machine is a common control model for hardware design. We can use it to model the conditional execution of expressions. In that case, we will use the outputs of the FSM to control the execution of expressions. Similarly, we will use the inputs to feed runtime conditions into the FSM, so that conditional sequencing can be implemented. In the next section, we will use a modified form of a FSM, called FSMD, which is the combination of an FSM and a datapath (modeled using expressions).

## 5.4  Finite State Machines with Datapath

A Finite State Machine with Datapath combines a hardware control model (an FSM) with a datapath. The datapath is described as cycle-based bit-parallel hardware using expressions, as discussed in Sect. 5.1. However, in contrast to datapaths using only `always` blocks, FSMD datapaths define also one or more *instructions*: conditionally executed 'always' blocks.

### 5.4.1  Modeling

Listing 5.12 shows an example of a datapath with an `always` block and three instructions: `inc`, `dec`, and `clr`. The meaning of the `always` block is the same as before: it contains expressions that will execute every clock cycle. In contrast to an `always` block, instructions will only execute when told to do so by a controller. Thus, the three instructions of the datapath in Listing 5.12 can increment, decrement, or clear register a, depending on what the controller tells the datapath to do.

**Listing 5.12** Datapath for an up-down counter with three instructions

```
1  dp updown(out c   : ns(3)) {
```

```
2    reg a : ns(3);
3    always { c = a; }
4    sfg inc { a = a + 1; } // instruction inc
5    sfg dec { a = a - 1; } // instruction dec
6    sfg clr { a = 0;      } // instruction clr
7 }
```

An instruction has the same meaning for expressions as an `always` block: it specifies operations in combinational logic and describes register-updates. An instruction has a name such as `inc`, `dec` and `clr` so it can be referenced by a controller. The keyword `sfg` precedes the instruction name, and this keyword is an acronym for *signal flow graph*: An instruction is a snippet of a dataflow graph of a hardware description during one clock cycle of processing.

A datapath with instructions needs a controller to select what instruction should execute in each clock cycle. The FSMD model uses an FSM for this. Listing 5.13 shows a Finite State Machine controller for the datapath of Listing 5.12. In this case, the controller will steer the datapath such that it first counts up from 0 to 3, and next counts down from 3 to 0. The finite state machine is a textual format for a state transition diagram. Going through the listing, we can identify the following features.

- Line 1: The `fsm` keyword defines a finite state machine with name `ctl_updown`, and tied to the datapath `updown`.
- Lines 2–3: The FSM contains three states. One state, called `s0`, will be the initial state. Two other states, called `s1` and `s2`, are regular states. The current state of the FSM will always be one of `s0`, `s1` or `s2`.
- Line 4: When the current state is `s0`, the FSM will unconditionally transition to state `s1`. State transitions will be taken at each clock edge. At the same time, the datapath will receive the instruction `clr` from the FSM, which will cause the register a to be cleared (See Listing 5.9 on Page 121).
- Lines 5–6: When the current state of the FSM equals `s1`, a conditional state transition will be taken. Depending on the value of the condition, the FSM will transition either to state `s1` (Line 5), or else to state `s2` (Line 6), and the datapath will receive either the instruction `inc` or else `dec`. This will either increment or else decrement register a. The state transition condition is given by the expression a < 3. Thus, the FSM will remain in state `s1` as long as register a is below three. When a equals three, the FSM will transition to `s2` and a decrementing sequence is initiated.
- Lines 7–8: When the current state of the FSM equals `s2`, a conditional state transition to either `s1` or else `s2` will be taken. These two state transitions will issue a decrement instruction to the datapath as long as register a is above zero. When the register equals zero, the controller will transition to `s1` and restart the incrementing sequence.

**Listing 5.13** Controller for the up-down counter

```
1 fsm ctl_updown(updown) {
```

**Table 5.4** Behavior of the
FSMD in Listing 5.9

| Cycle | FSM curr/next | DP instr | DP expr | a curr/next |
|---|---|---|---|---|
| 0 | s0/s1 | clr | a = 0; | 0/0 |
| 1 | s1/s1 | inc | a = a + 1; | 0/1 |
| 2 | s1/s1 | inc | a = a + 1; | 1/2 |
| 3 | s1/s1 | inc | a = a + 1; | 2/3 |
| 4 | s1/s2 | dec | a = a − 1; | 3/2 |
| 5 | s2/s2 | dec | a = a − 1; | 2/1 |
| 6 | s2/s2 | dec | a = a − 1; | 1/0 |
| 7 | s2/s1 | inc | a = a + 1; | 0/1 |
| 9 | s1/s1 | inc | a = a + 1; | 1/2 |
| 9 | s1/s1 | inc | a = a + 1; | 2/3 |

```
2    initial s0;
3    state s1, s2;
4    @s0 (clr) -> s1;
5    @s1 if (a < 3) then (inc) -> s1;
6                   else (dec) -> s2;
7    @s2 if (a > 0) then (dec) -> s2;
8                   else (inc) -> s1;
9  }
```

Thus, the FSM controller determines the schedule of instructions on the datapath. There are many possible schedules, and the example shown in Listing 5.13 is one of them. However, since an FSM is not programmable, the implementation of the FSM will fix the schedule of instructions for the datapath. Table 5.4 illustrates the first ten clock cycles of operation for this FSMD. Each row shows the clock cycle, the current and next FSM state, the datapath instruction selected by the controller, the datapath expression, and the current and next value of the register a.

In the up/down counter example, each datapath instruction contains a single expression, and the FSM selects a single datapath instruction for execution during each clock cycle. This is not a strict requirement. An instruction may contain as many expressions as needed, and the FSM controller may select multiple instructions for execution during any clock cycle. You can think of a group of scheduled instructions and the always block as a single, large always block that is active for a single clock cycle. Of course, not all combinations will work in each case. For example, in the datapath shown in Listing 5.12, the instructions clr, inc and dec are all exclusive, since all of them modify register a. The set of expressions that execute during a given clock cycle (as a result of the always block and the scheduled instructions) have to be conform to the same rules as if there were only a single always block. We will define these rules precisely in Sect. 5.6.

Listing 5.14 shows the implementation of Euclid's algorithm as an FSMD. In this case, several datapath instructions contain multiple expressions. In addition, the controller selects multiple datapath instructions during one state transition (line 21). The body of the reduce instruction was presented earlier as the computational core

**Fig. 5.7** An FSMD consists
of two stacked FSMs

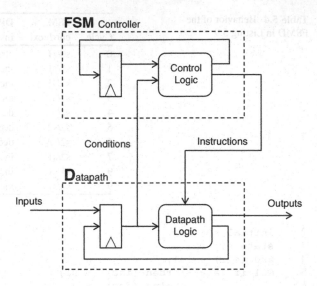

of GCD (Listing 5.14). The additional functionality provided by the controller is the
initialization of the registers m and n, and the detection of the algorithm completion.

**Listing 5.14** Euclid's GCD as an FSMD

```
1  dp euclid(in  m_in, n_in : ns(16);
2              out gcd        : ns(16)) {
3    reg m, n                  : ns(16);
4    reg done                  : ns(1);
5    sfg init     { m    = m_in;
6                   n    = n_in;
7                   done = 0;
8                   gcd  = 0; }
9    sfg reduce   { m    = (m >= n) ? m - n : m;
10                  n    = (n >  m) ? n - m : n; }
11   sfg outidle { gcd   = 0;
12                  done = ((m == 0) | (n == 0)); }
13   sfg complete{ gcd   = ((m > n) ? m : n);
14                  $display(``gcd = '', gcd); }
15 }
16 fsm euclid_ctl(euclid) {
17   initial s0;
18   state s1, s2;
19   @s0 (init) -> s1;
20   @s1 if (done) then (complete)           -> s2;
21                 else (reduce, outidle) -> s1;
22   @s2 (outidle) -> s2;
23 }
```

## 5.4.2 The FSMD Model As Two Stacked FSM

Here is another way to look upon and FSMD and its associated execution model. An FSMD consists of two stacked FSMs, as illustrated in Fig. 5.7. The top FSM contains the controller, and it is specified using a state transition diagram. The bottom FSM contains the datapath, and it is specified using expressions. The top FSM send instructions to the bottom FSM, and receives status information in return. Both FSM operate synchronously, and are connected to the same clock.

Each clock cycle, the two FSM go through the following activities.

1. Just after the clock edge, the state variables of both FSM are updated. For the controller, this means that a state transition is completed and the state register holds the new current-state value. For the datapath, this means that the register variables are updated as a result of assigning expressions to them.
2. The control FSM combines the control-state and datapath-state to evaluate the new next-state for the control FSM. At the same time, it will also select what instructions should be executed by the datapath.
3. The datapath FSM will evaluate the next-state for the state variables in the datapath, using the updated datapath state as well as the instructions received from the control FSM.
4. Just before the next clock edge, both the control FSM and the datapath FSM have evaluated and prepared the next-state value for the control state as well as the datapath state.

What makes a controller FSM different from a datapath FSM? Indeed, as illustrated in Fig. 5.7, both the datapath and the controller are sequential digital machines. Yet, from a designers' viewpoint, the creation of datapath logic and control logic is very different.

- Control logic tends to have an irregular structure. Datapath logic tends to have a regular structure (especially once you work with multi-bit words).
- Control logic is easy to describe using a finite state transition diagram, and hard to describe using expressions. Datapath logic is just the opposite: easy to describe using expressions, but hard to capture in state transition diagrams.
- The registers (state) in a controller have a different purpose than those in the datapath. Datapath registers contain algorithmic state. Control registers contain sequencing state.

In conclusion, FSMDs are useful because they capture control flow as well as data flow in hardware. Recall that C programs are also a combination of control flow and dataflow (Chap. 4). We relied on this commonality, for example, to convert a C program into a hardware FSMD.

### 5.4.3  An FSMD Is Not Unique

Figure 5.7 demonstrated how an FSMD actually consists of two stacked FSM. This has an interesting implication. From an implementation perspective, the partitioning between control logic and datapath logic is not unique. To illustrate this on Fig. 5.7, assume that we would merge the control logic and datapath logic into a single logic module, and assume that we would combine the registers in the controller with those in the datapath. The resulting design would still look as a FSM. The implementation does not make a clear distinction between the datapath part and FSM part.

In the previous subsection, we showed that the modeling of the FSM controller and the datapath is very different: using state transition graphs and expressions respectively. Since the partitioning between a controller and the datapath is not unique, this implies that we should be able to describe an FSM using expressions. This is illustrated in Listing 5.15, which shows the datapath-version of the controller in Listing 5.13. Apart from the difference in notation (state transition diagrams versus expressions), there is another important difference between both listings. In Listing 5.15, we have chosen the encoding for controller states. In Listing 5.13, the state encoding is symbolic (s0, s1, ...).

**Listing 5.15**  FSM controller for updown counter using expressions

```
1   dp updown_ctl(in   a_sm_3, a_gt_0 : ns(1);
2                  out instruction    : ns(2)) {
3     reg state_reg : ns(2);
4     // state encoding: s0 = 0, s1 = 1, s2 = 2
5     // instruction encoding: clr = 0, inc = 1, dec = 2
6     always {
7       state_reg   = (state_reg == 0) ? 1 :
8                     ((state_reg == 1) &  a_sm_3) ? 1 :
9                     ((state_reg == 1) & ~a_sm_3) ? 2 :
10                    ((state_reg == 2) &  a_gt_0) ? 2 : 1;
11      instruction = (state_reg == 0) ? 0 :
12                    ((state_reg == 1) &  a_sm_3) ? 1 :
13                    ((state_reg == 1) & ~a_sm_3) ? 2 :
14                    ((state_reg == 2) &  a_gt_0) ? 2 : 1;
15    }
16  }
```

If we can model an FSM with expressions, we can also merge it with the datapath controlled by this FSM. The resulting design for the up-down counter is shown in Listing 5.16. In this case, an entire FSMD is captured in a single datapath. How then, would you choose between developing separate FSM and datapath descriptions, versus capturing all of them in a single datapath? The following are some of the considerations you could make.

- State transition conditions of an FSM need to be stored in registers, which introduces a latency of one clock cycle for each conditional state transition. When capturing an FSM with datapath expressions, state transition conditions can be generated and evaluated in the same clock cycle.

- When capturing the FSMD in a single datapath, the expressions in the datapath include scheduling as well as data processing. On the other hand, in an FSMD with separate FSM and datapath descriptions, the datapath expressions only represent data processing. Listing 5.16 shows that, merging a controller into a datapath description tends to make the overall design more complicated to understand.
- When a datapath includes scheduling as well as data processing, it becomes harder to reuse it in a different schedule. Using an FSMD with separate FSM and datapath description on the other hand, will allow changes to the scheduling (the FSM) while reusing most of the datapath description.
- When capturing the FSMD in a single datapath, the state assignment is chosen by the designer, and may be optimized for specific applications. In an FSMD with separate FSM and datapath description, the state assignment is left to the logic synthesis tool.
- An FSMD captured as a single datapath is good for designs with simple or no control scheduling, such as designs with high-throughput requirements. An FSMD with separate FSM and datapath description is good for more complicated, structured designs.

**Listing 5.16** Updown counter using expressions

```
1   dp updown_ctl(out c : ns(3)) {
2     reg a        : ns(3);
3     reg state    : ns(2);
4     sig a_sm_3 : ns(1);
5     sig a_gt_0 : ns(1);
6     // state encoding: s0 = 0, s1 = 1, s2 = 2
7     always {
8       state = (state == 0) ? 1 :
9                ((state == 1) &  a_sm_3) ? 1 :
10               ((state == 1) & ~a_sm_3) ? 2 :
11               ((state == 2) &  a_gt_0) ? 2 : 1;
12      a_sm_3 = (a < 3);
13      a_gt_0 = (a > 0);
14      a = (state == 0) ? 0 :
15               ((state == 1) &  a_sm_3) ? a + 1 :
16               ((state == 1) & ~a_sm_3) ? a - 1 :
17               ((state == 2) &  a_gt_0) ? a + 1 : a - 1;
18      c = a;
19    }
20  }
```

In the above examples, we showed how an FSM can be modeled using datapath expressions, and how this allowed to capture an entire FSMD in a single datapath. The opposite case (modeling a datapath as a state transition diagram) is very uncommon. Problem 5.10 investigates some of the reasons for this.

**Fig. 5.8** Implementation of the up-down counter FSMD

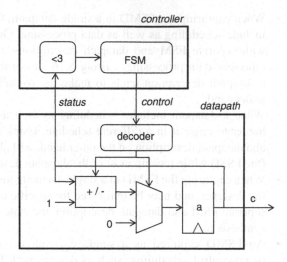

### 5.4.4  Implementation

**Listing 5.17** Datapath for an up-down counter with three instructions

```
1   dp updown(out c    : ns(3)) {
2     reg a : ns(3);
3     always  { c = a;       }
4     sfg inc { a = a + 1; }
5     sfg dec { a = a - 1; }
6     sfg clr { a = 0;       }
7   }
8   fsm ctl_updown(updown) {
9     initial s0;
10    state s1, s2;
11    @s0 (clr) -> s1;
12    @s1 if (a < 3) then (inc) -> s1;
13                   else (dec) -> s2;
14    @s2 if (a > 0) then (dec) -> s2;
15                   else (inc) -> s1;
16  }
```

How to determine the hardware implementation of an FSMD? The basic rules for mapping expressions on registers/signals into synchronous digital hardware are the same as with `always` blocks. However, there is also an important difference. When an expression occurs inside of a datapath instruction, it should execute only when that instruction is selected by the controller. Thus, the final datapath structure will depend on the schedule of instructions selected by the FSMD controller.

To clarify this point, consider again the up-down counter in Listing 5.17. From the FSM model, it's easy to see that `clr`, `inc`, and `dec` will always execute in different clock cycles. Therefore, the datapath operators used to implement each of `clr`, `inc`, and `dec` can be shared. A possible implementation of the FSMD is shown in Fig. 5.8. The datapath includes an adder/subtractor and a multiplexer, which implement the instruction-set (clr, inc, dec). A local decoder is needed to convert the encoding used for these instructions into local control signals for the multiplexer and the adder subtractor. The datapath is attached to a controller, which is implemented with an FSM and a local decoder for datapath status information. In this case, the value in register a is datapath status.

The resulting implementation looks as shown in Fig. 5.8. Many of the detailed implementation of Fig. 5.8 will be generated using automatic design tools (see next section). A designer who develops an FSMD will make the following overall decisions.

- The designer determines the amount of work done in a single clock cycle. This is simply the set of all datapath instructions which will execute concurrently in a single clock cycle. Hence, in order to obtain the best possible sharing, a designer must distribute similar operations over multiple clock cycles. For example, if there are 16 multiplications to perform with a clock cycle budget of four clock cycles, then an implementation with 4 multiplies each clock cycle will most likely be smaller than one which performs 16 multiplies in the first clock cycle and nothing in the next three cycles.
- The designer can also influence, indirectly, the maximum clock frequency attainable by the hardware design. This frequency is determined by the complexity of the expressions given in the datapath instructions. Obviously, small and short expressions will result in smaller and faster logic in the datapath. If the expressions used to capture the datapath become too complex, the resulting design may be too slow for the intended system clock period.

GEZEL descriptions can be simulated directly, as source code, or they can be converted into VHDL and simulated by a VHDL simulator. Refer to Appendix A for a brief overview on the installation and use of GEZEL tools.


## 5.5    FSMD Design Example: A Median Processor

So far, we discussed all building blocks to design a complete hardware implementation. In this section, we will apply these modeling techniques to create a full hardware implementation of a reference design in software.


### 5.5.1    Design Specification: Calculating the Median

The target design is a processor to compute the median of a list of numbers. The median operation selects the middle-ranked element of a list (for a list with an odd number of elements), or the average of the two middle-ranked elements (for a list with an even number of elements). For example, in the list

L = {4, 56, 2, 10, 32}

the median would be 10, since the ranked list corresponds to {2, 4, 10, 32, 56}. For our design, we will assume a fixed-length list with an odd number of elements. A median operation has an application when used as a filter in image processing applications. In that case, a stream of pixels is fed through the median operation, and each pixel is replaced with the median of the surrounding pixels. The effect of this operation is to reduce noise, which appears as speckle in the input image. Indeed, the median operation is very effective at removing outlier values within a list of pixels. This example also demonstrates the need for a high-speed implementation of the median, in particular when considering moving images.

Calculation of the median of a list requires one to sort the elements of the list, in order to find the middle element. A more efficient algorithm is the following: the median of a list L is the element e for which there are ($\sharp$L - 1)/2 elements in the list that are smaller than e. For example, to find the median of a list of five elements, we would identify the element for which there are two elements in L smaller than that element. In the example above, you can see that the value 10 meets this condition.

**Listing 5.18** C function to compute the median of five numbers

```
1   int median(int a1, int a2, int a3, int a4, int a5) {
2       int z1, z2, z3, z4, z5, z6, z7, z8, z9, z10;
3       int s1, s2, s3, s4;
4       z1  = (a1 < a2);
5       z2  = (a1 < a3);
6       z3  = (a1 < a4);
7       z4  = (a1 < a5);
8       z5  = (a2 < a3);
9       z6  = (a2 < a4);
10      z7  = (a2 < a5);
11      z8  = (a3 < a4);
12      z9  = (a3 < a5);
13      z10 = (a4 < a5);
14      s1 = ((    z1  +      z2  +      z3  +       z4)  == 2);
15      s2 = (((1-z1) +      z5  +      z6  +       z7)  == 2);
16      s3 = (((1-z2) + (1-z5) +      z8  +       z9)  == 2);
17      s4 = (((1-z3) + (1-z6) + (1-z8) +      z10)  == 2);
18      return ( s1 ? a1 : s2 ? a2 : s3 ? a3 : s4 ? a4 : a5);
19  }
```

Listing 5.18 shows a reference implementation of a C function to compute the median of five inputs. The function implements the algorithm described earlier. For each input, it calculates how many other inputs are smaller than that input. Since a<b implies b>a, ten comparisons are needed for five inputs. Finally, the return value of the function is the element for which the sum of comparisons equals 2.

## 5.5.2 Mapping the Median in Hardware

Next we design an FSMD for this function, and illustrate some of the issues that one must handle when writing a hardware implementation from a software specification. We will start, at first, by designing a datapath that accepts five values, and that produces the median as the output. Based on Listing 5.18, such a design is straightforward in GEZEL. Some wordlength optimization is possible, since the result of a comparison is only a single bit. Listing 5.19 shows an equivalent datapath design in GEZEL.

**Listing 5.19** GEZEL Datapath of a median calculation of five numbers

```
1   dp median(in a1, a2, a3, a4, a5 : ns(32); out q1 : ns(32)) {
2     sig z1, z2, z3, z4, z5, z6, z7, z8, z9, z10 : ns(3);
3     sig s1, s2, s3, s4, s5 : ns(1);
4     always {
5       z1  = (a1 < a2);
6       z2  = (a1 < a3);
7       z3  = (a1 < a4);
8       z4  = (a1 < a5);
9       z5  = (a2 < a3);
10      z6  = (a2 < a4);
11      z7  = (a2 < a5);
12      z8  = (a3 < a4);
13      z9  = (a3 < a5);
14      z10 = (a4 < a5);
15      s1 = ((    z1  +     z2  +     z3  +      z4) == 2);
16      s2 = (((1-z1) +     z5  +     z6  +      z7) == 2);
17      s3 = (((1-z2) + (1-z5) +     z8  +      z9) == 2);
18      s4 = (((1-z3) + (1-z6) + (1-z8) +     z10) == 2);
19      q1  = s1 ? a1 : s2 ? a2 : s3 ? a3 : s4 ? a4 : a5;
20    }
21  }
```

Compared to the software implementation, Listing 5.19 completes in a single clock cycle. However, this function makes an assumption which is not present in the C implementation. The GEZEL version assumes that all inputs of median are available at the same moment! In C, variables are stored in memory, and memory access are sequential with one another. Hence, although Listing 5.18 and Listing 5.19 are syntactically almost identical, they correspond to a very different behavior. In this case, the difference in implementation affects the communication between the median function and the rest of the system. The hardware design requires a bus of 6 times 32 bits (5 inputs and 1 output), or 192 bits, to carry all of the data from and to the median calculation unit.

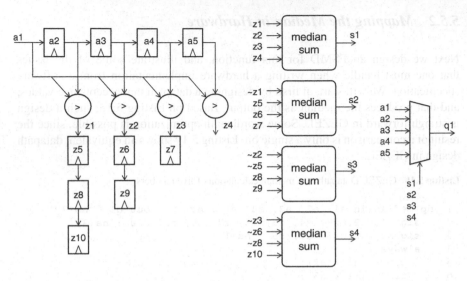

**Fig. 5.9** Median-calculation datapath for a stream of values

### 5.5.3 Sequentializing the Data Input

Let's consider an alternative: a design with a single input, 32 bit wide, and a single output, 32 bit wide. For each input, the design will evaluate the median over the last five values provided to the system. For each new input, a new median can be calculated. Clearly, we can simply pipe the input into a chain of registers. This will store a window of values. The median computation requires ten comparisons for five arbitrarily chosen inputs. In this case, however, we will recompute the median over every input value. This allows reusing some of the comparisons from previous iterations. Indeed, the number of new comparisons required per median value is 4, with 6 comparisons reusable from previous iterations. Reusing comparisons is advantageous because these involve 32-bit comparators. The resulting design is illustrated in Fig. 5.9. An equivalent GEZEL description of this design is shown in Listing 5.20.

**Listing 5.20** GEZEL Datapath of a median calculation of five numbers with sequentialized data input

```
1  dp median(in a1 : ns(32); out q1 : ns(32)) {
2    reg a2, a3, a4, a5 : ns(32);
3    sig z1, z2, z3, z4;
4    reg z5, z6, z7, z8, z9, z10 : ns(3);
5    sig s1, s2, s3, s4, s5 : ns(1);
6    always {
7      a2  = a1;
8      a3  = a2;
```

```
 9         a4   = a3;
10         a5   = a4;
11         z1   = (a1 < a2);
12         z2   = (a1 < a3);
13         z3   = (a1 < a4);
14         z4   = (a1 < a5);
15         z5   = z1;
16         z6   = z2;
17         z7   = z3;
18         z8   = z5;
19         z9   = z6;
20         z10  = z8;
21         s1 = ((    z1  +     z2  +     z3  +     z4)  == 2);
22         s2 = (((1-z1)  +     z5  +     z6  +     z7)  == 2);
23         s3 = (((1-z2)  + (1-z5)  +     z8  +     z9)  == 2);
24         s4 = (((1-z3)  + (1-z6)  + (1-z8)  +     z10) == 2);
25         q1   = s1 ? a1 : s2 ? a2 : s3 ? a3 : s4 ? a4 : a5;
26     }
27  }
```

### 5.5.4  Fully Sequentialized Computation

In Listing 5.20, there are still four parallel 32-bit comparators required, to compute a single median value per clock cycle. In addition, the computation of the terms s1, s2, s3 and s4 is very similar. By spreading the computation of a median value over multiple clock cycles, the number of operators in the datapath can be reduced, at the expense of adding registers and multiplexers. This section will demonstrate this by building a fully sequentialized version of the median computation unit, implemented as an FSMD. The FSM is particularly useful to capture the controller required for such a sequentialized design.

The first step to design a sequentialized hardware module is to create a schedule – a time-sequence for the computations involved in the hardware design. This could be done directly on a C program, but in this case we'll demonstrate the idea directly on the hardware of Fig. 5.10. The white-on-black labels indicate the clock cycles of our schedule. Each clock cycle, a different set of operations will be evaluated, until the final output can be computed, in clock cycle 12. Allocating similar operations in different clock cycles (for example, the comparison in clock cycles 2, 3, 4, and 5) will enable reuse of datapath operators.

Once the schedule is created, the resulting design can be coded. Listing 5.21 shows such a fully sequentialized version of the median computation. It is functionally identical to the previous designs, but uses 12 cycles to compute a median. It looks very different as a result. The design consists of the following major features.

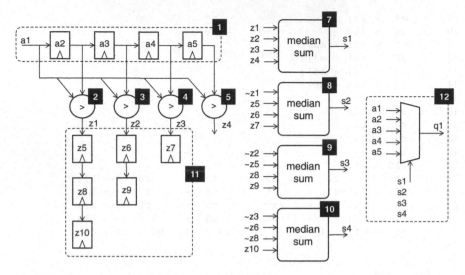

**Fig. 5.10** Sequential schedule median-calculation datapath for a stream of values

- The module `hastwo` checks if exactly two input bits are one. If so, it returns true. This module is a simplified formulation of the expressions used to compute `s1, s2, s3` and `s4`. The module is instantiated in `median` (line 27), and used in four datapath instructions: `c_s1, c_s2, c_s3, c_s4` (lines 44–47).
- The module `comp` is a 32-bit comparator. It is instantiated in `median` (line 31), and used in four datapath instructions: `c_z1, c_z2, c_z3` and `c_z4` (lines 40–43).
- Because of the sequentialized computation, all intermediate signals from Listing 5.20 have become registers in Listing 5.21. Also, care is taken to schedule the datapath instructions in a sequence so as never to overwrite a register that would still be needed (lines 67–80).
- The datapath uses an input data-strobe and an output data-strobe. The strobe is asserted (1) when the input is read or the output is produced. Such strobes are common in multi-cycle hardware designs to support the integration of this hardware module in a larger system. In this case, the strobes are used by the testbench (lines 83–108) to decide when to provide a new input. The testbench generates a pseudorandom sequence of 16-bit values, and the sequence is advanced each time a strobe is detected.

**Listing 5.21** Fully sequentialized median calculation with testbench

```
1  dp hastwo(in a, b, c, d : ns(1); out q : ns(1)) {
2    always {
3      q = ( a &  b & ~c & ~d) |
4          ( a & ~b & ~c &  d) |
5          (~a & ~b &  c &  d) |
6          ( a & ~b &  c & ~d) |
```

```
 7              (~a &  b & ~c &  d) |
 8              (~a &  b &  c & ~d);
 9     }
10  }
11
12  dp comp(in a, b : ns(32); out q : ns(1)) {
13     always { q = (a > b); }
14  }
15
16  dp median(in  istr  : ns(1);
17            in  a1_in : ns(32);
18            out ostr  : ns(1);
19            out q1    : ns(32)) {
20     reg a1, a2, a3, a4, a5 : ns(32);
21     reg z1, z2, z3, z4, z5, z6, z7, z8, z9, z10 : ns(3);
22     reg s1, s2, s3, s4 : ns(1);
23
24     reg ristr : ns(1);
25
26     sig h1, h2, h3, h4, qh : ns(1);
27     use hastwo(h1, h2, h3, h4, qh);
28
29     sig m1, m2 : ns(32);
30     sig mq      : ns(1);
31     use comp(m1, m2, mq);
32
33     always        { ristr = istr; }
34     sfg getinput  { a1 = istr ? a1_in : a1;
35                     a2 = istr ? a1 : a2;
36                     a3 = istr ? a2 : a3;
37                     a4 = istr ? a3 : a4;
38                     a5 = istr ? a4 : a5;
39                   }
40     sfg c_z1      { m1 =  a1; m2 = a2; z1 = mq; }
41     sfg c_z2      { m1 =  a1; m2 = a3; z2 = mq; }
42     sfg c_z3      { m1 =  a1; m2 = a4; z3 = mq; }
43     sfg c_z4      { m1 =  a1; m2 = a5; z4 = mq; }
44     sfg c_s1      { h1 =  z1; h2 = z2; h3 = z3; h4 =  z4;
                       s1 = qh; }
45     sfg c_s2      { h1 = ~z1; h2 = z5; h3 = z6; h4 =  z7;
                       s2 = qh; }
46     sfg c_s3      { h1 = ~z2; h2 =~z5; h3 = z8; h4 =  z9;
                       s3 = qh; }
47     sfg c_s4      { h1 = ~z3; h2 =~z6; h3 =~z8; h4 = z10;
                       s4 = qh; }
48     sfg c_z5z10   { z5 =  z1; z6 = z2; z7 = z3; z8 =  z5;
                       z9 = z6; z10 = z8; }
49     sfg c_notwo   { h1   =  0; h2 = 0; h3 =  0; h4 =  0;}
50     sfg c_nocomp  { m1   =  0; m2 = 0; }
51     sfg c_noout   { ostr =  0; q1 = 0; }
52     sfg putoutput { q1 = s1 ? a1 : s2 ? a2 : s3 ? a3 : s4 ?
                       a4 : a5;
53                     ostr =  1;
```

```
54                    $display($cycle, `` a1 '', a1, `` q1 '', q1
                      );
55                    }
56  }
57  fsm ctl_median(median) {
58   initial s0;
59   state s1, s2, s3, s4, s5, s6, s7, s8, s9, s10;
60   state s11, s12, s13, s14, s15, s16;
61
62   @s0 if (ristr) then (          c_notwo, c_nocomp, c_noout)
                             ->s1;
63                  else (getinput, c_notwo, c_nocomp, c_noout)
                             ->s0;
64   @s1  (c_notwo, c_z1, c_noout) -> s2;
65   @s2  (c_notwo, c_z2, c_noout) -> s3;
66   @s3  (c_notwo, c_z3, c_noout) -> s4;
67   @s4  (c_notwo, c_z4, c_noout) -> s5;
68   @s5  (c_s1,    c_nocomp, c_noout) -> s6;
69   @s6  (c_s2,    c_nocomp, c_noout) -> s7;
70   @s7  (c_s3,    c_nocomp, c_noout) -> s8;
71   @s8  (c_s4,    c_nocomp, c_noout) -> s9;
72   @s9  (c_notwo, c_nocomp, c_z5z10, c_noout) -> s10;
73   @s10 (c_notwo, c_nocomp, putoutput)        -> s0;
74  }
75
76  dp t_median {
77   sig istr, ostr : ns(1);
78   sig a1_in, q1 : ns(32);
79   use median(istr, a1_in, ostr, q1);
80   reg r : ns(1);
81   reg c : ns(16);
82   always    { r     = ostr;    }
83   sfg init  { c     - 0x1234;  }
84   sfg sendin { a1_in = c;
85              c     = (c[0] ^ c[2] ^ c[3] ^ c[5]) #
                             c[15:1];
86              istr  = 1;        }
87   sfg noin  { a1_in = 0;
88              istr  = 0;        }
89  }
90  fsm ctl_t_median(t_median) {
91   initial s0;
92   state s1, s2;
93   @s0 (init, noin)     -> s1;
94   @s1 (sendin)         -> s2;
95   @s2 if (r) then (noin) -> s1;
96           else (noin) -> s2;
97  }
98
99  system S {
100    t_median;
101  }
```

The simulation output of the 200 first clock cycles of Listing 5.21 is shown below. Each line of output shows the clock cycle at which a computation completes, followed by the input arguments a1, followed by the computed median q1. The output is shown in hexadecimal notation, and register values use the notation output/input. At a given output, the value of q1 is the median of the last five values entered. For example, at cycle 116, q1 has value b848, which is the median of the list ( 6e12, dc24, b848, 7091, e123). Refer to Appendix A for further details on installing GEZEL on a computer and running simulations.

```
/opt/gezel/bin/fdlsim m2.fdl 200
12 a1 1234/1234 q1 0
25 a1 91a/91a q1 0
38 a1 848d/848d q1 91a
51 a1 c246/c246 q1 1234
64 a1 e123/e123 q1 848d
77 a1 7091/7091 q1 848d
90 a1 b848/b848 q1 b848
103 a1 dc24/dc24 q1 c246
116 a1 6e12/6e12 q1 b848
129 a1 3709/3709 q1 7091
142 a1 1b84/1b84 q1 6e12
155 a1 8dc2/8dc2 q1 6e12
168 a1 46e1/46e1 q1 46e1
181 a1 2370/2370 q1 3709
194 a1 91b8/91b8 q1 46e1
```

## 5.6 Proper FSMD

A *proper FSMD* is one which has deterministic behavior. In general, a model with deterministic behavior is one which will always show the same response given the same initial state and the same input stimuli. In Chap. 2, we discussed how determinacy ensured that SDF graphs will compute the same result regardless of the schedule. Deterministic behavior is a desirable feature of most applications, including hardware/software codesign applications.

For a hardware FSMD implementation, deterministic behavior means that the hardware is free of race conditions. Without determinacy, a hardware model may end up in an unknown state (often represented using an 'X' in multi-valued logic hardware simulation) . A proper FSMD has no such race condition. It is obtained by enforcing four properties in the FSMD model. These properties are easy to check, both by the FSMD developer as well as by the simulation tools. The four properties are the following.

1. Neither registers nor signals can be assigned more than once during a clock cycle.
2. No circular definition exists between signals (wires).
3. If a signal is used as an operand of an expression, it must have a known value in the same clock cycle.
4. All datapath outputs must be defined (assigned) during all clock cycles.

The first rule is obvious, and ensures that there will be at most a single assignment per register/signal and per clock cycle. Recall from our earlier discussion that in a synchronous hardware model, all expressions are evaluated simultaneously according to the data dependencies of the expressions. If we allow multiple assignments per register/signal, the resulting value in the register or signal will become ambiguous.

The second rule ensures that any signal will carry a single, stable value during a clock cycle. Indeed, a circular definition between signals (e.g. as shown in Listing 5.4) may result in more then a single valid value. For example, circular definitions would occur when you try to model flip-flops with combinational logic (state). A proper FSMD model forces you to use reg for all state variables. Another case where you would end up with circular definition between signals is when you create ring-oscillators by wiring inverters in a loop. In a cycle-based hardware description language, all events happen at the pace of the global clock, and free-running oscillators cannot be modeled.

The third rule ensures that no signal can be used as an operand when the signal value would be undefined. Indeed, when an undefined signal is used as an operand in an expression, the result of the expression may become unknown. Such unknown values propagate in the model and introduce ambiguity on the outputs.

The fourth rule deals with hierarchy, and makes sure that rules 2 and 3 will hold even across the boundaries of datapaths. Datapath inputs and outputs have the same semantics as wires. The value of a datapath input will be defined by the datapath output connected to it. Rule 4 says that this datapath output will always have a known and stable value. Rule 4 is stricter than required. For example, if we don't read a datapath input during a certain clock cycle, the corresponding connected datapath output could remain undefined without causing trouble. However, requiring all outputs to be always defined is much easier to remember for the FSMD designer.

All of the above rules are enforced by the GEZEL simulation tools, either at runtime (through an error message), or else when the model is parsed. The resulting hardware created by these modeling rules is determinate and race-free.

## 5.7  Language Mapping for FSMD by Example

Even though we will be using the GEZEL language throughout this book for modeling of FSMD, all concepts covered so far are equally valid in other modeling languages including Verilog, VHDL or SystemC. We use GEZEL because of the following reasons.

- It is easier to set up cosimulation experiments in GEZEL. We will cover different types of hardware-software interfaces, and all of these are directly covered using GEZEL primitives.
- More traditional modeling languages include additional concepts (such as multi-valued logic and event-driven simulation), which, even though important by themselves, are less relevant in the context of a practical introduction to hardware-software codesign.
- GEZEL designs can be expanded into Verilog, VHDL or SystemC, as will be illustrated in this section. In fact, the implementation path of GEZEL works by converting these GEZEL designs into VHDL, and then using hardware synthesis on the resulting design.

The example we will discuss is the *binary* greatest common divisor (GCD) algorithm, a lightly optimized version of the classic GCD that makes use of the odd-even parity of the GCD operands.

### 5.7.1  GCD in GEZEL

Listing 5.22  Binary GCD in GEZEL

```
1  dp euclid(in  m_in, n_in : ns(16);
2             out gcd        : ns(16)) {
3    reg m, n               : ns(16);
4    reg done               : ns(1);
5    reg factor             : ns(16);
6
7    sfg init     { m = m_in; n = n_in; factor = 0; done = 0;
                      gcd = 0;
8                    $display("cycle=", $cycle, " m=", m_in,    " n
                      =", n_in); }
9    sfg shiftm   { m = m >> 1; }
10   sfg shiftn   { n = n >> 1; }
11   sfg reduce   { m = (m >= n) ? m - n : m;
12                  n = (n > m) ? n - m : n; }
13   sfg shiftf   { factor = factor + 1; }
14   sfg outidle  { gcd = 0; done = ((m == 0) | (n == 0)); }
15   sfg complete { gcd = ((m > n) ? m : n) << factor;
16                  $display("cycle=", $cycle, " gcd=",
                      gcd); }
17 }
```

```
18
19   fsm euclid_ctl(euclid) {
20     initial s0;
21     state s1, s2;
22
23     @s0 (init) -> s1;
24     @s1 if (done)                       then (complete)              -> s2;
25         else if ( m[0] &  n[0]) then (reduce, outidle)    -> s1;
26         else if ( m[0] & ~n[0]) then (shiftn, outidle)    -> s1;
27         else if (~m[0] &  n[0]) then (shiftm, outidle)    -> s1;
28                                         else (shiftn, shiftm,
29                                              shiftf, outidle)  -> s1;
30     @s2 (outidle) -> s2;
31   }
```

## 5.7.2  GCD in Verilog

**Listing 5.23** Binary GCD in Verilog

```verilog
1    module euclid(m_in, n_in, gcd, clk, rst);
2    input   [15:0] m_in;
3    input   [15:0] n_in;
4    output  [15:0] gcd;
5    reg     [15:0] gcd;
6    input   clk;
7    input   rst;
8
9    reg [15:0] m, m_next;
10   reg [15:0] n, n_next;
11   reg done, done_next;
12   reg [15:0] factor, factor_next;
13   reg [1:0] state, state_next;
14
15   parameter s0 = 2'd0, s1 = 2'd1, s2 = 2'd2;
16
17   always @(posedge clk)
18     if (rst) begin
19       n     <= 16'd0;
20       m     <= 16'd0;
21       done  <= 1'd0;
22       factor <= 16'd0;
23       state <= s0;
24     end else begin
25       n     <= n_next;
26       m     <= m_next;
27       done  <= done_next;
28       factor <= factor_next;
29       state <= state_next;
30     end
31
32   always @(*) begin
33     n_next     <= n;          // default reg assignment
34     m_next     <= m;          // default reg assignment
```

```
35        done_next    <= done;    // default reg assignment
36        factor_next  <= factor;  // default reg assignment
37        gcd          <= 16'd0;   // default output assignment
38
39        case (state)
40
41          s0: begin
42                m_next       <= m_in;
43                n_next       <= n_in;
44                factor_next  <= 16'd0;
45                done_next    <= 1'd0;
46                gcd          <= 16'd0;
47                state_next <= s1;
48                end
49
50          s1: if (done) begin
51                   gcd <= ((m > n) ? m : n) << factor;
52                   state_next <= s2;
53                end else if (m[0] & n[0]) begin
54                   m_next       <= (m >= n) ? m - n : m;
55                   n_next       <= (n > m) ? n - m : n;
56                   gcd          <= 16'd0;
57                   done_next <= ((m == 0) | (n == 0));
58                   state_next <= s1;
59                end else if (m[0] & ~n[0]) begin
60                   n_next <= n >> 1;
61                   gcd          <= 16'd0;
62                   done_next <= ((m == 0) | (n == 0));
63                   state_next <= s1;
64                end else if (~m[0] & n[0]) begin
65                   m_next <= m >> 1;
66                   gcd          <= 16'd0;
67                   done_next <= ((m == 0) | (n == 0));
68                   state_next <= s1;
69                end else begin
70                   n_next <= n >> 1;
71                   m_next <= m >> 1;
72                   factor_next <= factor + 1;
73                   gcd          <= 16'd0;
74                   done_next <= ((m == 0) | (n == 0));
75                   state_next <= s1;
76                end
77
78          s2: begin
79                gcd          <= 16'd0;
80                done_next <= ((m == 0) | (n == 0));
81                state_next<= s2;
82                end
83
84          default: begin
85                   state_next <= s0;  // jump back to init
86                   end
87        endcase
```

```
88    end
89
90  endmodule
```

## 5.7.3   GCD in VHDL

**Listing 5.24**  Binary GCD in VHDL

```vhdl
 1  library ieee;
 2  use ieee.std_logic_1164.all;
 3  use ieee.std_logic_arith.all;
 4
 5  entity gcd is
 6    port( m_in, n_in : in  std_logic_vector(15 downto 0);
 7          gcd          : out std_logic_vector(15 downto 0);
 8          clk, rst   : in  std_logic
 9        );
10  end gcd;
11
12  architecture behavior of gcd is
13    type statetype is (s0, s1, s2);
14    signal state, state_next : statetype;
15    signal m, m_next : std_logic_vector(15 downto 0);
16    signal n, n_next : std_logic_vector(15 downto 0);
17    signal done, done_next : std_logic;
18    signal factor, factor_next : std_logic_vector(15 downto 0)
       ;
19    begin
20
21    update_regs: process(clk, rst)
22    begin
23      if (rst='1') then
24          m       <= (others => '0');
25          n       <= (others => '0');
26          done  <= '0';
27          factor <= (others => '0');
28          state <= s0;
29        elsif (clk='1' and clk'event) then
30          state  <= state_next;
31          m       <= m_next;
32          n       <= n_next;
33          done  <= done_next;
34          factor <= factor_next;
35          state  <= state_next;
36        end if;
37      end process;
38
39      eval_logic: process(m_in, n_in, state)
40      begin
41      n_next     <= n;
42      m_next     <= m;
43      done_next  <= done;
44      factor_next<= factor;
```

```vhdl
45    gcd              <= (others => '0');
46
47    case state is
48
49      when s0 =>
50         m_next         <= m_in;
51         n_next         <= n_in;
52         factor_next <= (others => '0');
53         done_next      <= '0';
54         gcd            <= (others => '0');
55         state_next     <= s1;
56
57      when s1 =>
58         if (done = '1') then
59            if (m > n) then
60               gcd <= conv_std_logic_vector(shl(unsigned(m),
                     unsigned(factor)),16);
61            else
62               gcd <= conv_std_logic_vector(shl(unsigned(n),
                     unsigned(factor)),16);
63            end if;
64            state_next  <= s2;
65         elsif ((m(0) = '1') and (n(0) = '1')) then
66            if (m >= n) then
67               m_next <= unsigned(m) - unsigned(n);
68               n_next <= n;
69            else
70               m_next <= m;
71               n_next <= unsigned(n) - unsigned(m);
72            end if;
73            gcd         <= (others => '0');
74            if ((m = "0000000000000000") or
                (n = "0000000000000000")) then
75               done_next <= '1';
76            else
77               done_next <= '0';
78            end if;
79            state_next <= s1;
80         elsif ((m(0) = '1') and (n(0) = '0')) then
81            n_next <= '0' & n(15 downto 1);
82            gcd         <= (others => '0');
83            if ((m = "0000000000000000") or
                (n = "0000000000000000")) then
84               done_next <= '1';
85            else
86               done_next <= '0';
87            end if;
88            state_next <= s1;
89         elsif ((m(0) = '0') and (n(0) = '1')) then
90            m_next <= '0' & m(15 downto 1);
91            gcd         <= (others => '0');
92            if ((m = "0000000000000000") or
                (n = "0000000000000000")) then
```

```
93                        done_next <= '1';
94                 else
95                        done_next <= '0';
96                 end if;
97                 state_next <= s1;
98            else
99                 n_next <= '0' & n(15 downto 1);
100                m_next <= '0' & m(15 downto 1);
101                factor_next <= conv_std_logic_vector(unsigned
                                        (factor) +
102                            conv_unsigned(1,16),16);
103                gcd        <= (others => '0');
104                if ((m = "0000000000000000") or
                       (n = "0000000000000000")) then
105                    done_next <= '1';
106                else
107                    done_next <= '0';
108                end if;
109                state_next <= s1;
110            end if;
111
112        when s2 =>
113            gcd <= (others => '0');
114            if ((m = "0000000000000000") or
                   (n = "0000000000000000")) then
115                done_next <= '1';
116            else
117                done_next <= '0';
118            end if;
119            state_next<= s2;
120
121        when others =>
122            state_next <= s0;
123
124        end case;
125    end process;
126 end behavior;
```

### 5.7.4   GCD in SystemC

Listing 5.25  Binary GCD in SystemC

```
1    #include ``systemc.h''
2
3    enum statetype {s0, s1, s2};
4
5    SC_MODULE(gcd_fsmd) {
6        sc_in <bool>          clk;
7        sc_in <bool>          rst;
8        sc_in <sc_uint<16> >  m_in, n_in;
9        sc_out <sc_uint<16> > gcd;
10
```

```
11    sc_signal<statetype>   state, state_next;
12    sc_uint<16>            m,     m_next;
13    sc_uint<16>            n,     n_next;
14    sc_uint<16>            factor,factor_next;
15    sc_uint< 1>            done,  done_next;
16
17    void update_regs();
18    void eval_logic();
19    SC_CTOR(gcd_fsmd) {
20      SC_METHOD(eval_logic);
21      sensitive << m_in << n_in << state;
22      SC_METHOD(update_regs);
23      sensitive_pos << rst << clk;
24    }
25  };
26
27  void gcd_fsmd::update_regs() {
28    if (rst.read() == 1) {
29      state  = s0;
30      m      = 0;
31      n      = 0;
32      factor = 0;
33      done   = 0;
34    } else {
35      state  = state_next;
36      m      = m_next;
37      n      = n_next;
38      factor = factor_next;
39      done   = done_next;
40    }
41  }
42
43  void gcd_fsmd::eval_logic() {
44
45    n_next      = n;
46    m_next      = m;
47    done_next   = done;
48    factor_next = factor;
49    gcd         = 0;
50
51    switch(state) {
52      case s0:
53        m_next      = m_in;
54        n_next      = n_in;
55        factor_next = 0;
56        done_next   = 0;
57        gcd         = 0;
58        state_next  = s1;
59        break;
60      case s1:
61        if (done == 1) {
62          gcd = ((m > n) ? m : n) << factor;
63          state_next = s2;
```

```
64         } else if (m[0] & n[0]) {
65             m_next      = (m >= n) ? m - n : m;
66             n_next      = (n >  m) ? n - m : n;
67             gcd         = 0;
68             done_next   = ((m == 0) | (n == 0));
69             state_next = s1;
70         } else if (m[0] & ~n[0]) {
71             n_next      = (n >> 1);
72             gcd         = 0;
73             done_next   = ((m == 0) | (n == 0));
74             state_next = s1;
75         } else if (~m[0] & n[0]) {
76             m_next      = m >> 1;
77             gcd         = 0;
78             done_next   = ((m == 0) | (n == 0));
79             state_next = s1;
80         } else {
81             n_next      = n >> 1;
82             m_next      = m >> 1;
83             factor_next= factor + 1;
84             gcd         = 0;
85             done_next   = ((m == 0) | (n == 0));
86             state_next = s1;
87         }
88       break;
89     case s2:
90       gcd = 0;
91       done_next   = ((m == 0) | (n == 0));
92       break;
93     default:
94       state_next = s0;
95     }
96 }
```

## 5.8   Summary

In this section, we discussed a synchronous hardware modeling mechanism, consisting of a datapath in combination with an FSM controller. The resulting model is called FSMD (Finite State Machine with Datapath). An FSMD models datapath instructions with expressions, and control with a state transition graph. Datapath expressions are created in terms of register variables and signals (wires). Register variables are implicitly attached to the global clock signal. Datapath instructions (groups of datapath expressions) form the connection between the controller and the datapath.

A given FSMD design is not unique. A given design can be decomposed into many different, equivalent FSMD descriptions. It is up to designer to pick a modeling style that feels natural and that is useful for the problem at hand.

We discussed a modeling syntax for FSMD called GEZEL. GEZEL models can be simulated and converted into synthesizable VHDL code. However, the

FSMD model is generic and can be captured into any suitable hardware description language. We demonstrated equivalent synthesizable implementations of an FSMD in GEZEL, Verilog, VHDL and SystemC.


## 5.9 Further Reading

The FSMD model has been recognized as a universal model for RTL modeling of hardware. See Vahid (2007a) for a textbook that starts from combinational and sequential logic, and gradually works up to FSMD based design. FSMD were popularized by Gajski, and are briefly covered in Gajski et al. (2009). Earlier, one can find an excellent development of the FSMD model in Davio et al. (1983).

Further details on installation and use of the GEZEL toolset are covered in Appendix A.

By using four easy-to-remember rules, one can construct *proper* FSMD, FSMD that will not have race conditions. A mathematical proof that these four rules are sufficient can be found in Schaumont et al. (2006).

There is an ongoing discussion on how to improve the productivity of hardware design. Some researchers believe that *high-level synthesis*, the automatic generation of RTL starting from high-level descriptions, is essential. Several academic and commercial design tools that support such high level synthesis are described in Gajski et al. (2009). See Gupta et al. (2004) for a detailed description of one such an environment. On the other hand, the nature of hardware design is such that designers like to think about clock cycles when they think about architecture. Hence, abstraction should be applied with utmost care. See Bluespec (Hoe 2000) and MADL (Qin 2004) for examples of such carefully abstracted hardware design and modeling paradigms.


## 5.10 Problems

**Problem 5.1.** Which of the circuits (a, b, c, d) in Fig. 5.11 can be simulated using a cycle-based simulator?

**Problem 5.2.** Design a high-speed sorter for four 32-bit registers (Fig. 5.12). Show how to create a sorting network for four numbers, using only simple two-input comparator modules. The comparator modules are built with combinational logic, and have a constant critical path. Optimize the critical path of the overall design, and create a maximally parallel implementation. You may make use of comparator modules, registers, and wiring. The input of the sorter comes from four registers marked 'input', the output of the sorter needs to be stored in four registers marked 'output'.

**Fig. 5.11** Sample circuits for
Problem 5.1

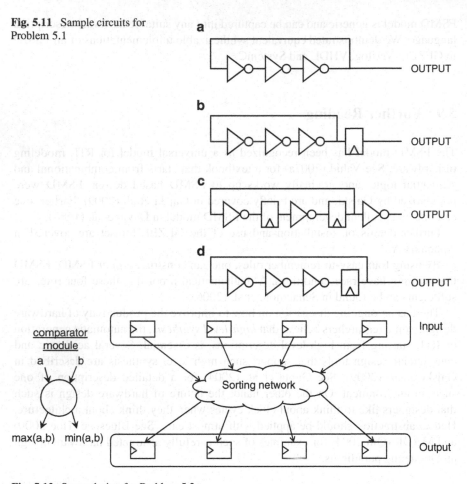

**Fig. 5.12** Sorter design for Problem 5.2

**Problem 5.3.** Design a Finite State Machine that recognizes the pattern '10' and
the pattern '01' in an infinite stream of bits. Make sure that the machine recognizes
only one pattern at a time, and that it is not triggered by overlapping patterns.
Figure 5.13 shows and example of the behavior of this FSM.

1. Draw a Mealy-type state diagram of this FSM.
2. Draw an RTL schematic of an implementation for this machine. Draw your
   implementation using registers and logic gates (AND, OR, NOT, and XOR).
   Make your implementation as compact as possible.

**Problem 5.4.** Design a Mealy-type finite state machine that recognizes either of
the following two patterns: 1101 or 0111. The patterns should be read left to right
(i.e. the leftmost bit is seen first), and they are to be matched into a stream of single
bits.

**Fig. 5.13** Sorter design for
Problem 5.2

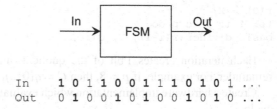

```
In    1 0 1 1 0 0 1 1 1 0 1 0 1 ...
Out   0 1 0 0 1 0 1 0 0 1 0 1 0 ...
```

**Problem 5.5.** Design an FSMD to divide natural numbers. The dividend and the
divider each have 8 bits of resolution. The quotient must have 10 bits of resolution,
and the remainder must have 8 bits of resolution. The divider has the following
interface

```
dp divider (in   x     : ns(8);
            in   y     : ns(8);
            in   start : ns(1);
            out  q     : ns(10);
            out  r     : ns(8);
            out  done  : ns(1)) {

  // Define the internals of the FSMD here ..

}
```

Given a dividend $X$ and a divider $Y$, the divider will evaluate a quotient $Q$ on $p$
bits of precision and a remainder $R$ such that

$$X.2^p = Q.Y + R \tag{5.1}$$

For example, if $p = 8, X = 12, Y = 15$, then a solution for $Q$ and $R$ is $Q = 204$
and $R = 12$ because $12 . 2^8 = 204.15 + 12$.

Your implementation must obtain the quotient and remainder within 32 clock
cycles. To implement the divider, you can use the restoring division algorithm as
follows. The basic operation evaluates a single bit of the quotient according to the
following pseudo-code:

```
basic_divider (input a, b;
               output q, r) {
  z := 2 * a - b;
  if (z < 0) then
      q = 0;
      r = 2 * a;
  else
      q = 1;
      r = z;
}
```

To evaluate the quotient over p bits, you repeat the basic 1-bit divider p times as
follows.

```
r(0) = X;
for i is 1 to p do
basic_divider(r(i-1), Y, q(i), r(i));
```

Each iteration creates 1 bit of the quotient, and the last iteration returns the remainder. For example, if $p = 8$, then $Q = q(0), q(1), q(2), \ldots, q(7)$ and $R = r(8)$.

Create a hardware implementation which evaluates 1 bit of the quotient per clock cycle.

**Problem 5.6.** How many flip-flops and how many adders do you need to implement the FSMD description in Listing 5.26? Count each single bit in each register, and assume binary encoding of the FSM state, to determine the flip-flop count.

**Listing 5.26** Program for Problem 5.6

```
1   dp mydp(in i : ns(5); out o : ns(5)) {
2     reg a1, a2, a3, a4 : ns(5);
3     sfg f1 { a1 = i;
4              a2 = 0;
5              a3 = 0;
6              a4 = 0; }
7     sfg f2 { a1 = a2 ? (a1 + a3) : (a1 + a4); }
8     sfg f3 { a3 = a3 + 1; }
9     sfg f4 { a4 = a4 + 1; }
10    sfg f5 { a2 = a2 + a1; }
11  }
12  fsm mydp_ctl(mydp) {
13    initial s0;
14    state s1, s2;
15    @s0 (f1) -> s1;
16    @s1 if (a1) then (f2, f3) -> s2;
17            else (f4)     -> s1;
18    @s2 if (a3) then (f2)     -> s1;
19            else (f5)     -> s2;
20  }
```

**Problem 5.7.** FSMD models provide modeling of control (conditional execution) as well as data processing in hardware. Therefore, it is easy to mimic the behavior of a C program and build and FSMD that reflects the same control flow as the C program. Write an FSMD model for the C function shown in Listing 5.27. Assume that the arguments of the function are the inputs of the FSMD, and that the result of the function is the FSMD output. Develop your model so that you need no more then a single multiplier.

**Listing 5.27** Program for Problem 5.7

```
1   int filter(int a) {
2     static int taps[5];
3     int c[] = {-1, 5, 10, 5, -1};
```

```
4     int r;
5
6     for (i=0; i<4; i++)
7       taps[i] = taps[i+1];
8     taps[4] = a;
9
10    r = 0;
11    for (i=0; i<5; i++)
12      r = r + taps[i] * c[i];
13
14    return r;
15  }
```

To model an array of constants in GEZEL, you can make use of the lookup table construct as follows:

```
dp lookup_example {

  lookup T : ns(8) = {5, 4, 3, 2, 1, 1, 1, 1};

  sig a, b : ns(3);

  always {
    a = 3;
    b = T[a]; // this assigns the fourth element of T to b
  }
}
```

**Problem 5.8.** Repeat Problem 5.7, but develop your FSMD so that the entire function completes in a single clock cycle.

**Problem 5.9.** Write the FSMD of Listing 5.28 in a single always block. This FSMD presents a Galois Field multiplier.

**Listing 5.28** Program for Problem 5.9

```
1   dp D( in fp, i1, i2 : ns(4); out mul: ns(4);
2         in mul_st: ns(1);
3         out mul_done : ns(1)) {
4     reg acc, sr2, fpr, r1 : ns(4);
5     reg mul_st_cmd : ns(1);
6     sfg ini { // initialization
7       fpr         = fp;
8       r1          = i1;
9       sr2         = i2;
10      acc         = 0;
11      mul_st_cmd = mul_st;
12    }
13    sfg calc { // calculation
14      sr2 = (sr2 << 1);
15      acc = (acc << 1) ^ (r1 & (tc(1)) sr2[3])   // add a if b=1
16              ^ (fpr & (tc(1)) acc[3]); // reduction if carry
17    }
```

```
18    sfg omul { // output inactive
19      mul      = acc;
20      mul_done = 1;
21      $display(``done. mul='', mul);
22    }
23    sfg noout { // output active
24      mul      = 0;
25      mul_done = 0;
26    }
27  }
28  fsm F(D) {
29    state s1, s2, s3, s4, s5;
30    initial   s0;
31    @s0 (ini,  noout) -> s1;
32    @s1 if (mul_st_cmd) then (calc, noout) -> s2;
33                        else (ini, noout)  -> s1;
34    @s2 (calc, noout) -> s3;
35    @s3 (calc, noout) -> s4;
36    @s4 (calc, noout) -> s5;
37    @s5 (ini,  omul ) -> s1;
38  }
```

**Problem 5.10.** In this chapter, we discussed how FSM can be expressed as datapath expressions (See Sect. 5.4.3 and Problem 5.8). It is also possible to go the opposite way, and model datapaths in terms of finite state machines. Write an FSM for the datapath shown in Listing 5.29.

**Listing 5.29** Program for Problem 5.10

```
1  dp tester(out o: ns(2)) {
2    reg a1 : ns(1);
3    reg a2 : ns(2);
4    always {
5      a1 = a1 + 1;
6      a2 = a2 + a1;
7      o  = a2;
8    }
9  }
```

# Chapter 6
# Microprogrammed Architectures

## 6.1 Limitations of Finite State Machines

Finite State Machines are well suited to capture the control flow and decision-making of algorithms. FSM state transition diagrams even resemble control dependency graphs (CDG). Nevertheless, FSM are no universal solution for control modeling. They suffer from several modeling weaknesses, especially when dealing with complex controllers.

A key issue is that FSMs are a flat model, without any hierarchy. A *flat* control model is like a C program consisting of just one single main function. Real systems do not use a flat control model: they need a control hierarchy. Many of the limitations of FSMs stem from their lack of hierarchy.

### 6.1.1 State Explosion

A flat FSM suffers from *state explosion*, which occurs when multiple independent activities interfere in a single model. Assume that a FSM has to capture two independent activities, each of which can be in one of three states. The resulting FSM, called a *product state-machine*, needs nine states to capture the control flow of the overall model. The product state-machine needs to keep track of the current state from two independent state machines at the same time. Figure 6.1 illustrates the effect of state explosion in a product state-machine. Two state machines, FSM1 and FSM2, need to be merged into a single product state-machine FSM1xFSM2. Due to conditional state transitions, one state machine can remain in a single state while the other state machine proceeds to the next state. This results in multiple intermediate states such as A1, A2, and A3. In order to represent all individual states, nine states are needed in total. The resulting number of state transitions (and state transition conditions) is even higher. Indeed, if we have $n$ independent state transition conditions in the individual state machines, the resulting product

P.R. Schaumont, *A Practical Introduction to Hardware/Software Codesign,*       157
DOI 10.1007/978-1-4614-3737-6_6, © Springer Science+Business Media New York 2013

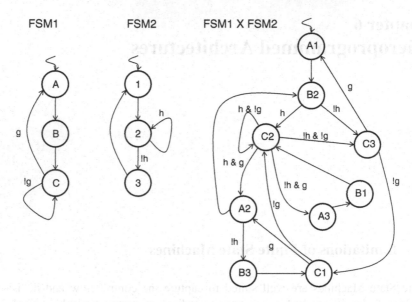

**Fig. 6.1** State explosion in FSM when creating a product state-machine

state-machine can have up to $2^n$ state transition conditions. Of course, there have been several proposals for hierarchical modeling extensions for FSMs, such as the Statecharts from David Harel. Currently, however, none of these are widely used for hardware design.

## 6.1.2  Exception Handling

A second issue with a flat FSM is the problematic handling of exceptions. An exception is a condition which may cause an immediate state transition, regardless of the current state of the finite state machine. The purpose of an exception is to abort the regular flow of control and to transfer control to a dedicated exception-handler. An exception may have internal causes, such as an overflow condition in a datapath, or external causes, such as an interrupt. Regardless of the cause, the effect of an exception on a finite state machine model is dramatic: an additional state transition needs to be added to every state of the finite state machine. For example, assume that the product state-machine in Fig. 6.1 needs to include an exception input called exc, and that the assertion of that input requires immediate transition to state A1. The resulting FSM, shown in Fig. 6.2, shows how exceptions degrade the clarity of the FSM state transition graph.

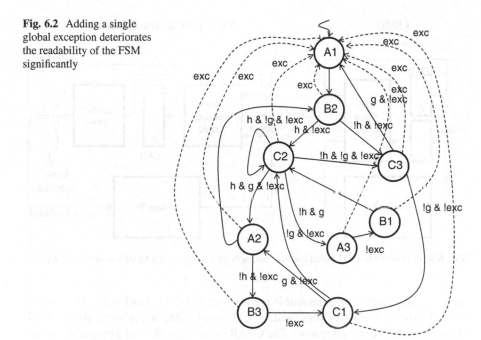

**Fig. 6.2** Adding a single global exception deteriorates the readability of the FSM significantly

### 6.1.3 Runtime Flexibility

A major concern, from the viewpoint of hardware-software codesign, is that a finite state machine is a non-flexible model. Once the states and the state transitions are defined, the control flow of the FSM is fixed. The hardware implementation of an FSM leads to a hardwired controller that cannot be modified after implementation.

As a result, designers have proposed improved techniques for specifying and implementing control, in order to deal with flexibility, exceptions, and hierarchical modeling. Microprogramming is one such a technique. Originally introduced in the 1950s by Maurice Wilkes as a means to create a programmable instruction-set for mainframe computers, it became very popular in the 1970s and throughout the 1980s. Microprogramming was found to be very useful to develop complex microprocessors with flexible instruction-sets. Nowadays, microprogramming is no longer popular; microprocessors are so cheap that they do no longer need a reprogrammable instruction-set. However, microprogramming is still very useful as a design technique to introduce flexibility in a hardware design.

## 6.2 Microprogrammed Control

Figure 6.3 shows a micro-programmed machine next to an FSMD. The fundamental idea in microprogramming is to replace the next-state logic of the finite

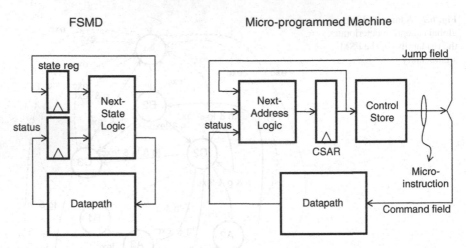

**Fig. 6.3** In contrast to FSM-based control, microprogramming uses a flexible control scheme

state-machine with a *programmable* memory, called the control store. The control store holds micro-instructions, and is addressed using a register called CSAR (Control Store Address Register). The CSAR is the equivalent of a program counter in microprocessors. The next-value of CSAR is determined by the next-address logic, using the current value of CSAR, the current micro-instruction and the value of status flags evaluated by the datapath. The default next-value of the CSAR corresponds to the previous CSAR value incremented by one. However, the next-address logic can also implement conditional jumps or immediate jumps.

Thus, the next-address logic, the CSAR, and the control store implement the equivalent of an instruction-fetch cycle in a microprocessor. In the design of Fig. 6.3, each micro-instruction takes a single clock cycle to execute. Within a single clock cycle, the following activities occur.

- The CSAR provides an address to the control store which retrieves a micro-instruction. The micro-instruction is split in two parts: a command-field and a jump-field. The command-field serves as a command for the datapath. The jump-field goes to the next-address logic.
- The datapath executes the command encoded in the micro-instruction, and returns status information to the next-address logic.
- The next-address logic combines the micro-instruction jump-field, the previous value of the CSAR, and the datapath status. The next-address logic updates the CSAR. Consequently, the critical path of the micro-programmed machine in Fig. 6.3 is determined by the combined logic delay through the control store, the next-address logic, and the datapath.

While the micro-programmed controller is more complicated than a finite state machine, it also addresses the problems of FSMs:

1. The micro-programmed controller scales well with complexity. For example, a 12-bit CSAR enables a control store with up the 4096 locations, and therefore a micro-program with 4096 steps. An equivalent FSM diagram with 4096 states, on the other hand, would be horrible to draw!
2. A micro-programmed machine deals very well with control hierarchy. With small modifications to the microprogrammed machine in Fig. 6.3, we can save the CSAR in a separate register or on a stack memory, and later restore it. This requires the definition of a separate micro-instruction to call a subroutine as well as a second micro-instruction to return from it.
3. A micro-programmed machine can deal efficiently with exception handling, since global exceptions are managed directly by the next-address logic, independently from the control store. For example, the presence of a global exception can feed a hard-coded value into the CSAR, immediately transferring the micro-programmed machine to an exception-handling routine. Exception handling in a micro-programmed machine is similar to a jump instruction, but it does not affect every instruction of the micro-program in the same way as it affects every state of a finite state machine.
4. Finally, micro-programs are flexible and very easy to change after the micro-programmed machine is implemented. Simply changing the contents of the control store is sufficient to change the program of the machine. In a micro-programmed machine, there is a clear distinction between the architecture of the machine and the functionality implemented using that architecture.

## 6.3 Micro-instruction Encoding

How should we define the encoding used by micro-instructions? In this section, we will discuss the design trade-offs that determine the micro-instruction format.

### 6.3.1 Jump Field

Figure 6.4 shows a 32-bit micro-instruction word, with 16 bits reserved for the datapath, and 16 bits reserved for the next-address logic. Let's first consider the part for the next-address logic. The address field holds an absolute target address, pointing to a location in the control store. In this case, the address is 12 bit, which means that this micro-instruction format would be suited for a control store with 4096 locations. The next field encodes the operation that will lead to the next value of CSAR. The default operation is, as discussed earlier, to increment CSAR. For such instructions, the address field remains unused. Otherwise, next will be used to encode various jump instructions. An absolute jump will transfer the value of the address field into CSAR. A conditional jump will use the value of a flag to conditionally update the CSAR or else increment it.

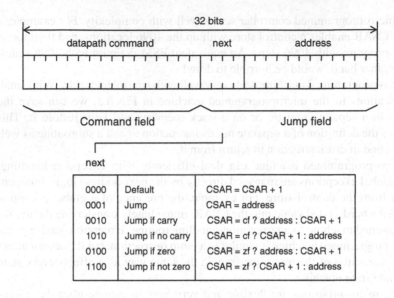

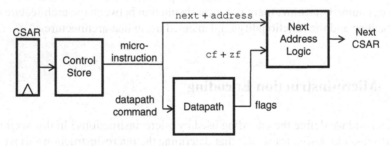

**Fig. 6.4** Sample format for a 32-bit micro-instruction word

Obviously, the format as shown is quite bulky, and may consume a large amount of storage. The `address` field, for example, is only used for jump instructions. If the micro-program contains only a few jump instructions, then the storage for the address field would be wasted. To avoid this, we will need to optimize the encoding format for micro-instructions. In micro-instructions other than jumps, the bits used for the `address` field could be given a different purpose.

## 6.3.2   Command Field

The design of the datapath command format reveals another interesting trade-off: we can either opt for a very wide micro-instruction word, or else we can prefer a

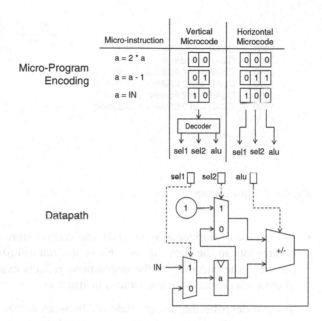

**Fig. 6.5** Example of vertical versus horizontal micro-programming

narrow micro-instruction word. A wide micro-instruction word allows each control bit of the data path to be stored separately. A narrow micro-instruction word, on the other hand, will require the creation of 'symbolic instructions', which are encoded groups of control-bits for the datapath. The FSMD model relies on such symbolic instructions. Each of the above approaches has a specific name. *Horizontal* micro-instructions use no encoding at all. It represents each control bit in the datapath with a separate bit in the micro-instruction format. *Vertical* micro-instructions on the other hand encode the control bits for the datapath as much as possible. A few bits of the micro-instruction can define the value of many more control bits in the data-path.

Figure 6.5 demonstrates an example of vertical and horizontal micro-instructions in the datapath. We wish to create a micro-programmed machine with three instructions on a single register a. The three instructions do one of the following: double the value in a, decrement the value in a, or initialize the value in a. The datapath shown on the bottom of Fig. 6.5 contains two multiplexers and a programmable adder/subtractor. It can be easily verified that each of the instructions enumerated above can be implemented as a combination of control bit values for each multiplexer and for the adder/subtractor. The controller on top shows two possible encodings for the three instructions: a horizontal encoding, and a vertical encoding.

- In the case of vertical microcode, the micro-instructions will be encoded. Since there are three different instructions, we can implement this machine with a 2-bit micro-instruction word. To generate the control bits for the datapath, we will have to decode each of the micro-instruction words into local control signals on the datapath.

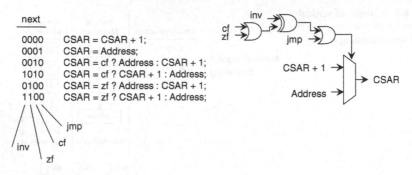

**Fig. 6.6**  CSAR encoding

- In the case of horizontal microcode, the control store will include each of the control bits in the datapath as a bit in the micro-instruction word. Hence, in this case, the encoding of the instructions reflects exactly the required setting of datapath elements for each micro-instruction.

We can describe the design trade-off between horizontal and vertical micro-programs as follows. Vertical micro-programs have a better code density, which is beneficial for the size of the control store. In Fig. 6.5, the vertically-encoded version of the microprogram will be only 2/3 of the size of the horizontally-encoded version. On the other hand, vertical micro-programs use an additional level of encoding, so that each micro-instruction needs to be decoded before it can drive the control bits of the datapath. Thus, the machine with the vertically encoded micro-program may be slower.

Obviously, the choice between a vertical and horizontal encoding needs to be made carefully. In practice, designers use a combination of vertical and horizontal encoding concepts, so that the resulting architecture is compact yet efficient. Consider for example the value of the next field of the micro-instruction word in Fig. 6.4. There are six different types of jump instructions, which would imply that a vertical micro-instruction would need no more then 3 bits to encode these six jumps. Yet, 4 bits have been used, indicating that there is some redundancy left. The encoding was chosen to simplify the design of the next-address logic, which is shown in Fig. 6.6. Such spare room in the micro-instruction encoding also supports future upgrades.

## 6.4  The Micro-programmed Datapath

The datapath of a micro-programmed machine consists of three elements: computation units, storage, and communication buses. Each of these may contribute a few control bits to the micro-instruction word. For example, multi-function computation units have function-selection bits, storage units have address bits and read/write

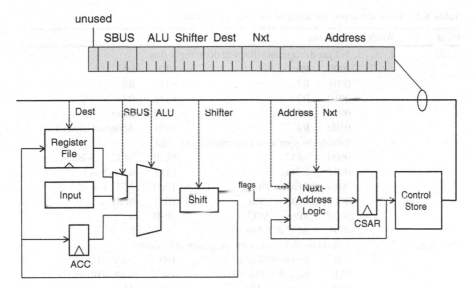

**Fig. 6.7** A micro-programmed datapath

command bits, and communication busses have source/destination control bits. The datapath may also generate status flags for the micro-programmed controller.

## 6.4.1 Datapath Architecture

Figure 6.7 illustrates a micro-programmed controller with a datapath attached. The datapath includes an ALU with shifter unit, a register file with eight entries, an accumulator register, and an input port. The micro-instruction word is shown on top of Fig. 6.7, and contains six fields. Two fields, Nxt and Address, are used by the micro-programmed controller. The other are used by the datapath. The type of encoding is mixed horizontal/vertical. The overall machine uses a horizontal encoding: each module of the machine is controlled independently. The sub-modules within the machine on the other hand use a vertical encoding. For example, the ALU field contains 4 bits. In this case, the ALU component in the datapath will execute up to 16 different commands.

The machine completes a single instruction per clock cycle. The ALU combines an operand from the accumulator register with an operand from the register file or the input port. The result of the operation is returned to the register file or the accumulator register. The communication used by datapath operations is controlled by two fields in the micro-instruction word. The *SBUS* field and the *Dest* field indicate the source and destination respectively.

**Table 6.1** Micro-instruction encoding of the example machine

| Field | Width | Encoding | | | |
|---|---|---|---|---|---|
| SBUS | 4 | Selects the operand that will drive the S-Bus | | | |
| | | 0000 | R0 | 0101 | R5 |
| | | 0001 | R1 | 0110 | R6 |
| | | 0010 | R2 | 0111 | R7 |
| | | 0011 | R3 | 1000 | Input |
| | | 0100 | R4 | 1001 | Address/Constant |
| ALU | 4 | Selects the operation performed by the ALU | | | |
| | | 0000 | ACC | 0110 | ACC — S-Bus |
| | | 0001 | S-Bus | 0111 | not S-Bus |
| | | 0010 | ACC + SBus | 1000 | ACC + 1 |
| | | 0011 | ACC − SBus | 1001 | SBus − 1 |
| | | 0100 | SBus − ACC | 1010 | 0 |
| | | 0101 | ACC & S-Bus | 1011 | 1 |
| Shifter | 3 | Selects the function of the programmable shifter | | | |
| | | 000 | logical SHL(ALU) | 100 | arith SHL(ALU) |
| | | 001 | logical SHR(ALU) | 101 | arith SHR(ALU) |
| | | 010 | rotate left ALU | 111 | ALU |
| | | 011 | rotate right ALU | | |
| Dest | 4 | Selects the target that will store S-Bus | | | |
| | | 0000 | R0 | 0101 | R5 |
| | | 0001 | R1 | 0110 | R6 |
| | | 0010 | R2 | 0111 | R7 |
| | | 0011 | R3 | 1000 | ACC |
| | | 0100 | R4 | 1111 | unconnected |
| Nxt | 4 | Selects next-value for CSAR | | | |
| | | 0000 | CSAR + 1 | 1010 | cf ? CSAR + 1 : Address |
| | | 0001 | Address | 0100 | zf ? Address : CSAR + 1 |
| | | 0010 | cf ? Address : CSAR + 1 | 1100 | zf ? CSAR + 1 : Address |

The Shifter module also generates flags, which are used by the micro-programmed controller to implement conditional jumps. Two flags are created: a zero-flag, which is high (1) when the output of the shifter is all-zero, and a carry-flag, which contains the bit shifted-out at the most-significant position.

### 6.4.2 Writing Micro-programs

Table 6.1 illustrates the encoding used by each module of the design from Fig. 6.7. A micro-instruction can be formed by selecting a module function for each module of the micro-programmed machine, including a next-address for the Address field. When a field remains unused during a particular instruction, a don't care value can be chosen. The don't care value should be carefully selected so that unwanted state changes in the datapath are avoided.

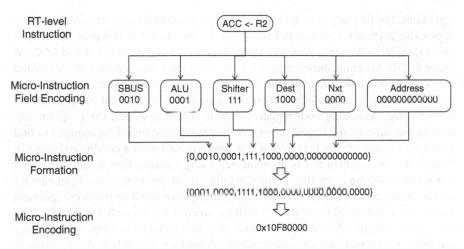

**Fig. 6.8**  Forming micro-instructions from register-transfer instructions

For example, an instruction to copy register R2 into the accumulator register ACC would be formed as illustrated in Fig. 6.8. The instruction should read out register R2 from the register file, pass the register contents over the SBus, through the ALU and the shifter, and write the result in the ACC register. This observation allows to determine the value of each field in the micro-instruction.

- The SBus needs to carry the value of R2. Using Table 6.1 we find SBUS equals 0010.
- The ALU needs to pass the SBus input to the output. Based on Table 6.1, ALU must equal 0001.
- The shifter passes the ALU output unmodified, hence Shifter must equal 111.
- The output of the shifter is used to update the accumulator register, so the Dest field equals 1000.
- Assuming that no jump or control transfer is executed by this instruction, the next micro-instruction will simply be one beyond the current CSAR location. This implies that Nxt should equal 0000 and Address is a don't-care, for example all-zeroes.
- Finally, we can find the overall micro-instruction code by putting all instruction fields together. Figure 6.8 illustrates this process. We conclude that a micro-instruction to copy R2 into ACC can be encoded as 0x10F80000 in the control store.

Writing a micro-program thus consists of formulating the desired behavior as a sequence of register transfers, and next encoding these register transfers as micro-instruction fields. More complex control operations, such as loops and if-then-else statements, can be expressed as a combination (or sequence) of register transfers.

As an example, let's develop a micro-program that reads two numbers from the input port and that evaluates their greatest common divisor (GCD) using Euclid's

algorithm. The first step is to develop a micro-program in terms of register transfers.
A possible approach is shown in Listing 6.1. Lines 2 and 3 in this program read in
two values from the input port, and store these values in registers R0 and ACC. At
the end of the program, the resulting GCD will be available in either ACC or R0, and
the program will continue until both values are equal. The stop test is implemented
in line 4, using a subtraction of two registers and a conditional jump based on
the zero-flag. Assuming both registers contain different values, the program will
continue to subtract the largest register from the smallest one. This requires to find
which of R0 and ACC is bigger, and it is implemented with a conditional jump in
Line 5. The bigger-then test is implemented using a subtraction, a left-shift and a
test on the resulting carry-flag. If the carry-flag is set, then the most-significant bit
of the subtraction would be one, indicating a negative result in two's complement
logic. This conditional jump-if-carry will be taken if R0 is smaller then ACC. The
combination of lines 5–7 shows how an if-then-else statement can be created using
multiple conditional and unconditional jump instructions. When the program is
complete, in Line 8, it spins in an infinite loop.

**Listing 6.1**  Micro-program to evaluate a GCD

```
1  ;    Command Field          || Jump Field
2            IN -> R0
3            IN -> ACC
4  Lcheck: (R0 - ACC)          || JUMP_IF_Z Ldone
5          (R0 - ACC) << 1     || JUMP_IF_C LSmall
6           R0 - ACC -> R0     || JUMP Lcheck
7  Lsmall: ACC - R0 -> ACC     || JUMP Lcheck
8  Ldone:                         JUMP Ldone
```

## 6.5   Implementing a Micro-programmed Machine

In this section, we discuss a sample implementation of a micro-programmed
machine in the GEZEL language. It can be used as a template for other implementations.

### 6.5.1   Micro-instruction Word Definition

A convenient starting point in the design of a micro-programmed machine is the
definition of the micro-instruction. This includes the allocation of micro-instruction
control bits, and their encoding.

The individual control fields are defined as subvectors of the micro-instruction.
Listing 6.2 shows the GEZEL implementation of the micro-programmed design
discussed in the previous section. The possible values for each micro-instruction

field are shown in Lines 5–71. The use of C macro's simplifies the writing of micro-programs.

The formation of a single micro-instruction is done using a C macro as well, shown in Lines 74–81. Lines 83–136 show the micro-programmed controller, which includes a control store with a micro-program and the next-address CSAR logic. The control store is a lookup table with a sequence of micro-instructions (lines 90–105). On line 115, a micro- instruction is fetched from the control store, and broken down into individual fields which form the output of the microprogrammed controller (lines 116–122). The next-address logic uses the next-address control field to find a new value for CSAR each clock cycle (lines 124–134).

The micro-programmed machine includes several data-paths, including a register file (lines 137–168), an ALU (lines 170–192), a shifter (lines 194–221). Each of the data-paths is crafted along a similar principle: based on the control field input, the data-input is transformed into a corresponding data-output. The decoding process of control fields is visible as a sequence of ternary selection-operators.

The top-level cell for the micro-programmed machine is contained in lines 223–254. The top-level includes the controller, a register file, an ALU and a shifter. The top-level module also defines a data-input port and a data-output port, and each has a strobe control signal that indicates a data-transfer. The strobe signals are generated by the top-level module based decoding of micro-instruction fields. The input strobe is generated when the SBUS control field indicates that the SBUS will be reading an external input. The output strobe is generated by a separate, dedicated micro-instruction bit.

A simple testbench for the top-level cell is shown on lines 256–276. The testbench feeds in a sequence of data to the micro-programmed machine, and prints out each number appearing at the data output port. The micro-program for this machine evaluates the GCD of each tuple in the list of numbers shown on line 264.

**Listing 6.2**  Micro-programmed controller in GEZEL

```
1   // wordlength in the datapath
2   #define WLEN 16
3
4   /* encoding for data output */
5   #define O_NIL     0     /* OT <- 0 */
6   #define O_WR      1     /* OT <- SBUS */
7
8   /* encoding for SBUS multiplexer */
9   #define SBUS_R0   0       /* SBUS <- R0  */
10  #define SBUS_R1   1       /* SBUS <- R1  */
11  #define SBUS_R2   2       /* SBUS <- R2  */
12  #define SBUS_R3   3       /* SBUS <- R3  */
13  #define SBUS_R4   4       /* SBUS <- R4  */
14  #define SBUS_R5   5       /* SBUS <- R5  */
15  #define SBUS_R6   6       /* SBUS <- R6  */
16  #define SBUS_R7   7       /* SBUS <- R7  */
17  #define SBUS_IN   8       /* SBUS <- IN  */
18  #define SBUS_X SBUS_R0 /* don't care */
19
```

```
20   /* encoding for ALU */
21   #define ALU_ACC  0      /* ALU <- ACC */
22   #define ALU_PASS 1      /* ALU <- SBUS */
23   #define ALU_ADD  2      /* ALU <- ACC + SBUS */
24   #define ALU_SUBA 3      /* ALU <- ACC - SBUS */
25   #define ALU_SUBS 4      /* ALU <- SBUS - ACC */
26   #define ALU_AND  5      /* ALU <- ACC and SBUS */
27   #define ALU_OR   6      /* ALU <- ACC or SBUS */
28   #define ALU_NOT  7      /* ALU <- not SBUS */
29   #define ALU_INCS 8      /* ALU <- ACC + 1 */
30   #define ALU_INCA 9      /* ALU <- SBUS - 1 */
31   #define ALU_CLR  10     /* ALU <- 0 */
32   #define ALU_SET  11     /* ALU <- 1 */
33   #define ALU_X ALU_ACC /* don't care */
34
35   /* encoding for shifter */
36   #define SHFT_SHL 1      /* Shifter <- shiftleft(alu) */
37   #define SHFT_SHR 2      /* Shifter <- shiftright(alu) */
38   #define SHFT_ROL 3      /* Shifter <- rotateleft(alu) */
39   #define SHFT_ROR 4      /* Shifter <- rotateright(alu) */
40   #define SHFT_SLA 5      /* Shifter <- shiftleftarithmetical
                                          (alu) */
41   #define SHFT_SRA 6      /* Shifter <- shiftrightarithmetical
                                          (alu) */
42   #define SHFT_NIL 7      /* Shifter <- ALU */
43   #define SHFT_X SHFT_NIL /* don't care */
44
45   /* encoding for result destination */
46   #define DST_R0  0       /* R0  <- Shifter */
47   #define DST_R1  1       /* R1  <- Shifter */
48   #define DST_R2  2       /* R2  <- Shifter */
49   #define DST_R3  3       /* R3  <- Shifter */
50   #define DST_R4  4       /* R4  <- Shifter */
51   #define DST_R5  5       /* R5  <- Shifter */
52   #define DST_R6  6       /* R6  <- Shifter */
53   #define DST_R7  7       /* R7  <- Shifter */
54   #define DST_ACC 8       /* IR  <- Shifter */
55   #define DST_NIL 15      /* not connected <- shifter */
56   #define DST_X   DST_NIL /* don't care instruction */
57
58   /* encoding for command field */
59   #define NXT_NXT  0      /* CSAR <- CSAR + 1 */
60   #define NXT_JMP  1      /* CSAR <- Address */
61   #define NXT_JC   2      /* CSAR <- (carry==1)? Address : CSAR
                                       + 1 */
62   #define NXT_JNC  10     /* CSAR <- (carry==0)? Address : CSAR
                                       + 1 */
63   #define NXT_JZ   4      /* CSAR <- (zero==1) ? Address : CSAR
                                       + 1 */
64   #define NXT_JNZ  12     /* CSAR <- (zero==0) ? Address : CSAR
                                       + 1 */
65   #define NXT_X NXT_NXT
66
```

```
67   /* encoding for the micro-instruction word */
68   #define MI(OUT, SBUS, ALU, SHFT, DEST, NXT, ADR) \
69        (OUT    << 31) | \
70        (SBUS   << 27) | \
71        (ALU    << 23) | \
72        (SHFT   << 20) | \
73        (DEST   << 16) | \
74        (NXT    << 12) | \
75        (ADR)
76
77   dp control(in  carry, zero : ns(1);
78                out ctl_ot       : ns(1);
79                out ctl_sbus     : ns(4);
80                out ctl_alu      : ns(4);
81                out ctl_shft     : ns(3);
82                out ctl_dest     : ns(4)) {
83
84      lookup cstore : ns(32) = {
85      // 0 Lstart: IN -> R0
86      MI(O_NIL, SBUS_IN, ALU_PASS, SHFT_NIL,DST_R0, NXT_NXT,0),
87      // 1          IN -> ACC
88      MI(O_NIL, SBUS_IN, ALU_PASS, SHFT_NIL,DST_ACC,NXT_NXT,0),
89      // 2 Lcheck: (R0 - ACC)        || JUMP_IF_Z Ldone
90      MI(O_NIL, SBUS_R0, ALU_SUBS, SHFT_NIL, DST_NIL,NXT_JZ,6),
91      // 3         (R0 - ACC) << 1   || JUMP_IF_C LSmall
92      MI(O_NIL, SBUS_R0, ALU_SUBS, SHFT_SHL, DST_NIL,NXT_JC,5),
93      // 4         R0 - ACC -> R0    || JUMP Lcheck
94      MI(O_NIL, SBUS_R0, ALU_SUBS,SHFT_NIL, DST_R0, NXT_JMP,2),
95      // 5 Lsmall: ACC - R0 -> ACC   || JUMP Lcheck
96      MI(O_NIL, SBUS_R0, ALU_SUBA, SHFT_NIL,DST_ACC,NXT_JMP,2),
97      // 6 Ldone:  R0 -> OUT         || JUMP Lstart
98      MI(O_WR,  SBUS_R0, ALU_X,    SHFT_X,   DST_X, NXT_JMP,0)
99      };
100
101     reg csar        : ns(12);
102     sig mir         : ns(32);
103     sig ctl_nxt     : ns(4);
104     sig csar_nxt    : ns(12);
105     sig ctl_address : ns(12);
106
107     always {
108
109        mir = cstore(csar);
110        ctl_ot      = mir[31];
111        ctl_sbus    = mir[27:30];
112        ctl_alu     = mir[23:26];
113        ctl_shft    = mir[20:22];
114        ctl_dest    = mir[16:19];
115        ctl_nxt     = mir[12:15];
116        ctl_address = mir[ 0:11];
117
118        csar_nxt = csar + 1;
119        csar = (ctl_nxt == NXT_NXT)  ? csar_nxt :
```

```
120                    (ctl_nxt == NXT_JMP)  ? ctl_address :
121                    (ctl_nxt == NXT_JC)   ? ((carry==1) ? ctl_address
                           : csar_nxt) :
122                    (ctl_nxt == NXT_JZ)   ? ((zero==1) ? ctl_address
                           : csar_nxt) :
123                    (ctl_nxt == NXT_JNC)  ? ((carry==0) ? ctl_address
                           : csar_nxt) :
124                    (ctl_nxt == NXT_JNZ)  ? ((zero==0) ? ctl_address
                           : csar_nxt) :
125              csar;
126       }
127    }
128
129    dp regfile (in  ctl_dest : ns(4);
130                in  ctl_sbus : ns(4);
131                in  data_in  : ns(WLEN);
132                out data_out : ns(WLEN)) {
133      reg r0 : ns(WLEN);
134      reg r1 : ns(WLEN);
135      reg r2 : ns(WLEN);
136      reg r3 : ns(WLEN);
137      reg r4 : ns(WLEN);
138      reg r5 : ns(WLEN);
139      reg r6 : ns(WLEN);
140      reg r7 : ns(WLEN);
141      always {
142       r0 = (ctl_dest == DST_R0) ? data_in : r0;
143       r1 = (ctl_dest == DST_R1) ? data_in : r1;
144       r2 = (ctl_dest == DST_R2) ? data_in : r2;
145       r3 = (ctl_dest == DST_R3) ? data_in : r3;
146       r4 = (ctl_dest == DST_R4) ? data_in : r4;
147       r5 = (ctl_dest == DST_R5) ? data_in : r5;
148       r6 = (ctl_dest == DST_R6) ? data_in : r6;
149       r7 = (ctl_dest == DST_R7) ? data_in : r7;
150       data_out = (ctl_sbus == SBUS_R0) ? r0 :
151                  (ctl_sbus == SBUS_R1) ? r1 :
152                  (ctl_sbus == SBUS_R2) ? r2 :
153                  (ctl_sbus == SBUS_R3) ? r3 :
154                  (ctl_sbus == SBUS_R4) ? r4 :
155                  (ctl_sbus == SBUS_R5) ? r5 :
156                  (ctl_sbus == SBUS_R6) ? r6 :
157                  (ctl_sbus == SBUS_R7) ? r7 :
158                  r0;
159      }
160    }
161
162    dp alu (in  ctl_dest : ns(4);
163            in  ctl_alu  : ns(4);
164            in  sbus     : ns(WLEN);
165            in  shift    : ns(WLEN);
166            out q        : ns(WLEN)) {
167      reg acc : ns(WLEN);
168      always {
```

```
169       q = (ctl_alu == ALU_ACC)  ? acc :
170           (ctl_alu == ALU_PASS) ? sbus :
171           (ctl_alu == ALU_ADD)  ? acc + sbus :
172           (ctl_alu == ALU_SUBA) ? acc - sbus :
173           (ctl_alu == ALU_SUBS) ? sbus - acc :
174           (ctl_alu == ALU_AND)  ? acc & sbus :
175           (ctl_alu == ALU_OR)   ? acc | sbus :
176           (ctl_alu == ALU_NOT)  ? ~ sbus       :
177           (ctl_alu == ALU_INCS) ? sbus + 1    :
178           (ctl_alu == ALU_INCA) ? acc + 1     :
179           (ctl_alu == ALU_CLR)  ? 0           :
180           (ctl_alu == ALU_SET)  ? 1           :
181             0;
182       acc = (ctl_dest == DST_ACC) ? shift : acc;
183     }
184   }
185
186   dp shifter(in  ctl     : ns(3);
187              out zero    : ns(1);
188              out cy      : ns(1);
189              in  shft_in : ns(WLEN);
190              out so      : ns(WLEN)) {
191     always {
192       so = (ctl == SHFT_NIL) ? shft_in :
193            (ctl == SHFT_SHL) ? (ns(WLEN)) (shft_in << 1) :
194            (ctl == SHFT_SHR) ? (ns(WLEN)) (shft_in >> 1) :
195            (ctl == SHFT_ROL) ? (ns(WLEN)) (shft_in #
196                shft_in[WLEN-1]) :
                (ctl == SHFT_ROR) ? (ns(WLEN)) (shft_in[0] #
                    (shft_in >> 1)):
197            (ctl == SHFT_SLA) ? (ns(WLEN)) (shft_in << 1) :
198            (ctl == SHFT_SRA) ? (ns(WLEN))
                        (((tc(WLEN)) shft_in) >> 1) :
199            0;
200       zero = (shft_out == 0);
201       cy   = (ctl == SHFT_NIL) ? 0 :
202            (ctl == SHFT_SHL) ? shft_in[WLEN-1] :
203            (ctl == SHFT_SHR) ? 0 :
204            (ctl == SHFT_ROL) ? shft_in[WLEN-1] :
205            (ctl == SHFT_ROR) ? shft_in[0] :
206            (ctl == SHFT_SLA) ? shft_in[WLEN-1] :
207            (ctl == SHFT_SRA) ? 0 :
208            0;
209     }
210   }
211
212   dp hmm(in  din  : ns(WLEN); out din_strb  : ns(1);
213          out dout : ns(WLEN); out dout_strb : ns(1)) {
214     sig carry, zero : ns(1);
215     sig ctl_ot      : ns(1);
216     sig ctl_sbus    : ns(4);
217     sig ctl_alu     : ns(4);
218     sig ctl_shft    : ns(3);
```

```
219    sig ctl_acc      : ns(1);
220    sig ctl_dest     : ns(4);
221
222    sig rf_out, rf_in : ns(WLEN);
223    sig sbus         : ns(WLEN);
224    sig alu_in       : ns(WLEN);
225    sig alu_out      : ns(WLEN);
226    sig shft_in      : ns(WLEN);
227    sig shft_out     : ns(WLEN);
228    use control(carry,    zero,
229                   ctl_ot, ctl_sbus, ctl_alu, ctl_shft, ctl_dest)
                      ;
230    use regfile(ctl_dest, ctl_sbus, rf_in,    rf_out);
231    use alu    (ctl_dest, ctl_alu,  sbus,    alu_in,  alu_out);
232    use shifter(ctl_shft, zero, carry, shft_in, shft_out);
233
234    always {
235      sbus       = (ctl_sbus == SBUS_IN) ? din   : rf_out;
236      din_strb   = (ctl_sbus == SBUS_IN) ? 1     : 0;
237      dout       = sbus;
238      dout_strb  = (ctl_ot == O_WR) ? 1 : 0;
239      rf_in      = shft_out;
240      alu_in     = shft_out;
241      shft_in    = alu_out;
242    }
243  }
244
245  dp hmmtest {
246    sig din         : ns(WLEN);
247    sig din_strb    : ns(1);
248    sig dout        : ns(WLEN);
249    sig dout_strb   : ns(1);
250    use hmm(din, din_strb, dout, dout_strb);
251
252    reg dcnt        : ns(5);
253    lookup stim     : ns(WLEN) = {14,32,87, 12, 23, 99, 32, 22};
254
255    always {
256      dcnt = (din_strb) ? dcnt + 1 : dcnt;
257      din  = stim(dcnt & 7);
258      $display($cycle, " IO ", din_strb, " ", dout_strb, " ",
259      $dec, din, " ", dout);
260    }
261  }
262
263  system S {
264    hmmtest;
265  }
```

This design can be simulated with the `fdlsim` GEZEL simulator. Because of the C macro's included in the source, the program first needs to be processed using the C preprocessor. The following command line illustrates how to simulate the first 100 cycles of this design.

```
>cpp -P hmm2.fdl | fdlsim 100
```

The first few lines of output look as follows.

```
0 IO 1 0 14 14
1 IO 1 0 32 32
2 IO 0 0 87 14
3 IO 0 0 87 14
4 IO 0 0 87 14
. . .
```

The micro-programmed machine reads the numbers 14 and 32 in clock cycles 0 and 1 respectively, and starts the GCD calculation. To find the corresponding GCD, we look for a '1' in the fourth column (output strobe). Around cycle 21, the first one appears. We can find that GCD(32,14) = 2. The testbench then proceeds with the next two inputs in cycles 23 and 24.

```
. . .
18 IO 0 0 87 2
19 IO 0 0 87 2
20 IO 0 0 87 2
21 IO 0 1 87 2
22 IO 1 0 87 87
23 IO 1 0 12 12
24 IO 0 0 23 87
. . .
```

A quick command to filter out the valid outputs during simulation is the following.

```
> cpp -P hmm2.fdl | fdlsim 200 | awk '{if ($4 == "1") print $0}'
21 IO 0 1 87 2
55 IO 0 1 23 3
92 IO 0 1 32 1
117 IO 0 1 14 2
139 IO 0 1 87 2
173 IO 0 1 23 3
```

This design demonstrates how the FSMD model can be applied to create a more complex, micro-programmed machine. In the following, we show how this can be used to create programming concepts at even higher levels of abstraction, using micro-program interpreters.

## 6.6 Micro-program Interpreters

A micro-program is a highly-optimized sequence of commands for a datapath. The sequence of register transfers is optimized for parallelism. Writing efficient micro-programs is not easy, and requires in-depth understanding of the machine architecture. An obvious question is if a programming language, such as a pseudo-assembly language, would be of help in developing micro-programs. Certainly, the writing process itself could be made more convenient.

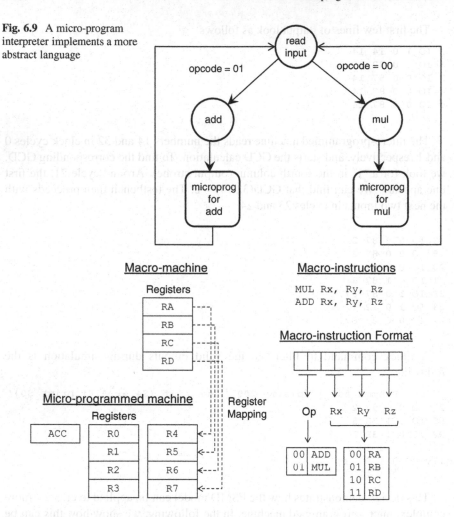

**Fig. 6.9** A micro-program interpreter implements a more abstract language

**Fig. 6.10** Programmer's model for the macro-machine example

A common use for micro-programs is therefore not to encode complete applications, but instead to work as *interpreters* for other programs, developed at a high abstraction level. An interpreter is a machine that decodes and executes instruction sequences of an abstract high-level machine, which we will call the macro-machine. The instructions from the macro-machine will be implemented in terms of micro-programs for a micro-programmed machine. Such a construct is illustrated in Fig. 6.9, and is called a *micro-program interpreter*. We create a micro-program in the form of an infinite loop, which reads a macro-instruction byte and breaks down a byte in opcode and operand fields. It then takes specific actions depending on the values of the opcode.

We discuss, by means of an example, how to implement such a macro-machine. Figure 6.10 shows the programmers' model of the macro-machine. It is a very simple machine, with four registers RA through RD, and two instructions for adding and multiplying those registers. The macro-machine will have the same wordlength as the micro-programmed machine, but it has fewer register than the original micro-programmed machine. To implement the macro-machine, we will map the macro-register set directly onto the micro-register set. In this case, we will map register RA to RD onto register R4 to R7 respectively. This leaves register R0 to R3, as well as the accumulator, available to implement macro-instructions. The macro-machine has two instructions: add and mul. Each of these instructions takes two source operands and generates one destination operand. The operands are macro machine registers. Because the micro machine has to decode the macro-instructions, we also need to choose the instruction-encoding of the macro-instructions. This is illustrated on the right of Fig. 6.10. Each macro-instruction is a single byte, with 2 bits for the macro-opcode, and 2 bits for each of the macro-instruction operands.

Listing 6.3 shows a sample implementation for each of the ADD and MUL instructions. We have assumed that single-level subroutines are supported at the level of the micro-machine. See Problem 6.3 how such a subroutine call can be implemented in the micro-programmed machine.

The micro-interpreter loop, on line 21–29, reads one macro-instruction from the input, and determines the macro-instruction opcode with a couple of shift instructions. Depending on the value of the opcode field, the micro-program will then jump to a routine to implement the appropriate macro-instruction, add or mul.

The implementation of ADD is shown in lines 35–40. The micro-instructions use fixed source operands and a fixed destination operand. Since the macro-instructions can use one of four possible operand registers, an additional register-move operation is needed to prepare the micro-instruction operands. This is done by the putarg and getarg subroutines, starting on line 62. The getarg subroutine copies data from the macro-machine source registers (RA through RD) to the micro-machine source working registers (R1 and R2). The putarg subroutine moves data from the micro-machine destination working register (R1) back to the destination macro-machine register (RA through RD).

The implementation of the add instruction starts on line 35. At the start of this section of code, the accumulator contains the macro-instruction. The accumulator value is passed on to the getarg routine, which decodes the two source operand registers can copies them into micro-machine register R1 and R2. Next, the add macro-instruction performs the addition, and stores the result in R1 (line 36–39). The putarg and getarg routines assume that the opcode of the macro-instruction is stored in the accumulator. Since the body of the add instruction changes the accumulator, it needs to be preserved before putarg is called. This is the purpose of the register-copy instructions on lines 36 and 39.

The implementation of the mul macro-instruction starts on line 46, and follows the same principles as the add instruction. In this case, the body of the instruction is more complex since the multiply operation needs to be performed using an add-and-shift loop. A loop counter is created in register R3 to perform eight iterations of add-and-shift. Because the accumulator register is only 8 bit, the multiply instruction cannot capture all 16 output bits of an 8-by-8 bit multiply. The implementation of mul preserves only the least significant byte.

**Listing 6.3** Implementation of the macro-instructions ADD and MUL

```
 1  //-------------------------------------------------
 2  // Macro-machine for the instructions
 3  //
 4  //    ADD Rx, Ry, Rz
 5  //    MUL Rx, Ry, Rz
 6  //
 7  //    Macro-instruction encoding:
 8  //    +----+----+----+----+
 9  //    | ii + Rx + Ry + Rz +
10  //    +----+----+----+----+
11  //
12  //    where ii = 00 for ADD
13  //               01 for MUL
14  //    where Rx, Ry and Rz are encoded as follows:
15  //               00 for RA (mapped to R4)
16  //               01 for RB (mapped to R5)
17  //               10 for RC (mapped to R6)
18  //               11 for RD (mapped to R7)
19  //
20  // Interpreter loop reads instructions from input
21  macro:    IN -> ACC
22            (ACC & 0xC0) >> 1 -> R0
23            R0 >> 1 -> R0
24            R0 >> 1 -> R0
25            R0 >> 1 -> R0
26            R0 >> 1 -> R0
27            R0 >> 1 -> R0              || JUMP_IF_NZ mul
28            (no_op)                    || JUMP add
29  macro_done: (no_op)                  || JUMP macro
30
31  //-------------------------------------------------
32  //
33  // Rx = Ry + Rz
34  //
35  add:      (no_op)                    || CALL getarg
36            ACC -> R0
37            R2  -> ACC
38            (R1 + ACC) -> R1
39            R0 -> ACC                  || CALL putarg
40            (no_op)                    || JUMP macro_done
41
42  //-------------------------------------------------
43  //
44  // Rx = Ry * Rz
45  //
46  mul:      (no_op)                    || CALL getarg
47            ACC -> R0
```

```
48              0    -> ACC
49              8    -> R3
50  loopmul:    (R1 << 1) -> R1            || JUMP_IF_NC nopartial
51              (ACC << 1) -> ACC
52              (R2 + ACC) -> ACC
53  nopartial:  (R3 - 1) -> R3             || JUMP_IF_NZ loopmul
54              ACC -> R1
55              R0 -> ACC                  || CALL putarg
56              (no_op)                    || JUMP macro_done
57
58  //------------------------------------------------------------
59  //
60  // GETARG
61  //
62  getarg:     (ACC & 0x03) -> R0         || JUMP_IF_Z Rz_is_R4
63              (R0 - 0x1)                 || JUMP_IF_Z Rz_is_R5
64              (R0 - 0x2)                 || JUMP_IF_Z Rz_is_R6
65  Rz_is_R7:   R7 -> R1                   || JUMP get_Ry
66  Rz_is_R6:   R6 -> R1                   || JUMP get_Ry
67  Rz_is_R5:   R5 -> R1                   || JUMP get_Ry
68  Rz_is_R4:   R4 -> R1                   || JUMP get_Ry
69  get_Ry:     (ACC & 0x0C) >> 1 -> R0
70              R0 >> 1 -> R0              || JUMP_IF_Z Ry_is_R4
71              (R0 - 0x1)                 || JUMP_IF_Z Ry_is_R5
72              (R0 - 0x2)                 || JUMP_IF_Z Ry_is_R6
73  Ry_is_R7:   R7 -> R2                   || RETURN
74  Ry_is_R6:   R6 -> R2                   || RETURN
75  Ry_is_R5:   R5 -> R2                   || RETURN
76  Ry_is_R4:   R4 -> R2                   || RETURN
77
78  //------------------------------------------------------------
79  //
80  // PUTARG
81  //
82  putarg:     (ACC & 0x30) >> 1 -> R0
83              R0 >> 1 -> R0
84              R0 >> 1 -> R0
85              R0 >> 1 -> R0              || JUMP_IF_Z Rx_is_R4
86              (R0 - 0x1)                 || JUMP_IF_Z Rx_is_R5
87              (R0 - 0x2)                 || JUMP_IF_Z Rx_is_R6
88  Rx_is_R7:   R1 -> R7                   || RETURN
89  Rx_is_R6:   R1 -> R6                   || RETURN
90  Rx_is_R5:   R1 -> R5                   || RETURN
91  Rx_is_R4:   R1 -> R4                   || RETURN
```

A micro-programmed interpreter creates the illusion of a machine that has more powerful instructions than the original micro-programmed architecture. The trade-off made with such an interpreter is ease-of-programming versus performance: each instruction of the macro-machine may need many micro-machine instructions. The concept of microprogram interpreters has been used extensively to design processors with configurable instruction sets, and was originally used to enhance the flexibility of expensive hardware. Today, the technique of micro-program interpreter design is still very useful to create an additional level of abstraction on top of a micro-programmed architecture.

## 6.7  Micro-program Pipelining

As can be observed from Fig. 6.11, the micro-program controller may be part of a long chain of combinational logic. Pipeline registers can be used to break these long chains. However, the introduction of pipeline registers has a large impact on the design of micro-programs. This section will study these effects in more detail.

First, observe in Fig. 6.11 that the CSAR register is part of possibly three loops with logic. The first loop runs through the next-address logic. The second loop runs through the control store and the next-address logic. The third loop runs through the control store, the data path, and the next-address logic. These combinational paths may limit the maximum clock frequency of the micro-programmed machine. There are three common places where additional pipeline registers may be inserted in the design of this machine, and they are marked with shaded boxes in Fig. 6.11.

- A common location to insert a pipeline register is at the output of the control store. A register at that location is called the micro-instruction register, and it enables overlap of the datapath evaluation, the next address evaluation, and the micro-instruction fetch.

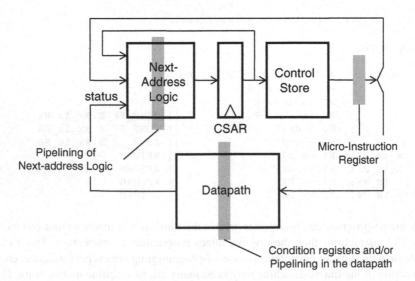

**Fig. 6.11** Typical placement of pipeline registers in a micro-program interpreter

**Table 6.2** Effect of the micro-instruction register on jump instructions

| Cycle | CSAR | Micro-instruction register |
| --- | --- | --- |
| N | 4 | |
| N+1 | 5 | CSTORE(4) = JUMP 10 |
| N+2 | 10 | CSTORE(5) *need to cancel* |
| N+3 | 11 | CSTORE(10) *execute* |

- Another location for pipeline registers is the datapath. Besides pipeline register inside of the data path, additional condition-code registers can be placed at the datapath outputs.
- Finally, the next-address logic may be pipelined as well, in case high-speed operation is required and the target CSAR address cannot be evaluated within a single clock cycle.

Each of these register cuts through a different update-loop of the CSAR register, and therefore each of them has a different effect on the micro-program.

### 6.7.1 Micro-instruction Register

Let's first consider the effect of adding the micro-instruction register. Due to this register, the micro-instruction fetch (i.e. addressing the CSTORE and retrieving the next micro-instruction) is offset by one cycle from the evaluation of that micro-instruction. For example, when the CSAR is fetching instruction $i$ from a sequence of instructions, the datapath and next-address logic will be executing instruction $i - 1$.

Table 6.2 illustrates the effect of this offset on the instruction stream, when that stream contains a jump instruction. The micro-programmer entered a JUMP 10 instruction in CSTORE location 4, and that instruction will be fetched in clock cycle N. In clock cycle N+1, the micro-instruction will appear at the output of the micro-instruction register. The execution of that instruction will complete in cycle N+2. For a JUMP, this means that the value of CSAR will be affected in cycle N+2. As a result, a JUMP instruction cannot modify the value of CSAR within a single clock cycle. If CSTORE(4) contains a JUMP, then the instruction located in CSTORE(5) will be fetched as well. The micro-programmer needs to be aware of this. The possible strategies are (a) take into account that a JUMP will be executed with one cycle of delay (so-called 'delayed branch'), or (b) include support in the micro-programmed machine to cancel the execution of an instruction in case of a jump.

**Table 6.3** Effect of the micro-instruction register and condition-code register on conditional jump instructions

| Cycle | CSAR | Micro-instruction register |
|---|---|---|
| N | 3 | |
| N+1 | 4 | CSTORE(3) = TEST R0 sets Z-flag |
| N+2 | 5 | CSTORE(4) = JZ 10 |
| N+3 | 10 | CSTORE(5) *need to cancel* |
| N+4 | 11 | CSTORE(10) *execute* |

**Table 6.4** Effect of additional pipeline registers in the CSAR update loop

| Cycle | CSAR_pipe | CSAR | Micro-instruction register |
|---|---|---|---|
| 0 | 0 | 0 | CSTORE(0) |
| 1 | 1 | 0 | CSTORE(0) twice? |
| 2 | 1 | 1 | CSTORE(1) |
| 3 | 2 | 1 | CSTORE(1) twice? |

## 6.7.2   Datapath Condition-Code Register

As a second case, assume that we have a condition-code register in the data-path, in addition to a micro-instruction register. The net effect of a condition code register is that a condition value will only be available one clock cycle after the corresponding datapath operation. As a result, a conditional-jump instruction can only operate on datapath conditions from the previous clock cycle. Table 6.3 illustrates this effect. The branch instruction in CSTORE(4) is a conditional jump. When the condition is true, the jump will be executed with one clock cycle delay, as was discussed before. However, the JZ is evaluated in cycle N+2 based on a condition code generated in cycle N+1. Thus, the micro-programmer needs to be aware that conditions need to be available one clock cycle before they will be used in conditional jumps.

## 6.7.3   Pipelined Next-Address Logic

Finally, assume that there is a third level of pipelining available inside of the next-address update loop. For simplicity, we will assume there are two CSAR registers back-to-back in the next-address loop. The output of the next-address-logic is fed into a register CSAR_pipe, and the output of CSAR_pipe is connected to CSAR. Table 6.4 shows the operation of this micro-programmed machine, assuming all registers are initially zero. As shown in the table, the two CSAR registers in the next-address loop result in two (independent) address sequences. When all registers start out at 0, then each instruction of the micro-program will be executed twice. Solving this problem is not easy. While one can do a careful initialization of CSAR_pipe and CSAR such that they start out at different values (e.g. 1 and 0), this re-initialization will need to be done at each jump instruction. This makes the design and the programming of pipelined next-address logic very hard.

These examples show that a micro-programmer must be aware of the implementation details of the micro-architecture, and in particular of all the delay effects present in the controller. This can significantly increase the complexity of the development of micro-programs.

## 6.8 Microprogramming with Microcontrollers

Although microprogramming originated many years ago, its ideas are still very useful. When complex systems are created in hardware, the design of an adequate control architecture often becomes a challenge. In this section, we illustrate a possible solution based on the use of a micro-controller.

### 6.8.1 System Architecture

A micro-controller has minimal computational capabilities, such as an ALU with basic logical and arithmetic operations. However, they are pervasively used for all sorts of control applications. In this section, we will discuss using them as micro-program controllers.

Figure 6.12 shows an example of a micro-programmed architecture that uses a microcontroller. We assume a device which has 8-bit digital I/O ports. The microcontroller has dedicated instructions to read from, and write to, such ports.

In the design of Fig. 6.12, the I/O ports are used to control a microcoded datapath. There are three ports involved.

- A digital input port is used to carry data or status from the microcoded datapath to the microcontroller.
- A first digital output port is used to carry data from the microcontroller to the microcoded datapath.
- A second digital output port is used to carry control or micro-instructions from the microcontroller to the microcoded datapath.

The machine works as follows. For each micro-instruction, the microcontroller will combine a micro-instruction with an optional argument, and send that to the microcoded datapath. The microcoded datapath will then return any result to the microcontroller. Care needs to be taken to keep the microcoded datapath and the microcontroller synchronized. If the execution of a microcoded instruction takes multiple cycles, the microcontroller will need to delay reading data output and/or status over an appropriate amount of cycles. The next subsection demonstrates an example of this technique.

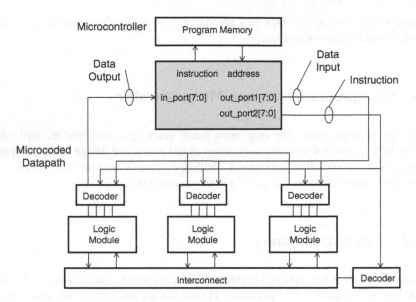

**Fig. 6.12** Using a microcontroller as a microprogram controller

## 6.8.2  Example: Bresenham Line Drawing

We will discuss an example microcoded machine, designed using an 8051 micro-controller. The application is that of Line Drawing on a grid. Figure 6.13 illustrates the concept. A continuous line is defined between (0,0) and (tx,ty). The line is drawn using a discrete grid of pixels, and the problem to solve is to decide what pixels should be turned on. We assume that the grid has unit spacing, so that tx and ty are positive integers.

An algorithm to solve this problem was proposed by Bresenham. His solution starts with the following observation. If a line segment lies in the first octant (or, tx > ty > 0), then the line can be drawn, pixel per pixel, by taking only horizontal and diagonal steps. Thus, if the pixel (x1, y1) is turned on, then the next pixel to turn on will be either (x1+1,y1) or else (x1+1,y1+1). The pixel that should be selected is the one that lies 'closest' to the true line. Bresenham's insight was to show that the closest pixel can be obtained using integer arithmetic only.

Indeed, assume that pixel (x1, y1) is already turned on, and that its center has a distance e from the true line. If the center of pixel (x1,y1) is above the true line, then e is negative. If the center of pixel (x1, y1) is below the true line, then e is positive. The line drawing algorithm can be controlled based on the sign of e. As long as e is positive, we need to take diagonal steps. When e becomes negative, we should take a horizontal step. In this manner, the error e is minimized over the entire line.

To compute e at every pixel, we proceed as follows. When taking a horizontal step, the true line will move up following the slope ty/tx. The error will increase

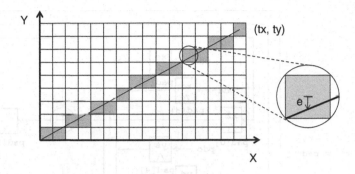

**Fig. 6.13** Bresenham line drawing algorithm

by es = ty/tx. If we would take a diagonal step, then the error will increase by
ed = ty/tx - 1. The factor 1 represents the step in the y direction, assuming
a unit-spaced grid. Both quantities, es and ed, can be scaled to integer values by
multiplying them with tx, which is a constant over the entire line.

Listing 6.4 shows a C function to draw pixels using the Bresenham algorithm.
This function can draw lines in the first quadrant, and hence distinguishes between
the cases ty>tx and tx>ty. The algorithm is written using unsigned char,
and is limited in precision to byte-size coordinates. The sign test for the error
accumulator is written as a test on the most significant bit of e, i.e. (e & 0x80).
The algorithm can be compiled and executed on an 8-bit microcontroller.

**Listing 6.4** C program for line drawing in the first quadrant

```
1   void bresen(unsigned tx, unsigned ty) {
2       unsigned char  x, y;     // pixel coordinates
3       unsigned char  e;        // error accumulator
4       unsigned char  es, ed;   // error inc for straight/diag
5       unsigned char  xs, xd;   // x inc for straight/diag
6       unsigned char  ys, yd;   // y inc for straight/diag
7       unsigned i;
8
9       x  = 0; y = 0;
10      e  = 0;
11      ed = (tx > ty) ? (ty - tx) : (tx - ty);
12      es = (tx > ty) ? ty         : tx;
13      xd = 1;
14      xs = (tx > ty) ? 1 : 0;
15      yd = 1;
16      ys = (tx > ty) ? 0 : 1;
17
18      for (i=0; i<64; i++) {  // plot 64 pixels
19          // plot(x, y);
20          x = (e & 0x80) ? x + xs : x + xd;
21          y = (e & 0x80) ? y + ys : y + yd;
22          e = (e & 0x80) ? e + es : e + ed;
23      }
24  }
```

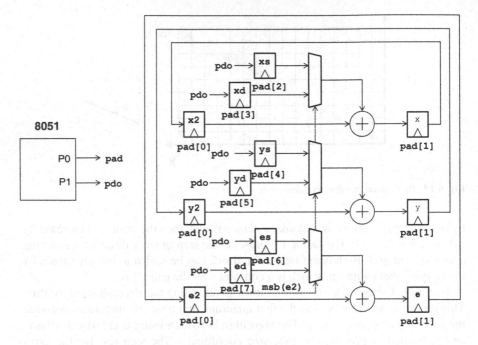

**Fig. 6.14** Bresenham line drawing microcoded datapath

Next, we design a microcoded datapath for the Bresenham line drawing algorithm. The most obvious candidate for such a datapath is the most intensively executed part of the code, the loop body of Listing 6.4 (lines 20–22). Figure 6.14 shows a hardware implementation for the microcoded datapath. Three adders work in parallel to update the x and y registers holding the current pixel coordinate, as well as the e register, holding the error accumulator. The values to add are stored in additional registers within the microcoded datapath: xs, xd, ys, yd, es, and ed. The microcoded datapath is fully controlled from the 8051 microcontroller through two 8-bit ports. These ports define the value of pad and pdo.

Listing 6.5 shows the GEZEL version of the microcoded datapath in Fig. 6.14. The listing also shows the inclusion of the 8051 microcontroller, and two ports. The C driver code for the microcoded datapath is shown in Listing 6.6.

The microinstructions generated from the 8051 will look as combinations of pad and pdo. For example, let's see how to program the step registers in the datapath. There are four step registers (xd, xs, yd, ys). Their update is tied to pad[3], pad[2], pad[5], pad[4] respectively. To program xd, the 8051 needs to write the desired value to pdo, then toggle bit pad[3]. Thus, in the GEZEL description (Listing 6.5) on line 40, we find:

```
xd = pad[3] ? pdo : xd;
```

The corresponding programming command is shown on Line 27 of the C driver (Listing 6.6):

```
P1 = xd; P0 = 0x08; P0 = 0x0;
```

As a second example, let's see how the 8051 will implement loop iterations. In Fig. 6.14, each of the loops that contains register x, y and e, also contains a second register x2, y2 and e2. The control bits in pad can now be steered to update either the set x, y and e, or else the set x2, y2 and e2. The update of x, y and e is tied to pad[1], while the update of x2, y2 and e2 is tied to pad[0]. Hence, in the GEZEL description (Listing 6.5) around line 30, we find:

```
xi = e2[7] ? xs : xd;
x2 = pad[0] ? x  : x2;
x  = pad[1] ? x2 + xi : x;
```

To implement one iteration of the loop, the 8051 controller first toggles pad[1], and next toggles pad[2]. This can be done simultaneously, so that the 8051 controller writes the values 0x1 and 0x2 to pad. The corresponding loop iteration command is shown on Lines 30 and 31 of the C driver (Listing 6.6):

```
P0 = 0x1;
P0 = 0x2;
```

**Listing 6.5** GEZEL Microcoded Datapath

```
1   ipblock my8051 {
2     iptype "i8051system";
3     ipparm "exec=bresen.ihx";
4   }
5
6   ipblock mi_8051(out data : ns(8)) {
7     iptype "i8051systemsource";
8     ipparm "core=my8051";
9     ipparm "port=P0";
10  }
11
12  ipblock dout_8051(out data : ns(8)) {
13    iptype "i8051systemsource";
14    ipparm "core=my8051";
15    ipparm "port=P1";
16  }
17
18  dp microdp(in  pad : ns(8);
19             in  pdo : ns(8)) {
20    reg xd, yd : ns(1);
21    reg xs, ys : ns(1);
22    reg x,  y  : ns(8);
23    reg ed, es : ns(8);
24    reg e      : ns(8);
25    reg x2, y2 : ns(8);
26    reg e2     : ns(8);
```

```
27    sig xi, yi : ns(8);
28    sig ei      : ns(8);
29    always {
30      xi  = e2[7]  ? xs       : xd;
31      yi  = e2[7]  ? ys       : yd;
32      ei  = e2[7]  ? es       : ed;
33      x2  = pad[0] ? x        : x2;
34      y2  = pad[0] ? y        : y2;
35      e2  = pad[0] ? e        : e2;
36      x   = pad[1] ? x2 + xi  : x;
37      y   = pad[1] ? y2 + yi  : y;
38      e   = pad[1] ? e2 + ei  : e;
39      xs  = pad[2] ? pdo      : xs;
40      xd  = pad[3] ? pdo      : xd;
41      ys  = pad[4] ? pdo      : ys;
42      yd  = pad[5] ? pdo      : yd;
43      es  = pad[6] ? pdo      : es;
44      ed  = pad[7] ? pdo      : ed;
45    }
46  }
47
48  dp top {
49    sig pad, pdo : ns(8);
50    use my8051;
51    use mi_8051(pad);
52    use dout_8051(pdo);
53    use microdp(pad, pdo);
54  }
55
56  system S {
57    top;
58  }
```

**Listing 6.6** 8051 Driver Code for GEZEL Microcoded Datapath

```
1   #include <8051.h>
2
3   void bresen_hw(unsigned tx, unsigned ty) {
4     unsigned char  x, y;   // pixel coordinates
5     unsigned char  e;      // error accumulator
6     unsigned char  es, ed; // error inc for straight/diag
7     unsigned char  xs, xd; // x inc for straight/diag
8     unsigned char  ys, yd; // y inc for straight/diag
9     unsigned i;
10
11    x  = 0;
12    y  = 0;
13    e  = 0;
14    ed = (tx > ty) ? (ty - tx) : (tx - ty);
15    es = (tx > ty) ? ty        : tx;
16    xd = 1;
17    xs = (tx > ty) ? 1 : 0;
18    yd = 1;
19    ys = (tx > ty) ? 0 : 1;
```

```
20      // P0 - MI    [ed es yd ys xd xs xye x2y2e2]
21      //              7   6   5   4   3   2   1       0
22      // P1 - DOUT
23      P1 = ed;   P0 = 0x80;   P0 = 0x0;
24      P1 = es;   P0 = 0x40;   P0 = 0x0;
25      P1 = yd;   P0 = 0x20;   P0 = 0x0;
26      P1 = ys;   P0 = 0x10;   P0 = 0x0;
27      P1 = xd;   P0 = 0x8;    P0 = 0x0;
28      P1 = xs;   P0 = 0x4;    P0 = 0x0;
29      for (i=0; i<64; i++) {
30          P0 = 0x1;
31          P0 = 0x2;
32      }
33  }
```

This concludes the example of Bresenham line drawing using a microcoded datapath and a microcontroller. This technique is very useful if one is coping with the design of control functions in hardware design.


## 6.9  Summary

In this section we introduced Microprogramming as means to deal with control design problems in hardware. Finite State Machines are good for small, compact specifications, but they result in a few issues. Finite State Machines cannot easily express hierarchy (FSMs calling other FSMs). Therefore control problems can easily blow up when specified as a finite-state-machine, yielding so-called 'state explosion'.

In a micro-programmed architecture, the hardcoded next-state logic of a finite state machine is replaced with a programmable control store and a program control (called CSAR or Control Store Address Register). This takes care of most problems: micro-programs can call other micro-programs using jump instructions or using the equivalent of subroutines. Micro-programs have a much higher scalability than finite state machines.

Writing micro-programs is more difficult than writing assembly code or C code. Therefore, instead of mapping a full application directly into micro-code, it may be easier to develop a micro-programmed interpreter. Such an interpreter implements an instruction-set for a language at a higher level of abstraction. Still, even single micro-instructions may be hard to write, and in particular the programmer has to be aware of all pipelining and delay effects inside of the micro-programmed architecture.

## 6.10   Further Reading

The limitations of FSM as a mechanism for control modeling have long been recognized in literature. While it has not been the topic of the chapter, there are several alternate control modeling mechanisms available. A key contribution to hierarchical modeling of FSM was defined by Harel in StateCharts (Harel 1987). Additionally, the development of so-called synchronous languages have support the specification of control as event-driven programs. See for example Esterel by Berry (2000) and Potop-Butucaru et al. (2007).

A nice introduction and historical review of Microprogramming can be found on-line on the pages of Smotherman (2009). Most of the work on micro-programming was done in the late 1980s and early 1990s. Conference proceedings and computer-architecture books from that period are an excellent source of design ideas. For example, a extensive description of micro-programming is found in the textbook by Lynch (1993). Control optimization issues of micro-programming are discussed by Davio et al. (1983).

## 6.11   Problems

**Problem 6.1.** Figure 6.15 shows a micro-programmed datapath. There are six control bits for the datapath: 2 bits for each of the multiplexers M1 and M2, and 2 bits for the ALU. The encoding of the control bits is indicated in the figure.

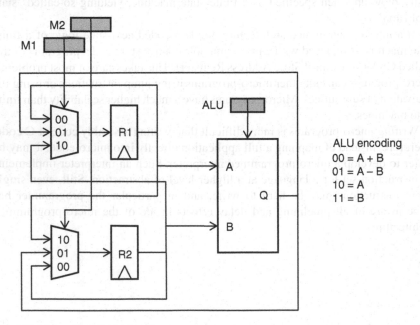

**Fig. 6.15**  Datapath for Problem 6.1

**Table 6.5** Micro-instructions for Problem 6.1

| SWAP | Interchange the content of R1 and R2. |
|---|---|
| ADD Rx | Add the contents of R1 and R2 and store the contents in Rx, which is equal to R1 or R2. There are two variants of this instruction depending on Rx. |
| COPY Rx | Copy the contents of Ry into Rx. (Rx, Ry) is either (R1, R2) or (R2, R1). There are two variants of this instruction depending on Rx. |
| NOP | Do nothing. |

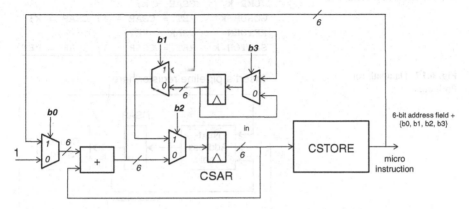

**Fig. 6.16** Datapath for Problem 6.3

(a) Develop a horizontal micro-instruction encoding for the list of micro-instructions shown in Table 6.5.
(b) Develop a vertical micro-instruction encoding for the same list of instructions. Use a reasonable encoding that results in a compact and efficient decoder for the datapath.

**Problem 6.2.** Using the micro-programmed machine discussed in Sect. 6.5, create a program that reads in a number from the input and that counts the number of non-zero bits in that number. The resulting bitcount must be stored in register R7.

**Problem 6.3.** Figure 6.16 shows the implementation of a next-address decoder. A total of 10 bits from the micro-instruction are used to control the next-address logic: a 6-bit address field, and four control bits, b0, b1, b2, and b3.

For each of the combinations of control bits shown in Table 6.6, find a good description of the instruction corresponding to the control bit values shown. Don't write generic descriptions (like 'CSAR register is incremented by one'), but give a high-level description of the instruction they implement. Use terms that a software programmer can understand

**Problem 6.4.** Design a next-address instruction decoder based on the set of micro-instructions shown in Table 6.7. Design your implementation in GEZEL or Verilog. An example IO definition is shown next. The CSAR has to be 10 bit wide, the

**Table 6.6** Next-address
instructions for Problem 6.3

| Combination | b0 | b1 | b2 | b3 |
|---|---|---|---|---|
| Instruction 1 | 1 | X | 0 | 0 |
| Instruction 2 | X | 1 | 1 | 0 |
| Instruction 3 | 0 | 1 | 1 | 1 |
| Instruction 4 | X | 0 | 1 | 0 |

**Table 6.7** Next-address
instructions for Problem 6.4

```
NXT        CSAR = CSAR + 1;
JUMP k     CSAR = k;
GOSUB k    RET = CSAR + 1; CSAR = k;
RETURN     CSAR = RET;
SWITCH k   RET = CSAR + 1; CSAR = RET;
```

**Fig. 6.17** Datapath for
Problem 6.5

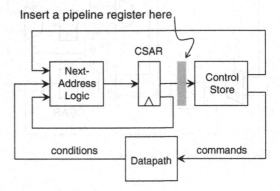

width of the Address field and the width of the next-address field must be chosen
accordingly.

```
dp nextaddress_decoder (in csar    : ns(10);
                        out address : ns(x);
                        out next    : ns(y)) {
    // ...
}
```

**Problem 6.5.** Your colleague asks you to evaluate an enhancement for a micro-
programmed architecture, as illustrated in Fig. 6.17. The enhancement is to insert a
pipeline register just in from of the control store.

(a) Does this additional register reduce the critical path of the overall architecture?
(b) Your colleague calls this a dual-thread architecture, and claims this enhance-
    ment allows the micro-control engine to run two completely independent
    programs in an interleaved fashion. Do you agree with this or not?

# Chapter 7
# General-Purpose Embedded Cores

## 7.1 Processors

The most successful programmable component over the past decades is, without doubt, the microprocessor. The microprocessor is truly ubiquitous. Any modern electronic device more complicated than a pushbutton seems fitted with a microprocessor; a modern car contains 50–100 embedded microprocessors. There are several reasons for the universal success of the microprocessor.

- Microprocessors, or the stored-program concept in general, separate software from hardware through the definition of an instruction-set. No other hardware development technique has ever been able to uncouple hardware and software in a similar way. Think about microprogramming. Microprograms are really shorthand notations for the control specification of a specialized datapath. The notion of a micro-program as an architecture-independent concept makes no sense: micro-instructions are architecture-specific.
- Microprocessors come with tools (compilers and assemblers), that help a designer to create applications. The availability of a compiler to translate a programming language into a program for a microprocessor is an enormous advantage. An embedded software designer can write applications in a high-level programming language, independently from the specific microprocessor. Compilers help embedded software designers to quickly migrate from one processor family to the next one.
- No other device has been able to cope as efficiently with reuse as microprocessors did. *Reuse*, in general, is the ability to save design effort over multiple applications. A general-purpose embedded core is an excellent example of reuse in itself. However, microprocessors have also a large impact on reuse in electronic system design. Microprocessors come with bus interfaces that support the physical integration of an electronic system consisting of multiple components. Their compilers have enabled the development of standard software libraries as well as the logical integration of a system.

P.R. Schaumont, *A Practical Introduction to Hardware/Software Codesign*,
DOI 10.1007/978-1-4614-3737-6_7, © Springer Science+Business Media New York 2013

- Fourth, no other programmable component has the same scalability as a microprocessor. The concept of the stored-program computer has been implemented across a wide range of word-lengths (4-bit ... 64-bit). Microprocessors have also gained significant traction as central unit in complex integrated circuits. In this approach, called System-on-Chip (SoC), a microprocessor controls the collaboration of one or more complex peripherals. We will discuss SoC in the next chapter.

In summary, the combination of instruction-set, tools, reuse, and scalability have turned the microprocessor into a dominant component in electronic systems. In fact, very often hardware/software codesign starts with a program on a general-purpose microprocessor, and specializes that design by adding dedicated hardware to the microprocessor.

### 7.1.1   The Toolchain of a Typical Micro-processor

Figure 7.1 illustrates the typical design flow to convert software source code into instructions for a processor. The Figure introduces the following terminology used in this Chapter. A compiler or an assembler is used to convert source code into object code. An object code file contains opcodes (instructions) and constants, along with supporting information to organize these instructions and constants in memory. A linker is used to combine several object code files into a single, standalone executable file. A linker will also resolve all unknown elements in the object code, such as the address of external variables or the entry point of library routines. Finally, a loader program determines how the information in an executable file is organized in memory. Typically, there will be a part of the memory space reserved for instructions, another part for constant data, another part for global data with read/write access, and so on. A very simple microprocessor system requires at least two elements: a processor, and a memory holding instructions for the processor. The memory is initialized with processor instructions by the loader. The processor will fetch these instructions from memory and execute them on the processor datapath.

### 7.1.2   From C to Assembly Instructions

Listing 7.1  A C program to find the maximum of greatest common divisors

```
1   int gcd(int a[5], int b[5]) {
2     int i, m, n, max;
3     max = 0;
4     for (i=0; i<5; i++) {
5       m = a[i];
6       n = b[i];
```

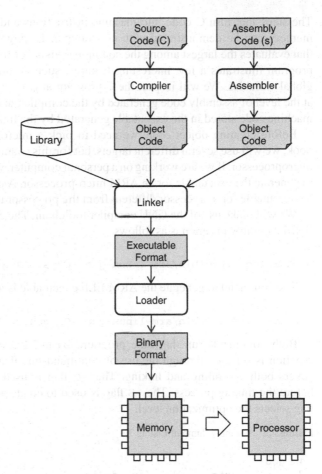

**Fig. 7.1** Standard design flow of software source code to processor instruction

```
 7      while (m != n) {
 8        if (m > n)
 9          m = m - n;
10        else
11          n = n - m;
12      }
13      if (max > m)
14        max = m;
15    }
16    return max;
17  }
18
19  int a[] = {26, 3,33,56,11};
20  int b[] = {87,12,23,45,17};
21
22  int main() {
23    return gcd(a, b);
24  }
```

The steps that turn C code into machine instructions and, eventually, opcodes in memory can be demonstrated with an example. Listing 7.1 shows a C program that evaluates the largest among the common divisors of five pairs of numbers. The program illustrates a few interesting features, such as function calls, arrays, and global variables. We will inspect the C program at two levels of abstraction. First, at the level of assembly code generated by the compiler, and next, at the level of the machine code stored in the executable generated by the linker.

Before creating object code, we need to pick a microprocessor target. In this book, we will use several different targets, but for this example, we choose the ARM microprocessor. If we are working on a personal computer, we need a *cross-compiler* to generate the executable for an ARM microprocessor. A cross-compiler generates an executable for a processor different from the processor used to run the compiler.

We will make use of the GNU compiler toolchain. The command to generate the ARM assembly program is as follows.

```
> /usr/local/arm/bin/arm-linux-gcc -c -S -O2 gcd.c -o gcd.s
```

The command to generate the ARM ELF executable is as follows.

```
> /usr/local/arm/bin/arm-linux-gcc -O2 gcd.c -o gcd
```

Both commands run the same program, arm-linux-gcc, but the specific function is selected through the use of command-line flags. The default behavior covers both compiling and linking. The -c flag is used to end the compilation before the linking process. The -S flag is used to create assembly code. The -O2 flag selects the optimization level.

**Listing 7.2** Assembly dump of Listing 7.1

```
 1   gcd:
 2              str     lr, [sp, #-4]!
 3              mov     lr, #0
 4              mov     ip, lr
 5   .L13:
 6              ldr     r3, [r0, ip, asl #2]
 7              ldr     r2, [r1, ip, asl #2]
 8              cmp     r3, r2
 9              beq     .L17
10   .L11:
11              cmp     r3, r2
12              rsbgt   r3, r2, r3
13              rsble   r2, r3, r2
14              cmp     r3, r2
15              bne     .L11
16   .L17:
17              add     ip, ip, #1
18              cmp     lr, r3
19              movge   lr, r3
20              cmp     ip, #4
21              movgt   r0, lr
```

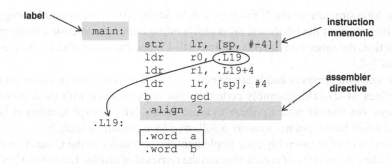

**Fig. 7.2** Elements of an assembly program produced by gcc

```
22              ldrgt    pc, [sp], #4
23              b        .L13
24   a:
25              .word    26, 3, 33, 56, 11
26   b:
27              .word    87, 12, 23, 45, 17
28   main:
29              str      lr, [sp, #-4]!
30              ldr      r0, .L19
31              ldr      r1, .L19+4
32              ldr      lr, [sp], #4
33              b        gcd
34              .align   2
35   .L19:
36              .word    a
37              .word    b
```

Listing 7.2 is the assembly program generated out of the C program in Listing 7.1. Figure 7.2 illustrates several noteworthy features of an assembly program. The program contains three elements: labels, instructions, and assembler directives. Labels are symbolic addresses. They are used as target locations for branch instructions, and as symbolic locations for variables. In Fig. 7.2, for example, variables a and b are addressed by the label .L19 and .L19+4 respectively. Assembler directives start with a dot; they do not make part of the assembled program, but are used by the assembler. The style of assembly source code shown in Listing 7.2 is common for gcc; only the instruction set will change from one cross-compiler to the next.

Understanding assembly programs is vital to understand the performance issues for many kinds of hardware/software codesign problems. In this book, we make use of C programs, and the gcc compiler, to create those assembly programs. You will find that it is easy to find a strong correspondence between a C program and its assembly version, even for an unknown processor. Let's compare the C program of Listing 7.1 with the assembly version of it in Listing 7.2.

- The overall *structure* of the assembly program preserves the structure of the C program. The gcd function is on lines 1–23, the main function is on lines 28–34.

The loop structure of the C program can be identified in the assembly program by inspection of the labels and the corresponding branch instructions. In the gcd function, the inner for loop is on lines 10–15, and the outer while loop is on lines 5–23.

- The constant arrays a and b are directly encoded as constants in the assembly, on lines 24–27. The assembly code does not directly work with these constant arrays, but instead with a *pointer* to these arrays. The storage location at label .L19 will hold a pointer to array a followed by a pointer to array b.
- *Function calls* in assembly code implement the semantics of the C function call, including the passing of parameters and the retrieval of results. Lines 30–32 of the assembly program show how this C function call is implemented. The assembly program copies the starting address of these arrays into r0 and r1. The gcd function in the assembly can make use of r0 and r1 as a pointer to array a and b respectively.

The micro-processor works with object code, binary opcodes generated out of assembly programs. Compiler tools can re-create the assembly code out of the executable format. This is achieved by the objdump program, another program in the gcc toolchain. The following command shows how to retrieve the opcodes for the gcd program.

```
> /usr/local/arm/bin/arm-linux-objdump -d gcd
```

Listing 7.3 shows the object code dump for the gcd program. The instructions are mapped to *sections* of memory, and the .text section holds the instructions of the program. Each function has a particular starting address, measured as an offset from the start of the executable. In this case, the gcd function starts at 0x8380 and the main functions starts at 0x83cc. Listing 7.3 also shows the *opcode* of each instruction, the binary representation of instructions handled by the microprocessor. As part of generating the executable, the address value of each label is encoded into each instruction. For example, the b .L13 instruction on line 23 of Listing 7.2 is encoded as a branch to address 0x838c on line 22 of Listing 7.3.

**Listing 7.3**  Object dump of Listing 7.2

```
 1  Disassembly of section .text:
 2
 3  00008380 <gcd>:
 4      8380:   e52de004   str  lr, [sp, -#4]!
 5      8384:   e3a0e000   mov  lr, #0   ; 0x0
 6      8388:   e1a0c00e   mov  ip, lr
 7      838c:   e790310c   ldr  r3, [r0, ip, lsl #2]
 8      8390:   e791210c   ldr  r2, [r1, ip, lsl #2]
 9      8394:   e1530002   cmp  r3, r2
10      8398:   0a000004   beq  83b0 <gcd+0x30>
11      839c:   e1530002   cmp  r3, r2
12      83a0:   c0623003   rsbgt  r3, r2, r3
13      83a4:   d0632002   rsble  r2, r3, r2
14      83a8:   e1530002   cmp  r3, r2
```

```
15        83ac:    1afffffa    bne 839c <gcd+0x1c>
16        83b0:    e28cc001    add ip, ip, #1   ; 0x1
17        83b4:    e15e0003    cmp lr, r3
18        83b8:    a1a0e003    movge   lr, r3
19        83bc:    e35c0004    cmp ip, #4   ; 0x4
20        83c0:    c1a0000e    movgt   r0, lr
21        83c4:    c49df004    ldrgt   pc, [sp], #4
22        83c8:    eaffffef    b   838c <gcd+0xc>
23   000083cc <main>:
24        83cc:    e52de004    str lr, [sp, -#4]!
25        83d0:    e59f0008    ldr r0, [pc, #8]    ;
                                        83e0 <main+0x14>
26        83d4:    e59f1008    ldr r1, [pc, #8]    ;
                                        83e4 <main+0x18>
27        83d8:    e49de004    ldr lr, [sp], #4
28        83dc:    eafffe7     b   8380 <gcd>
29        83e0:    00010444    andeq   r0, r1, r4, asr #8
30        83e4:    00010458    andeq   r0, r1, r8, asr r4
```

We will discuss the use of compiler tools further in Sect. 7.4. First, we will take
a closer look at the process of executing microprocessor instructions. In particular,
we will discuss the factors that affect the execution time of an instruction.

## 7.2 The RISC Pipeline

This section describes the internal architecture of a very common type of microprocessor, the Reduced Instruction Set Computer (RISC). We will review the basic
ideas in RISC architecture design. The material in this section is typically covered,
in far greater depth, in a computer-architecture course.

In a RISC processor, the execution of a single instruction is split in different
stages, which are chained together as a pipeline. Each instruction operates on a set
of registers contained within the processor. Processor registers are used as operands
or as targets for the processor instructions, and for control. For example, the ARM
processor contains 17 registers: data register r0 to r14, a program counter register
pc, and a processor status register cpsr. The Microblaze processor has 32 general-
purpose registers (r0 to r31) and up to 18 special-purpose registers (such as the
program counter, the status register, and more).

Each stage of a RISC pipeline takes one clock cycle to complete. A typical RISC
pipeline has three or five stages, and Fig. 7.3 illustrates a five-stage pipeline. The
five stages of the pipeline are called Instruction Fetch, Instruction Decode, Execute,
Buffer, and Write-back. As an instruction is executed, each of the stages performs
the following activities.

- *Instruction Fetch*: The processor retrieves the instruction addressed by the
  program counter register from the instruction memory.
- *Instruction Decode*: The processor examines the instruction opcode. For the case
  of a branch-instruction, the program counter will be updated. For the case of a

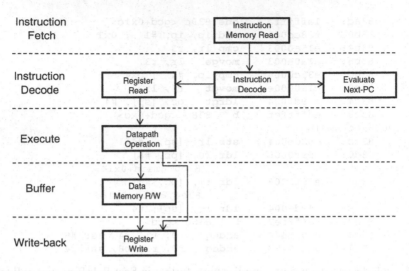

**Fig. 7.3** A five-stage RISC pipeline

compute-instruction, the processor will retrieve the processor data registers that are used as operands.

- *Execute*: The processor executes the computational part of the instruction on a datapath. In case the instruction will need to access data memory, the execute stage will prepare the address for the data memory.
- *Buffer*: In this stage, the processor may access the data memory, for reading or for writing. In case the instruction does not need to access data memory, the data will be forwarded to the next pipeline stage.
- *Write Back*: In the final stage of the pipeline, the processor registers are updated.

A three-stage RISC pipeline is similar to a five-stage RISC pipeline, but the Execute, Buffer, and Write-back stages are collapsed into a single stage.

Under ideal circumstances, the five-stage RISC pipeline is able to accept a new instruction every clock cycle. Thus, the instruction throughput in a RISC processor may be as high as one instruction every clock cycle. Because of the pipelining, each instruction may take up to five clock cycles to complete. The instruction latency therefore can be up to five clock cycles. A RISC pipeline improves instruction throughput at the expense of instruction latency. However, the increased instruction latency of a RISC processor is usually not a problem because the clock frequency of a pipelined processor is higher than that of a non-pipelined processor.

In some cases it is not possible for an instruction to finish within five clock cycles. A pipeline stall occurs when the progress of instructions through the pipeline is temporarily halted. The cause of such a stall is a pipeline hazard. In advanced RISC processors, pipeline interlock hardware can detect and resolve pipeline hazards automatically. Even when interlock hardware is present, pipeline hazards may still occur. We discuss three different categories of pipeline hazards, along with examples for an ARMv6 processor. The three categories are the following.

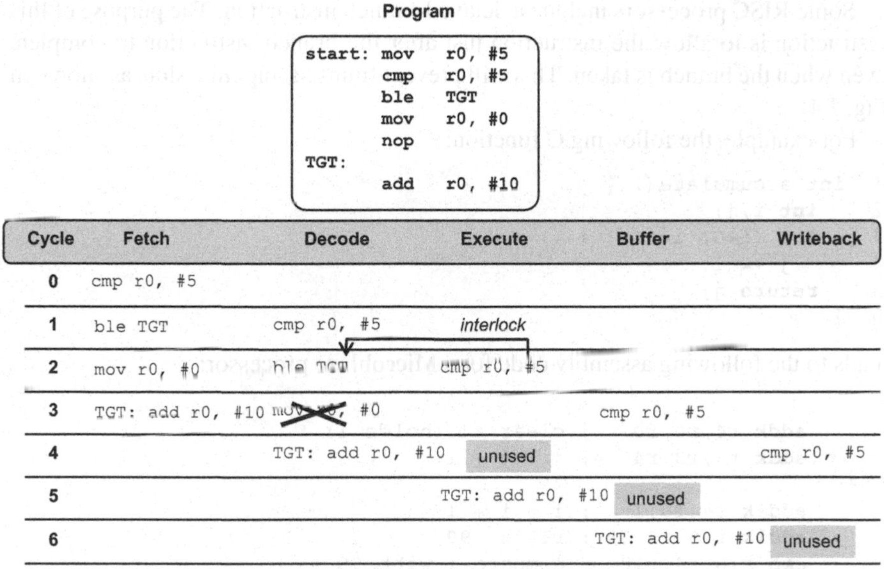

**Program**

```
start:  mov     r0, #5
        cmp     r0, #5
        ble     TGT
        mov     r0, #0
        nop
TGT:
        add     r0, #10
```

| Cycle | Fetch | Decode | Execute | Buffer | Writeback |
|-------|-------|--------|---------|--------|-----------|
| 0 | cmp r0, #5 | | | | |
| 1 | ble TGT | cmp r0, #5 | *interlock* | | |
| 2 | mov r0, #0 | ble TGT | cmp r0, #5 | | |
| 3 | TGT: add r0, #10 | mov r0, #0 | | cmp r0, #5 | |
| 4 | | TGT: add r0, #10 | unused | | cmp r0, #5 |
| 5 | | | TGT: add r0, #10 | unused | |
| 6 | | | | TGT: add r0, #10 | unused |

**Fig. 7.4** Example of a control hazard

- *Control hazards* are pipeline hazards caused by branches.
- *Data hazards* are pipeline hazards caused by unfulfilled data dependencies.
- *Structural hazards* are caused by resource conflicts and cache misses.

## 7.2.1   Control Hazards

Branch instructions are the most common form of pipeline stalls. As indicated in Fig. 7.3, a branch is only executed (i.e. it modifies the program counter register) in stage two of the pipeline. At that moment, another instruction has already entered the pipeline. As this instruction is located *after* the branch instruction, that instruction should be thrown away in order to preserve sequential execution semantics.

Figure 7.4 illustrates a control hazard. The pipeline is shown drawn on its side, running from left to right. Time runs down across the rows. A control hazard occurs because of the branch instruction ble TGT. In cycle 2, the new program counter value evaluates to the target address of the branch, TGT. Note that even though ble is a conditional branch that uses the result of the instruction just before that (cmp r0, #5), the branch condition is available in cycle 2 because of the interlock hardware in the pipeline. Starting in cycle 3, instructions from the target address TGT enter the pipeline. At the same time, the instruction just after the branch is canceled in the decode stage. This results in an unused instruction slot just after the branch instruction.

Some RISC processors include a delayed-branch instruction. The purpose of this instruction is to allow the instruction just after the branch instruction to complete even when the branch is taken. This will prevent 'unused' pipeline slots as shown in Fig. 7.4.

For example, the following C function:

```
1  int accumulate() {
2     int i,j;
3     for (i=0; i<100; i++)
4        j += i;
5     return j;
6  }
```

leads to the following assembly code for a Microblaze processor:

```
        addk  r4,r0,r0    ; clear r4 (holds i)
        addk  r3,r3,r4    ; j = j + i
$L9:
        addik r4,r4,1     ; i = i + 1
        addik r18,r0,99   ; r18 <- 99
        cmp   r18,r4,r18  ; compare i with 99
        bgeid r18,$L9     ; delayed branch if equal
        addk  r3,r3,r4    ; j = j + i -> branch delay slot
```

The delayed-branch instruction is bgeid, which is a 'branch if-greater-or-equal delayed'. The instruction just after the branch corresponds to the loop body j = j + i. Because it is a delayed-branch instruction, it will be executed regardless if the conditional branch is taken or not.

## 7.2.2  Data Hazards

A second cause of pipeline stalls are data hazards: pipeline delays caused by the unavailability of data. Processor registers are updated at the end of each instruction, during the write-back phase. But what if the data is required before it has updated a processor register? After all, as indicated in the pipeline diagram in Fig. 7.3, the write-back stage comes two cycles behind the execute stage. An instruction that reaches the write-back stage is two instructions after the instruction that is currently executing. In the following snippet, the add instruction will be in the buffer stage by the time the mov instruction reaches the write-back stage, and the addition would have already completed.

```
mov  r0, #5
add  r1, r0, r1
```

In a RISC pipeline, this is handled by pipeline interlock hardware. The pipeline interlock hardware observes the read/write patterns of all instructions currently flowing in the RISC pipeline, and makes sure they take data from the right source.

Program

```
start: mov  r0, #5
       ldr  r1, [r0]
       add  r2, r1, #3
       mov  r3, #0
       nop
```

| Cycle | Fetch | Decode | Execute | Buffer | Writeback |
|-------|-------|--------|---------|--------|-----------|
| 0 | mov r0, #5 | | | | |
| 1 | ldr r1,[r0] | mov r0, #5 | | | |
| 2 | add r2,r1,#3 | ldr r1,[r0] | mov r0, #5 | | |
| 3 | mov r3,#0 | add r2,r1,#3 | ldr r1,[r0] | mov r0, #5 | |
| 4 | wait | wait | wait | ldr r1,[r0] | mov r0, #5 |
| 5 | | mov r3,#0 | add r2,r1,#3 | unused | ldr r1,[r0] |
| 6 | | | mov r3,#0 | add r2,r1,#3 | unused |
| 7 | | | | mov r3,#0 | add r2,r1,#3 |
| 8 | | | | | mov r3,#0 |

**Fig. 7.5** Example of a data hazard

Consider the previous example again. When the add instruction is in the execute stage, it will use the result of the mov instruction as if flows through the buffer stage of the pipeline. This activity is called *forwarding*, and it is handled automatically by the processor.

In some cases, *forwarding* is not possible because the data is simply not yet available. This happens when a read-from-memory instruction is followed by an instruction that uses the data coming from memory. An example of this case is shown in Fig. 7.5. The second instruction fetches data from memory and stores it in register r1. The following add instruction uses the data from that register as an operand. In cycle 4, the add instruction reaches the execute stage. However, during the same clock cycle, the ldr instruction is still accessing the data memory. The new value of r1 is only available at the start of cycle 5. Therefore, the interlock hardware will stall all stages preceding the buffer stage in cycle 4. Starting in cycle 5, the entire pipeline moves forward again, but due to the stall in cycle 4, an unused pipeline slot flushes out in cycles 5 and 6.

Data hazards may lengthen the execution time of an instruction that would normally finish in just five clock cycles. For classic RISC processors, data hazards can be predicted statically, by examining the assembly program. When the execution time of a program needs to be estimated exactly, a programmer will need to identify all data hazards and their effects.

Program

```
mov    r0, #5
ldmia  r0, {r1,r2}
add    r4, r1, r2
add    r4, r4, r3
```

| Cycle | Fetch | Decode | Execute | Buffer | Writeback |
|---|---|---|---|---|---|
| 0 | mov r0, #5 | | | | |
| 1 | ldmia r0,{r1,r2} | mov r0, #5 | | | |
| 2 | add r4,r1,r2 | ldmia r0,{r1,r2} | mov r0, #5 | | |
| 3 | wait | wait | ldmia r0,{r1,r2} | mov r0, #5 | |
| 4 | add r4,r4,r3 | add r4,r1,r2 | ldmia r0,{r1,r2} | load r1 | mov r0, #5 |
| 5 | | add r4,r4,r3 | add r4,r1,r2 | load r2 | update r1 |
| 6 | | | add r4,r4,r3 | add r4,r1,r2 | update r2 |
| 7 | | | | add r4,r4,r3 | add r4,r1,r2 |
| 8 | | | | | add r4,r4,r3 |

**Fig. 7.6** Example of a structural hazard

## 7.2.3  Structural Hazards

The third class of hazards are structural hazards. These are hazards caused by instructions that require more resources from a processor than available. For example, a given instruction may require five concurrent additions while there is only a single ALU available. To implement such an instruction, the execution phase of the instruction will need to be extended over multiple clock cycles, while the pipeline stages before that will be stalled.

Another example of a structural hazard is illustrated in Fig. 7.6. The ldmia instruction (ARM) is a load-multiple instruction that reads consecutive memory locations and that stores the resulting values in memory. In the example shown, the value stored in address r0 will be copied to r1, while the value stored in address r0+4 will be copied to r2. When the ldmia instruction reaches the execute stage, the execute stage will be busy for two clock cycles in order to evaluate the memory addresses r0 and r0+4. Therefore, all pipeline stages before the execute stage are halted for a single clock cycle. After that, the pipeline proceeds normally.

A structural hazard is caused by the processor architecture, but it may have a broad number of causes: the width of memory ports, the number of execution units in the data-path, or the restrictions on communication busses. A programmer can only predict structural hazards through a solid understanding of the processor architecture. Furthermore, memory latency effects can also cause the execution time of the buffer stage to vary. A cache miss can extend the latency of a load-memory instruction with hundreds of cycles. While the memory-load instruction is waiting for data to be returned from memory, it will stall the pipeline in a manner similar to a structural hazard.

## 7.3  Program Organization

Efficient hardware/software codesign requires a simultaneous understanding of system architecture and software. This is different from traditional computer science, where a programmer is typically interested in running a C program 'as fast as possible', but without much concern for the computer hardware that runs the C program.

In this section, we consider the relation between the structure of a C program and its' implementation on a RISC processor. We cover the storage of C variables in memory, and the implementation of function calls. We are assuming a 32-bit architecture, and will provide examples based on ARM, Microblaze, and Nios-II RISC processors.

### 7.3.1  Data Types

A good starting point to discuss the mapping of C programs to RISC processors are the data types used by C programs. Table 7.1 shows how C maps to the native data types supported by 32-bit processors such as ARM, Nios-II, and Microblaze. All C data types, apart from `char`, are treated as signed (two's complement) numbers.

The hardware required for operations on two's complement signed numbers and unsigned numbers is almost identical. One exception is that signed-number arithmetic uses sign extension. Converting, for example, a signed integer to a signed long (64-bit) integer will replicate the most significant bit of the source integer into the upper 32 bits of the long integer. The second difference between signed and unsigned operations is that the comparison operation for signed and unsigned numbers has a different hardware implementation. Indeed, when comparing unsigned bytes, `0xff` is bigger then `0x01`. But, when comparing signed bytes, `0xff` is smaller then `0x01`.

| C data type | |
|---|---|
| Char | 8-bit |
| Short | Signed 16-bit |
| Int | Signed 32-bit |
| Long | Signed 32-bit |
| Long long | Signed 64-bit |

**Table 7.1** C compiler data types

The mapping of C data types to physical memory locations is affected by several factors.

The first is the *aligment* of data types in the memory. A typical memory organization for 32-bit RISC processors is word-based. Thus, a single memory transfer may be able to access any of the 4 bytes in a word, but a group of 4 bytes across a word boundary cannot be accessed in a single memory transfer. A word-oriented memory organization affects the mapping of data types in logical address space (Fig. 7.7a). A 32-bit integer, for example, cannot straddle a word-boundary address. The RISC processor uses only a single load instruction for an integer, and hence can use only a single memory-access.

A second factor to affect the mapping of data types is the storage order, illustrated in Fig. 7.7b. A little-endian storage order will map the lower-significant bytes of a word into lower memory locations. A big-endian storage order, on the other hand, will map the higher-significant bytes to lower memory locations. In a typical C program, the endianess is of no concern. In hardware/software codesign however, the physical representation of data types is important in the transition of software to hardware and back. Therefore, the endianess of a processor (and in some cases even the bit-ordering) is important. It is easy to check the endianess of a given processor using a small C program such as the following one.

```
int main() {
    char j[4];
    volatile int *pj;
    pj = (int *) j;

    j[0] = 0x12;
    j[1] = 0x34;
    j[2] = 0x56;
    j[3] = 0x78;

    printf("\%x\n", *pj);
}
```

For this program, a little-endian processor will print 78563412, while a big-endian processor will print 12345678. A Microblaze processor is big-endian, a Nios-II processor is little-endian, and an ARM is bi-endian (meaning it can work either way, and the endianess is a configuration of the processor).

## 7.3.2  Variables in the Memory Hierarchy

A second important implementation aspect of C programs is the relationship between the variables of a C program and the memory locations used to store those variables. The *memory hierarchy* creates the illusion of a uniform, very fast memory space. As illustrated in Fig. 7.8, a memory hierarchy includes the processor registers, the cache memory and the main memory. In embedded processors, cache memory is

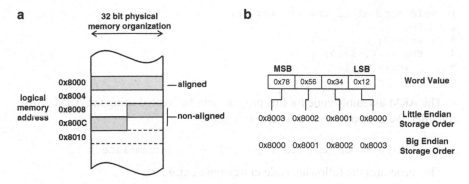

Fig. 7.7  (a) Alignment of data types. (b) Little-endian and Big-endian storage order

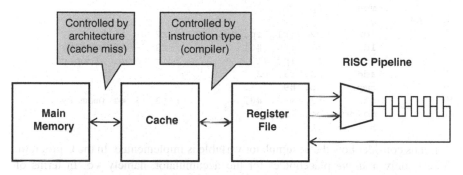

Fig. 7.8  Memory hierarchy

optional; in high-end processors on the other hand, multiple levels of cache memory are used.

The cache operates as a fast local memory which holds the most-frequently used main-memory locations. Whenever the processor requests data from a memory location, the cache may report a hit and return a locally-stored copy of the desired memory location. The cache may also report a miss and instead first fetch the requested memory location from main memory. A cache memory improves program efficiency when data (or instructions) are used more than once by a C program. The instructions of the body of a loop, for example, are good candidates for caching. Conversely, when every data element needs to be retrieved only once, a cache memory is of little help.

The memory hierarchy is transparent to a C programmer. Under normal circumstances, a C programmer will not worry what type of memory is used to store the variables from a program. It could be the processors' registers, the cache, or the main-memory. In reality, data travels up and down the memory hierarchy during program execution. This is illustrated by the following example, which shows a C function accumulating an array.

```
1   void accumulate(int *c, int a[10]) {
2       int i;
3       *c = 0;
4       for (i=0; i<10; i++)
5           *c += a[i];
6   }
```

The ARM assembly code for this program can be created using the ARM cross-compiler. We use optimization-level two for this.

```
/usr/local/arm/bin/arm-linux-gcc -O2 -c -S accumulate.c
```

This generates the following code in accumulate.s:

```
1               mov     r3, #0
2               str     r3, [r0, #0]
3               mov     ip, r3
4   .L6:
5               ldr     r2, [r1, ip, asl #2]   ; r2 <- a[i]
6               ldr     r3, [r0, #0]           ; r3 <- *c (memory)
7               add     ip, ip, #1             ; increment loop ctr
8               add     r3, r3, r2
9               cmp     ip, #9
10              str     r3, [r0, #0]           ; r3 -> *c (memory)
11              movgt   pc, lr
12              b       .L6
```

Let us consider how the accumulator variable is implemented. In the C program, there is only a *single* placeholder for the accumulator, namely *c. In terms of physical memory, there are at three different locations where a copy of *c may be found: the processor registers, the cache, and the main memory. In the assembly implementation, the *value* of the accumulator travels up in the memory hierarchy. According to the C function, the accumulator is provided through a pointer. This implies that the accumulator will be stored in main memory. On line 6 of the previous Listing, that variable is read from memory and stored in processor register r3. On line 10, the processor register r3 is written back to memory. Thus, depending on the nature and state of the cache memory, reading/writing processor registers from/to memory may trigger additional data transfers between the cache memory and the main memory. In the context of hardware/software codesign, the difference between the physical implementation of a C program and its logical design is important. For example, the physical mapping of variables is important when a communication link needs to be created between software and hardware.

A C programmer has a limited amount of control over the mapping of variables in the memory hierarchy. This control is offered through the use of *storage class specifiers* and *type qualifiers*. The most important ones are enumerated in Table 7.2. A few example declarations are shown below.

```
volatile int *c;     // c is a pointer to a volatile int
int * const y;       // y is a constant pointer to an int
register int x;      // x is preferably mapped into a register
```

**Table 7.2** C storage class specifiers and type qualifiers

| Keyword | Function |
|---|---|
| Storage specifier | |
| register | Indicates that the preferred mapping of the data type is in processor registers. This will keep the variable as class as possible to the RISC pipeline |
| static | Limits the scope (visibility) of the variable over multiple files. This specifier does not relate to the memory hierarchy, but to the functions where the variable may be accessed |
| extern | Extends the scope (visibility) of the variable to all files. This specifier does not relate to the memory hierarchy, but to the functions where the variable may be accessed |
| Type qualifier | |
| const | Indicates that the qualified variable cannot be changed |
| volatile | Indicates that the qualified variable can change its value at any time, even outside of the operations in the C program. As a result, the compiler will make sure to write the value always back to main memory after modification, and maintain a copy of it inside of the processor registers |

Type qualifiers will be important to access memory-mapped interfaces, which are hardware/software interfaces that appear as memory locations to software. We will discuss this in Chap. 11.

### 7.3.3 Function Calls

Behavioral hiearchy – C functions calling other functions – is key to mastering complexity with C programs. We briefly describe the concepts of C function calls in the context of RISC processors. We use the example C program in Listing 7.4.

**Listing 7.4** Sample program

```
1
2  int accumulate(int a[10]) {
3    int i;
4    int c = 0;
5    for (i=0; i<10; i++)
6      c += a[i];
7    return c;
8  }
9
10  int a[10];
11  int one = 1;
12
13  int main() {
14    return one + accumulate(a);
15  }
```

Let us assume that we have compiled this program for an ARM processor using the arm-linux-gcc cross compiler. It is possible to re-create the assembly listing

corresponding to the object file by *dis-assembling* the object code. The utility
`arm-linux-objdump` takes care of that. The `-d` flag on the command line
selects the dis-assembler functionality. The utility supports many other functions
(See Sect. 7.4 and Problem 7.8).

```
/usr/local/arm/bin/arm-linux-objdump -O2 -c accumulate.c -o
    accumulate.o
/usr/local/arm/bin/arm-linux-objdump -d accumulate
```

The ARM assembly listing of this program is shown in Listing 7.5.

**Listing 7.5** Sample program

```
00000000 <accumulate>:
   0:    e3a01000      mov     r1, #0
   4:    e1a02001      mov     r2, r1
   8:    e7903102      ldr     r3, [r0, r2, lsl #2]
   c:    e2822001      add     r2, r2, #1
  10:    e3520009      cmp     r2, #9
  14:    e0811003      add     r1, r1, r3
  18:    c1a00001      movgt   r0, r1
  1c:    c1a0f00e      movgt   pc, lr
  20:    ea000000      b       8 <accumulate+0x8>

00000024 <main>:
  24:    e52de004      str     lr, [sp, -#4]!
  28:    e59f0014      ldr     r0, [pc, #20]    ; 44 <main+0x20>
  2c:    ebfffffe      bl      0 <accumulate>
  30:    e59f2010      ldr     r2, [pc, #16]    ; 48 <main+0x24>
  34:    e5923000      ldr     r3, [r2]
  38:    e0833000      add     r3, r3, r0
  3c:    e1a00003      mov     r0, r3
  40:    e49df004      ldr     pc, [sp], #4
          . . .
```

Close inspection of the instructions reveals many practical aspects of the runtime
layout of this program, and in particular of the implementation of function calls.
The instruction that branches into `accumulate` is implemented at address `0x2c`
with a `bl` instruction – *branch with link*. This instruction will copy the program
counter in a separate link register `lr`, and load the address of the branch target into
the program counter. A return-from-subroutine can now be implemented by copying
the link register back into the program counter. This is shown at address `0x1c` in
`accumulate`. Obviously, care must be taken when making nested subroutine calls
so that `lr` is not overwritten. In the `main` function, this is solved at the entry, at
address `0x24`. There is an instruction that copies the current contents of `lr` into a
local area within the stack, and at the end of the main function the program counter
is directly read from the same location.

The arguments and return value of the `accumulate` function are passed
through register `r0` rather than main memory. This is obviously much faster when
only a few data elements need to be copied. The input argument of `accumulate` is

the base address from the array a. Indeed, the instruction on address 8 uses r0 as a base address and adds the loop counter multiplied by 4. This expression thus results in the effective address of element a[i] as shown on line 5 of the C program (Listing 7.4). The return argument from accumulate is register r0 as well. On address 0x18 of the assembly program, the accumulator value is passed from r1 to r0. For ARM processors, the full details of the procedure-calling convention are defined in the *ARM Procedure Call Standard* (APCS), a document used by compiler writers and software library developers.

In general, arguments are passed from function to function through a data structure known as a *stack frame*. A stack frame holds the return address, the local variables, the input and output arguments of the function, and the location of the calling stack frame. A full-fledged stack frame can be found when the accumulate function described earlier is compiled *without* optimizations. In that case, the C compiler takes a conservative approach and keeps all local variables in main memory, rather than in registers.

```
/usr/local/arm/bin/arm-linux-gcc -c -S accumulate.c
```

Listing 7.6 shows the resulting non-optimized assembly code of accumulate. Figure 7.9 illustrates the construction of the stack frame.

**Listing 7.6** Accumulate without compiler optimizations

```
 1   accumulate:
 2              mov      ip, sp
 3              stmfd    sp!, {fp, ip, lr, pc}
 4              sub      fp, ip, #4
 5              sub      sp, sp, #12
 6              str      r0, [fp, #-16]     ; base address a
 7              mov      r3, #0
 8              str      r3, [fp, #-24]     ; c
 9              mov      r3, #0
10              str      r3, [fp, #-20]     ; i
11   .L2:
12              ldr      r3, [fp, #-20]
13              cmp      r3, #9             ; i<10 ?
14              ble      .L5
15              b        .L3
16   .L5:
17              ldr      r3, [fp, #-20]     ; i * 4
18              mov      r2, r3, asl #2
19              ldr      r3, [fp, #-16]
20              add      r3, r2, r3         ; *a + 4 * i
21              ldr      r2, [fp, #-24]
22              ldr      r3, [r3, #0]
23              add      r3, r2, r3         ; c = c + a[i]
24              str      r3, [fp, #-24]     ; update c
25              ldr      r3, [fp, #-20]
26              add      r3, r3, #1
27              str      r3, [fp, #-20]     ; i = i + 1
28              b        .L2
```

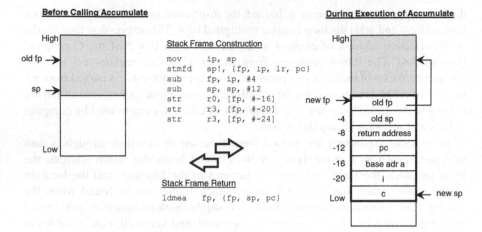

**Fig. 7.9** Stack frame construction

```
29    .L3:
30            ldr     r3, [fp, #-24]    ; return arg
31            mov     r0, r3
32            ldmea   fp, {fp, sp, pc}
```

The instructions on lines 2 and 3 are used to create the stack frame. On line 3, the frame pointer (fp), stack pointer (sp), link register or return address (lr) and current program counter (pc) are pushed onto the stack. The single instruction stmfd is able to perform multiple transfers (m), and it grows the stack downward (fd). These four elements take up 16 bytes of stack memory.

On line 3, the frame pointer is made to point to the first word of the stack frame. All variables stored in the stack frame will now be referenced based on the frame pointer fp. Since the first four words in the stack frame are already occupied, the first free word is at address fp - 16, the next free word is at address fp - 20, and so on. These addresses may be found back in Listing 7.6.

The following local variables of the function accumulate are stored within the stack frame: the base address of a, the variable i, and the variable c. Finally, on line 31, a return instruction is shown. With a single instruction, the frame pointer fp, the stack pointer sp, and the program counter pc are restored to the values just before calling the accumulate function.

### 7.3.4  Program Layout

Another aspect of C program implementation is the physical representation of the program and its data structures in the memory hierarchy. This leads to the *program layout*, the template that is used to organize instructions and data. A distinction must be made between the organization of a compiled C program in an executable file (or a program ROM), and the memory organization of that C program during

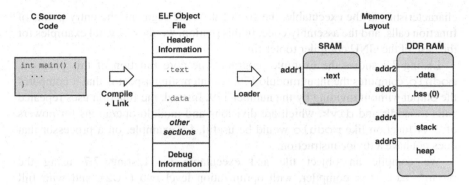

**Fig. 7.10** Static and dynamic program layout

execution. The former case is a static representation of all the instructions and constants defined by the C program. The latter case is a dynamic representation of all the instructions and the runtime data structures such as the stack and the heap.

Figure 7.10 shows how a C program is compiled into an executable file, which in turn is mapped into memory. There are several standards available for the organization of executable files. In the figure, the example of ELF (Executable Linkable Format) is shown. An ELF file is organized into sections, and each of these can take up a variable amount of space in the file. The sections commonly found in an ELF file are the `.text` section which contains the binary instructions of the C program and the `.data` section which contains initialized data (constants). The ELF file may also contain debugging information, such as the names of the variables in the C program. This debugging information is utilized by source level debuggers to relate the execution of a binary program to actions of the C source code.

When a compiled C program is first loaded into memory for execution, the ELF file is analyzed by the loader and the sections with relevant information are copied into memory locations. In contrast to a file, the resulting organization of instructions and data into memory does not need to be contiguous or even occupy the same physical memory. Each section of an ELF file can be mapped at a different address, and possibly map into a different memory module. The example in Fig. 7.10 shows how the `.text` segment maps into fast static RAM memory (SRAM) while the `.data`, stack and heap segments map into DDR RAM memory. During program execution, there may be sections of memory which do not appear in the ELF file, or which do not occupy any area within the ELF file. These sections include data storage areas: dynamic data (heap), local data (stack), and global data (bss).

## 7.4 Compiler Tools

A C compiler provides utilities to inspect the organization of an executable file or an object file. We can use those utilities to gain deeper insight into the low-level

characteristics of the executable: the size of the code segment, the entry points of
function calls, and the assembly code. In this section, we show several examples for
the case of the GNU compiler toolchain.

Listing 7.7 shows the example program. The core function of this program,
modulo, computes the input modulo 179. An interesting aspect is that it computes
the output without dividing by the number 179. Instead, the function uses repeated
calls to modk and divk, which are divisions and modulo-operations for powers
of 2. A function like modulo would be useful, for example, on a processor that
does not have a divide instruction.

We compile an object file and executable for Listing 7.7 using the
nios2-elf-gcc compiler, with optimization level two (-O2), and with full
debug information. This compiler targets the Nios-II core used in Altera FPGA.
The following examples illustrate some of the output formats produced by the
GNU Compiler Toolchain. The examples are not comprehensive, and you'll need
to consult the manual pages of these tools for further information. We will show
how to find the size, the symbols, the organization, and eventually the assembly
instructions for this program.

**Listing 7.7** Modulo-179 function

```
 1  unsigned modk(unsigned x, unsigned k) {
 2    return (x & ((1 << k) - 1));
 3  }
 4
 5  unsigned divk(unsigned x, unsigned k) {
 6    return (x >> k);
 7  }
 8
 9  unsigned modulo(unsigned x) {
10    unsigned r, q, k, a, m, z;
11    m = 0xB3; // 179
12    k = 8;
13    a = (1 << k) - m;
14    r = modk(x, k);
15    q = divk(x, k);
16    do {
17      do {
18        r = r + modk(q * a, k);
19        q = divk(q * a, k);
20      } while (q != 0);
21      q = divk(r, k);
22      r = modk(r, k);
23    } while (q != 0);
24    if (r >= m)
25      z = r - m;
26    else
27      z = r;
28    return z;
29  }
30
```

```
31  int num[4] = {221, 322, 768, 121};
32  int res[4];
33
34  int main() {
35    unsigned i;
36    for (i=0; i<4; i++)
37      res[i] = modulo(num[i]);
38    return 0;
39  }
```

In the following, we assume that we have compiled modulo.c using the nios2-elf-gcc design flow. We are concerned with examining the output of the compilation process.

## 7.4.1  Examining Size

A basic question of interest is: how large is this program? In resource-constrained applications, knowing the size of a program helps a designer to estimate the memory requirements. The size utility shows the static size of a program, that is, the amount of memory required to store instructions, constants, and global variables. The size utility is specific to the cross-compiler environment being used. For example, we use arm-linux-size, nios2-elf-size, and mb-size to analyze binary code compiler for ARM, Nios-II, and Microblaze, respectively.

The nios2-elf-size utility gives a summary printout of the amount of memory required for a given C program.

```
> nios2-elf-size modulo.o
   text      data       bss       dec      hex filename
    300        16         0       316      13c modulo.o
```

The output of the program on modulo.o shows that there are 300 bytes in the text section, 16 bytes in the initialized data-section data, and 0 bytes in the non-initialized data-section bss. The total amount of bytes required for this program is 316 (300 + 16) in decimal notation or 0xf4 in hexadecimal notation. Since Nios-II uses word-size instructions, a text segment of 300 bytes implies 75 instructions (228/4). These include the instructions contained within modk, divk, modulo and main. There are also 16 bytes of initialized data, used to store the global array num. After linking the compiler sections.o, we obtain an executable file modulo.elf. The utility nios2-elf-size can handle the executable as well. In this case, the amount of code and data increases significantly, due to the inclusion of C libraries into the program. The linked program uses 10,472 bytes for the code, 5,252 bytes of uninitialized data, and 488 bytes of constants. Libraries have great influence on the overall size of the program. In resource-constrained environments, a great deal of effort is done to reduce the size of the libraries.

```
> nios2-elf-size modulo.elf
   text    data    bss    dec     hex filename
  10472    5252    448  16172    3f2c modulo.elf
```

The size utility does not show the dynamic memory usage of a program. In fact, it cannot predict the amount of stack or the amount of heap that will be needed. The size of the stack and heap are not the result of compiling code and counting instructions or data bytes. Instead, they are a design parameter selected by the programmer.

## 7.4.2  Examining Sections

Code coming out of a compiler is organized in chunks of continuous memory, called sections. The text, data, and bss sections listed by size are the three most important ones, but a program can define additional sections, for example to store debugging information.

The objdump utility allows you to examine the relative size and position of each section, as well as the names of all the functions in it. The -h flag of objdump generates a printout of section headers. This reveals the following information for the modulo.elf program (with only partial information being shown).

```
modulo.elf:      file format elf32-littlenios2

Sections:
Idx Name           Size      VMA        LMA        File off  Algn
  0 .entry         00000020  09000000   09000000   00001000  2**5
                   CONTENTS, ALLOC, LOAD, READONLY, CODE
  1 .exceptions    00000198  09000020   09000020   00001020  2**2
                   CONTENTS, ALLOC, LOAD, READONLY, CODE
  2 .text          00002670  090001b8   090001b8   000011b8  2**2
                   CONTENTS, ALLOC, LOAD, READONLY, CODE
  3 .rodata        000000c0  09002828   09002828   00003828  2**2
                   CONTENTS, ALLOC, LOAD, READONLY, DATA
  4 .rwdata        00001484  090028e8   09003d6c   000038e8  2**2
                   CONTENTS, ALLOC, LOAD, DATA, SMALL_DATA
  5 .bss           000001c0  090051f0   090051f0   000051f0  2**2
                   ALLOC, SMALL_DATA
  ...
```

This listing illustrates the name of the sections, their size, the starting address (VMA and LMA), the offset within the ELF file and the alignment in bytes as a power of 2. VMA stands for *virtual memory address* and it reflects the address of the section during execution. LMA stands for *load memory address* and it reflects the address of the section when the program is first loaded into memory. In this case both numbers have the same value. They would be different in cases where the program is stored in a different memory than the one that holds the program at runtime. For example, a Flash memory can store the program sections at the address reflected by LMA. When the program executes, it is copied into RAM memory at the address reflected by VMA.

Another feature of the objdump program is to print the names of all the symbols included in an object file or executable. Symbols are the elements with external visibility, such as functions and variables. The -t flag on nios2-elf-objdump shows the symbol table. The sorted output of this command looks as follows.

```
> nios2-elf-objdump -t modulo.elf | sort

09000000 g       *ABS*      00000000 __alt_mem_Onchip_memory
...
090001b8 l     d .text      00000000 .text
090001b8 g     F .text      0000003c _start
090001f4 g     F .text      00000014 modk
09000208 g     F .text      00000008 divk
09000210 g     F .text      000000c0 modulo
090002d0 g     F .text      0000005c main
...
090028e8 l     d .rwdata    00000000 .rwdata
090028e8 g     O .rwdata    00000010 num
...
090051f0 l     d .bss       00000000 .bss
09005214 g     O .bss       00000010 res
```

The program indicates that the .text segment starts at address 0x90001b8. Next, it lists several symbols. The main function, for example, starts at address 0x90002d0. The symbol table also shows the variables num and res, and their position in the rwdata section and bss section.

Finally, it is also possible to extract symbol table information directly out the linker by inspecting the linker map file. The linker map can be generated as a byproduct of the compilation process. In the GNU C Compiler, a linker map file can be generated with the -Wl,-Map=.. command line option. For example, nios2-elf-gcc will generate the linker map file for modulo.elf through the following options.

```
nios2-elf-gcc  -Wl,-Map=modulo.map -O2 -g -o modulo.elf modulo.c
```

This linker map file comes the closest to the actual memory-map of the implementation; the linker map file will show the size and type of memories defined in the system, and how the different sections of the program map into these memories. The linker map will also show exactly what object files contribute to the final executable. The following listing shows the partial content of a linker map file produced for the modulo program.

```
...
.text          0x090001b8          0x3c hal_bsp/obj/HAL/src/crt0.o
               0x090001b8              _start
.text          0x090001f4          0x138 obj/modulo.o
               0x090001f4              modk
```

```
0x09000208                    divk
0x09000210                    modulo
0x090002d0                    main
```

## 7.4.3   Examining Assembly Code

The most detailed inspection of the compiler output is done by looking at the assembly output of the compiler. A straightforward way to do this is to run gcc with the -S option. The assembly output for modulo.c can be created as follows.

```
nios2-elf-gcc -S -O2 modulo.c
```

This leads to the file modulo.s. The modk function, for example, looks as follows.

```
modk:
    movi      r3,1
    sll       r2,r3,r5
    sub       r2,r2,r3
    and       r2,r2,r4
    ret
    .size     modk, .-modk
    .align    2
    .global   divk
    .type     divk, @function
```

Assembly code can also be recreated from an object file or an executable using the objdump function with the -D flag. For an executable, we will see the address as well as the opcode with each instruction. The same modk function will look as follows when generated by nios2-elf-objdump:

```
090001f4 <modk>:
 90001f4:   00c00044    movi    r3,1
 90001f8:   1944983a    sll     r2,r3,r5
 90001fc:   10c5c83a    sub     r2,r2,r3
 9000200:   1104703a    and     r2,r2,r4
 9000204:   f800283a    ret
```

There are many different ways to query the output of the compiler, using the compiler itself, the linker, and utilities such as size and objdump. All these utilities are very useful to analyze the detailed, low-level construction of a program. The next section gives an examples of such analysis.

## 7.5 Low-Level Program Analysis

Because the interaction of hardware and software is so fundamental to hardware/software codesign, understanding software at the lowest level is essential. Optimizing performance requires insight into the low-level construction of the program. For example, what would be the difference between using optimization level -O2 and -O3? How does the compiler generate address expressions from array indices? How does the compiler implement loop counters? These are all questions that can be solved by analyzing the assembly code.

We call this concept *Program Analysis*: interpreting and understanding the performance of a program based on observing the assembly code of that program. Program Analysis is useful for addressing many design activities, such as the following examples.

- Program analysis provides insight into the optimizations that a C compiler can or cannot offer. Many C programmers are hopeful about the capabilities of a C compiler to produce efficient assembly code. While this is generally true for a high-quality compiler, there are many cases where a compiler cannot help. For example, a compiler is unable to make optimizations that require specific knowledge in the statistical properties of the program data input. A compiler will not transform a program with double or float variables into one with int variables, even if an int would give sufficient precision for a particular application. A compiler cannot perform optimizations at high abstraction levels, such converting one type of sorting algorithm into another, equivalent, but more efficient, sorting algorithm. To understand what a compiler can and cannot do, it is helpful to compare C code and the corresponding assembly code.
- Program analysis enables a programmer to make quite accurate predictions on the execution time of a given program. In cases where you are addressing low-level issues, such as controlling hardware modules or controlling hardware interface signals from within software, these timing predictions may be very important.

Listing 7.8 A simple convolution function

```
1  int array[256];
2  int c[256];
3  int main() {
4    int i, a;
5    a = 0;
6    for (i=0; i<256; i++)
7      a += array[i] * c[256 - i];
8    return a;
9  }
```

We will discuss program analysis of Listing 7.8. This program illustrates the C implementation of a convolution operation: the cross-product of a data-array with a reversed data-array. We are interested in the efficiency of this program on a Microblaze processor. For this purpose, we generate the assembly listing of this

program using the Microblaze GNU compiler. The optimization level is set at O2.
The resulting assembly listing is shown in Listing 7.9.

**Listing 7.9**  Microblaze assembly for the convolution program

```
1           .text
2           .align 2
3           .globl main
4           .ent main
5    main:
6           .frame r1,44,r15
7           .mask 0x01c88000
8           addik  r1,r1,-44
9           swi    r22,r1,32
10          swi    r23,r1,36
11          addik  r22,r0,array
12          addik  r23,r0,c+1024
13          swi    r19,r1,28
14          swi    r24,r1,40
15          swi    r15,r1,0
16          addk   r24,r0,r0
17          addik  r19,r0,255
18   $L5:
19          lwi    r5,r22,0
20          lwi    r6,r23,0
21          brlid  r15,__mulsi3
22          addik  r19,r19,-1
23          addk   r24,r24,r3
24          addik  r22,r22,4
25          bgeid  r19,$L5
26          addik  r23,r23,-4
27          addk   r3,r24,r0
28          lwi    r15,r1,0
29          lwi    r19,r1,28
30          lwi    r22,r1,32
31          lwi    r23,r1,36
32          lwi    r24,r1,40
33          rtsd   r15,8
34          addik  r1,r1,44
35          .end   main
36   $Lfe1:
37          .size  main,$Lfe1-main
38          .bss
39          .comm  array,1024,4
40          .type  array, @object
41          .comm  c,1024,4
42          .type  c, @object
```

Did the C compiler a good job while compiling the C program in Listing 7.8? In
previous sections, we already discussed several of the elements to help us answer
this question, including the *stack frame*. Another concept is the *Application Binary
Interface* (ABI). The ABI defines how a processor will use its registers to implement
a C program. For the case of a Microblaze processor and the example in Listing 7.8,
the following aspects are relevant.

- The Microblaze has 32 registers
- Register r0 is always zero and used as a zero-constant.
- Register r1 is the stack pointer.
- Registers r19 through r31 are callee-saved registers: a function that whishes to use these registers must preserve their contents before returning to the caller.

Therefore, we can make the following observations on lines 8–17 from Listing 7.9.

- Line 8 grows the stack pointer by 44 bytes (11 words). Note that the Microblaze stack grows downwards.
- Lines 9, 10, 13, 14 save registers on the stack. These registers (r22, r23, r19, r24) will be used as temporary variables by the program. They are restored just before the main function terminates.

From the values loaded into the working registers, we can infer what they actually represent.

- Register r22 is initialized with array, the starting address of the array.
- Register r23 is initialized with c+1024, which is the start of the c variable plus 1,024. Since the c variable is an array with 256 integers, we conclude that r23 points to the end of the c variable.
- Register r19 is initialized to 255, which is the loop count minus 1.
- Register r24 is initialized to 0, and could be the loop counter or the accumulator.

We now know enough to the tackle the loop body in lines 18–26. Loops in assembly code are easy to find since they always start with a label (like $L5), and terminate with a branch instruction to that label. In this case, the last instruction of the loop body is on line 26 because the branch on line 25 is a delayed-branch (ends with a 'd'). The loop body reads an element from the variables array and c (lines 18 and 19) and stores the result in r5 and r6. The next instruction is a function call. It is implemented in a RISC processor as a branch which saves the return address on the stack. r15 is used to hold the return address. The function is called __mulsi3. From its name, this function hints to be a multiplication, indicating that the compiler generated code for a micro-processor without a built-in multiplier. The multiplication will support the implementation of the following C code.

```
a += array[i] * c[256 - i];
```

The result of the function __mulsi3 is provided in registers r3 and r4 (this is another convention of the ABI). Indeed we see that r3 is accumulated to r24 on line 21. This clears up the meaning of register r24: it is the accumulator. Note that there are three adjustments to counter values in the loop body: Register r19 is decremented by 1, register r22 is incremented by 4, and register r23 is decremented by 4. The adjustment to r23 is still part of the loop because this instruction is located in the branch delay slot after the bgeid branch. We already know that register r22 and r32 are pointers pointing to the variables array and c. Register r19 is the loop counter. Thus, we conclude that the compiler was able to find out that the address expressions for c and array are sequential, just as the loop counter i. It is as if the compiler has automatically performed the following very effective optimization:

```
int array[256];
int c[256];
int main() {
  int i, a;
  int *p1, *p2;
  p1 = array;
  p2 = &(c[255]);
  a = 0;
  for (i=0; i<256; i++)
    a += (*(p1++)) * (*(p2--));
  return a;
}
```

The C compiler is thus able to do fairly advanced dataflow analysis and optimization (when the optimization flag is turned on). Static program analysis does not reveal cycle counts and performance numbers. Rather, it provides a qualitative appreciation of a program. Being able to investigate assembly code, even for processors foreign to you, enables you to make accurate decisions on potential software performance.

## 7.6    Processor Simulation

When working with embedded processors, it is frequently needed to simulate a model of the processor before the actual design is made. The difference with the hardware simulations we discussed so far, is that embedded processor simulation needs a cross-compiled software binary to run.

### 7.6.1    Instruction-Set Simulation

Simulations with processor models are very common in hardware-software codesign; they are meant to test the executables created with a cross-compiler, and to evaluate the performance of the resulting program. Micro-processors such as ARM can be simulated with an instruction-set simulator, a simulation engine specialized at simulating the instruction-set for a particular micro-processor.

The GEZEL cosimulation environment integrates several instruction-simulation engines, including one for the ARM processor, one for the 8051 micro-controller, one for the picoblaze micro-controller, and one for AVR micro-controller. These simulation engines are open-source software projects by themselves. SimIt-ARM was developed by Wei Qin, the Dalton 8051 simulator was developed by the team

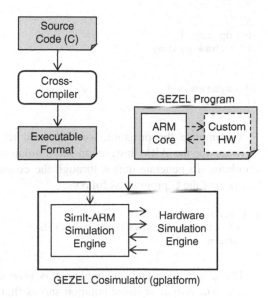

**Fig. 7.11** An instruction-set simulator, integrated into GEZEL, can simulate cross-compiled executables

of Frank Vahid, the Picoblaze simulator was developed by Mark Six, and the AVR simulator is derived from the SimulAVR project.

Figure 7.11 shows how instruction-set simulators are integrated into the GEZEL cosimulation engine, `gplatform`. In this figure, an instruction set simulator is used to combine two pieces of a cosimulation: a hardware description written in GEZEL, and a software application for an embedded core. The software application for the embedded core is written in C and compiled into executable code using a cross compiler. The hardware part of the application is written in GEZEL, and it specifies the platform architecture: the microprocessor, and its interaction with other hardware modules. The combination of the GEZEL program and the cross-compiled executable format is the input for the cosimulation.

The GEZEL cosimulator is cycle-accurate, including all the instruction-set simulators integrated within `gplatform`. Listing 7.10 shows a GEZEL program that simulates a stand-alone ARM core to execute the `gcd` program of Listing 7.1. Lines 1–4 define an ARM core which runs an executable program called `gcd`. The `ipblock` is a special type of GEZEL module which represents a black-box simulation model, a simulation model without internal details. This particular module does not have any input/output ports. We will introduce such input/output ports while discussing the various hardware/software interfaces (Chap. 11). Lines 6–12 of the GEZEL program simply configure the `myarm` module for execution.

**Listing 7.10** A GEZEL top-level module with a single ARM core

```
1  ipblock myarm {
2     iptype "armsystem";
3     ipparm "exec = gcd";
4  }
```

```
 5
 6  dp top {
 7     use myarm;
 8  }
 9
10  system S {
11     top;
12  }
```

To simulate the program, we will need to cross-compile the C application software for the ARM instruction-set simulator. Next, we run the instruction-set simulator. To generate output through the co-simulation, we modified the main function of the C program as follows:

```
int main() {
  printf("gcd(a,b)=%d\n", gcd(a,b));
  return 0;
}
```

The compilation and co-simulation is now done through the following commands. The output of the simulation shows that the program takes 14,338 cycles to execute.

```
> /usr/local/arm/bin/arm-linux-gcc -static gcd.c -o gcd
> gplatform top.fdl
core myarm
armsystem: loading executable [gcd]
gcd(a,b)=3
Total Cycles: 14338
```

Cycle-accurate simulation provides a detailed view on the execution of an embedded software program. Some instruction-set simulators provide feedback on the execution of individual instructions in the processor pipeline. The next subsection shows an example.

### 7.6.2 Analysis Based on Execution of Object Code

Processor features such as pipeline stalls and cache misses are not easy to determine using static program analysis alone. Using processor simulation, a designer can observe the dynamic effects of program execution.

SimIt-ARM, one of the instruction simulators integrated in GEZEL, is able to report the activities of each instruction as it flows through the processor pipeline. This includes quantities such as the value of the program counter, the simulation cycle-count, and the instruction completion time. Obviously, collecting this type of information will generate huge amounts of data, and a programmer needs to trace instructions selectively. SimIt-ARM provides the means to turn the instruction-tracing feature on or off, so that a designer can focus on a particular program area of interest.

Listing 7.11 shows the listing of a GCD function. Lines 11–13 illustrate a pseudo-systemcall that is used to turn the instruction-tracing feature of SimIt-ARM on and off. This pseudo-systemcall is simulator-specific, and will be implemented differently when a different processor or simulation environment is used. As shown in the `main` function on lines 17–19, the `gcd` function is called after turning the tracing feature on, and it is turned-off again after that.

**Listing 7.11** Microblaze assembly for the convolution program

```
1   int gcd (int a, int b) {
2     while (a != b) {
3       if (a > b)
4         a = a - b;
5       else
6         b = b - a;
7     }
8     return a;
9   }
10
11  void instructiontrace (unsigned a) {
12    asm("swi 514");
13  }
14
15  int main() {
16    int a, i;
17    instructiontrace (1);
18    a = gcd(6, 8);
19    instructiontrace (0);
20    printf("GCD = \%d\n", a);
21    return 0;
22  }
```

We will also generate the assembly code for the gcd function, which is useful as a guide during instruction tracing. Listing 7.12 shows the resulting code, annotated with the corresponding C statements in Listing 7.11. First, let's look at the execution without compiler optimization.

```
/usr/local/arm/bin/arm-linux-gcc -static -S gcd.c -o gcd.S
```

**Listing 7.12** ARM assembly code for gcd function

```
1   gcd:
2       mov  ip, sp                          ; set up stack frame
3       stmfd sp!, {fp, ip, lr, pc}
4       sub  fp, ip, #4
5       sub  sp, sp, #8
6       str  r0, [fp, #-16]                  ; storage for var_a
7       str  r1, [fp, #-20]                  ; storage for var_b
8   .L2:
9       ldr  r2, [fp, #-16]
10      ldr  r3, [fp, #-20]
11      cmp  r2, r3                          ; while (var_a != var_b)
```

```
12        bne   .L4
13        b     .L3
14  .L4:
15        ldr r2, [fp, #-16]                ; if (var_a > var_b)
16        ldr r3, [fp, #-20]
17        cmp r2, r3
18        ble .L5
19        ldr r3, [fp, #-16]
20        ldr r2, [fp, #-20]
21        rsb r3, r2, r3                    ;    var_a = var_a - var_b;
22        str r3, [fp, #-16]
23        b     .L2
24  .L5:                                    ; else
25        ldr r3, [fp, #-20]
26        ldr r2, [fp, #-16]
27        rsb r3, r2, r3                    ;    var_b = var_b - var_a;
28        str r3, [fp, #-20]
29        b     .L2
30  .L3:
31        mov r0, r3
32        ldmea   fp, {fp, sp, pc}
```

The Simit-ARM processor configuration uses the following parameters.

- D-cache of 16 KB, organized as a 32-set associative cache with a line size of 32-bytes.
- I-cache of 16 KB, organized as a 32-set associative cache with a line size of 32-bytes.
- Sixty-four-cycle memory-access latency, one-cycle cache-access latency.

The operation of a set-associative cache is as follows. Consider the address mapping used by a 16 KB set-associative cache with 32 sets and a line size of 32 bytes. Since the entire cache is 16 KB, each of the 32 sets in the cache contains 512 bytes or 16 lines. If we number the cache lines from 0 to 15, then address $n$ from the address space will map into line $(n/32) \bmod 16$. For example, assume that the processor performs an instruction fetch from address 0x8524. Figure 7.12 shows how this address maps into the second word of the tenth line of the cache. The cache will thus check each tenth line in each of the 32 sets before declaring a cache-miss.

We can now perform the simulation with instruction tracing on. The output is shown, in part, below. The columns in this listing have the following meaning.

- **Cycle**: The simulation cycle count at the instruction fetch
- **Addr**: The location of that instruction in program memory
- **Opcode**: The instruction opcode
- **P**: Pipeline mis-speculation. If a 1 appears in this column, then the instruction is not completed but removed from the pipeline.
- **I**: Instruction-cache miss. If a 1 appears in this column, then there is a cache miss when this instruction is fetched.

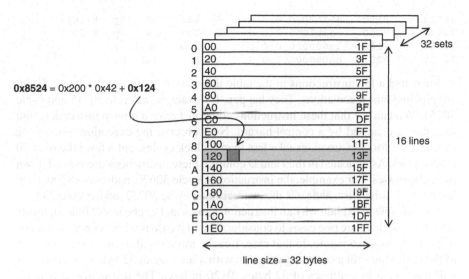

0x8524 = 0x200 * 0x42 + 0x124

**Fig. 7.12** Mapping of address 0x8524 in a 32-set, 16-line, 32-bytes-per-line set-associative cache

- **D**: Data-cache miss. If a 1 appears in this column, then there is a data cache miss when this instruction executes.
- **Time**: The total time that this instruction is active in the pipeline, from the cycle it is fetched to the cycle it is retired.
- **Mnemonic**: Assembly code for this instruction.

```
Cycle     Addr    Opcode      P I D  Time   Mnemonic
30601     81e4    e1a0c00d    0 1 0    70    mov    ip, sp;
30667     81e8    e92dd800    0 0 0     8    stmdb  sp!, {fp, ip, lr, pc};
30668     81ec    e24cb004    0 0 0     8    sub    fp, ip, #4;
30672     81f0    e24dd008    0 0 0     5    sub    sp, sp, #8;
30673     81f4    e50b0010    0 0 0     5    str    r0, [fp, #-16];
30674     81f8    e50b1014    0 0 0     5    str    r1, [fp, #-20];
30675     81fc    e51b2010    0 0 0     5    ldr    r2, [fp, #-16];
30676     8200    e51b3014    0 1 0    70    ldr    r3, [fp, #-20];
30742     8204    e1520003    0 0 0     6    cmp    r2, r3;
30743     8208    1a000000    0 0 0     3    bne    0x8210;
30745     820c    ea00000d    1 0 0     1    b      0x8248;
30746     8210    e51b2010    0 0 0     5    ldr    r2, [fp, #-16];
30747     8214    e51b3014    0 0 0     5    ldr    r3, [fp, #-20];
30748     8218    e1520003    0 0 0     6    cmp    r2, r3;
30749     821c    da000004    0 0 0     3    ble    0x8234;
30751     8220    e51b3010    1 1 0     1    ldr    r3, [fp, #-16];
30752     8234    e51b3014    0 1 0    69    ldr    r3, [fp, #-20];
30817     8238    e51b2010    0 0 0     5    ldr    r2, [fp, #-16];
30818     823c    e0623003    0 0 0     6    rsb    r3, r2, r3;
30819     8240    e50b3014    0 0 0     6    str    r3, [fp, #-20];
30821     8244    eaffffec    0 0 0     2    b      0x81fc;
30822     8248    e1a00003    1 0 0     1    mov    r0, r3;
```

```
30823      81fc    e51b2010 0 0 0     5  ldr    r2, [fp, #-16];
30824      8200    e51b3014 0 0 0     5  ldr    r3, [fp, #-20];
30826      8208    1a000000 0 0 0     3  bne    0x8210;
30828      820c    ea00000d 1 0 0     1  b      0x8248;
```

First, find a few instructions in the table that have a '1' in the P column. These are pipeline mis-speculations. They happen, for example, at cycle 30745 and cycle 30751. You can see that these instructions come just *after* a branch instruction, and thus they are caused by a control hazard. Next observe the execution time of the instructions. Most instructions take less than 6 clock cycles, but a few take over 50 clock cycles. As indicated in the I and D column, these instructions are slowed down by cache misses. For example, the instruction at cycle 30676, address 0x8200, is an instruction-cache miss, and so is the instruction at cycle 30752, address 0x8234.

It is possible to explain why an instruction causes an I-cache miss? Indeed, this is possible, and there are two cases to consider. The first case is when a linear sequence of instructions is executing. In that case, I-cache misses will occur at the boundary of the cache-lines. In a cache organization with a line size of 32 bytes, cache misses will thus occur at multiples of 32 bytes (0x20 in hex). The instruction at address 0x8200 is an example of this case. This is the first instruction of a cache line which is not in the cache. Therefore, the instruction-fetch stage stalls for 64 clock cycles in order to update the cache. The second case is when a jump instruction executes and moves to a program location which is not in the cache. In that case, the target address may be in the middle of a cache line, and a cache miss may still occur. The instruction at address 0x8234 is an example of that case. That instruction is executed as a result of jump. In fact, the instruction just before that (at address 0x8220) is also cache miss. That instruction does not complete, however, because it is part of a control hazard.

Finally, observe also that some regular instructions take five clock cycles to complete, while others take six clock cycles. Relevant examples are the instructions on addresses 0x8214 and 0x8218. The first of these instructions is a memory-fetch that loads the value of a local variable (b) into register r3. The following instruction is a compare instruction that uses the value of r3. As discussed earlier, this is an example of a data hazard, where the value of a register is only available after the buffer stage of the RISC pipeline. The compare-instruction at address 0x8218 cannot benefit from pipeline interlock hardware and it must be stalled for one clock cycle until the result is available from data-memory.

Using processor simulation at low abstraction level is therefore useful to understand the precise behavior of an embedded software program.

## 7.6.3  Simulation at Low Abstraction Level

Even more simulation detail is available at the level of a hardware simulation. Figure 7.13 illustrates how a cosimulation can be created using a hardware description language. By replacing the instruction-set simulator with a hardware

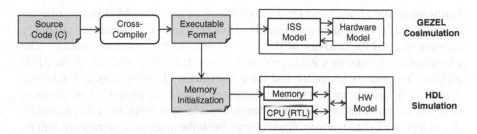

**Fig. 7.13** Processor simulation using a hardware simulator

model (in a HDL) of the actual processor, a standard hardware simulator can be used. This type of simulation is very detailed, but it may have a few caveats as opposed to the use of an instruction-set simulator.

- One needs a processor HDL model before such a simulation can be done. However, not all processors have a freely available Verilog or VHDL implementation. Although several processors (OpenRISC and Leon, for example) have freely available HDL, the models for typical commercial processors are closed. Indeed, a very detailed simulation model is considered to be intellectual property of the processor designer.
- The speed of a simulation decreases with the level of simulation detail. HDL-level simulations of full processor models are typically much slower than ISS based processor models. Furthermore, a co-design puts emphasis on the interaction between hardware and software. It may therefore be useful to abstract the internal details of the processor during simulation.
- The HDL based simulation of a processor needs special preparation. In particular, before the simulation can start, the program memory of the processor needs to be properly initialized. This is done by using a memory initialization file.
- The high level of detail may implies the loss of abstraction. For example, in an HDL based cosimulation, the distinction between hardware and software is effectively lost: everything is expressed in terms of Verilog or VHDL. In particular when analyzing the behavior of software, the loss of abstraction is problematic. Using only the HDL simulation, a designer can trace the program counter value and the instruction op-codes. The interpretation of those opcodes is still the task of the designer.

Hence, although HDL based simulation of processor models is attractive in the later stages of a design, it is not a panacea. Instruction-set simulation plays a very important role in the initial development of embedded software.

## 7.7 Summary

In this chapter, we discussed the organization and operation of typical RISC processors, using the ARM, the Nios-II and the Microblaze as examples. In hardware-software codesign, processors are the entry-point of software into the hardware world. Hence, to analyze the operation of a low-level hardware-software interface, it is very useful to understand the link between a C program, its' assembly instructions, and the behavior of these instructions in the processor pipeline. The execution of software by a RISC processor is affected by the behavior of the RISC pipeline, its memory hierarchy, and the organization of instructions and data into memory. Through the understanding of a limited set of concepts in C, these complex interactions can be understood and controlled to a fairly detailed level. For example, the mapping of data types to memory can be influenced with storage qualifiers, and detailed performance optimization is possible through careful rewriting of C code in combination with study of the resulting program through static and dynamic analysis. This chapter has prepared us for the next big step in a hardware/software codesigned system: the extension of a simple RISC processor into a system-on-chip architecture that integrates software, processors, and custom hardware functions. Clearly, the RISC processor will play a pivotal role in this story.

## 7.8 Further Reading

The classic work on RISC processor architectures is by Hennessy and Patterson (2006). It is essential reading if you want to delve into the internals of RISC processors. Good documentation on the low-level software tools such as size and objdump is not easy to find; the manual pages unfortunately are rather specialized. Books on Embedded Linux Programming, such as (Yaghmour et al. 2008), are the right place to start if the man pages do not help. The ELF format is described in detail in the Tool Interface Standard ELF format (ELF Committee 1995). Processor documentation can be found with the processor designers or processor vendors. For example, ARM has an extensive on-line library documenting all the features of ARM processors (ARM 2009b), and Xilinx provides a detailed specification of the Microblaze instruction-set (Xilinx 2009a).

An effective method to learn about the low-level implementation of a RISC core is to implement one, for example starting from open source code. The LEON series of processors by Gaisler Aeroflex, for example, provides a complete collection of HDL source code, compilers, and debuggers (Aeroflex 2009). The internals of a processor simulator, and of the SimIt-ARM instruction-set simulator, are described by Qin in several articles (D'Errico and Qin 2006; Qin and Malik 2003).

## 7.9 Problems

**Problem 7.1.** Write a short C program that helps you to determine if the stack grows upwards or downwards.

**Problem 7.2.** Write a short C program that helps you to determine the position of the stack segment, the text segment, the heap, and data segment (global variables).

**Problem 7.3.** Explain the difference between the following terms:

- Control hazard and data hazard
- Delayed branch and conditional branch
- Little Endian and Big Endian
- volatile int * a and int * const a
- Virtual Memory Address (VMA) and Load Memory Address (LMA)

**Problem 7.4.** This problem considers C Qualifiers and Specifiers.

(a) Correct or not: The volatile qualifier will prevent a processor from storing that variable in the cache?
(b) When writing a C program, you can create an integer variable a as follows: register int a. This specifier tells the compiler that a should be preferably kept in a register as much as possible, in the interest of program execution speed. Explain why this specifier cannot be used for the memory-mapped registers in a hardware coprocessor.

**Problem 7.5.** The following C function was compiled for Microblaze with optimization-level O2. It results in a sequence of four assembly instructions. Carefully examine the C code (Listing 7.13) and the assembly code (Listing 7.14), and answer the following questions. Note that register r5 holds the function argument and register r3 holds the function return value.

(a) Explain why the assembly code does not have a loop?
(b) Suppose line 5 of the C program reads a = a - 1 instead of a = a + 1. Determine how the assembly code would change.

**Listing 7.13** C listing for Problem 7.5

```
1   int dummy(int a) {
2     int i, j = a;
3     for (i=0; i<3; i++) {
4       j += a;
5       a = a + 1;
6     }
7     return a + j;
8   }
```

**Listing 7.14** Assembly listing for Problem 7.5

```
1       muli   r3, r5,  4
2       addk   r4, r3, r5
3       rtsd   r15, 8
4       addik  r3, r3, 6
```

**Problem 7.6.** The following C statement implements a pseudorandom generator. It translates to the sequence of assembly instructions as shown in Listing 7.15. The assembly instructions are those of a five-stage pipelined StrongARM processor.

```
unsigned rnstate;
rnstate = (rnstate >> 1) ^ (-(signed int)(rnstate &1)
                                    & 0xd0000001u);
```

Answer the following questions:

(a) What is the purpose of line 5 in Listing 7.15 (the rsb instruction) in the StrongArm Code? Point out exactly what part of the C expression it will implement.

(b) What types of hazard can be caused by line 3 in Listing 7.15?

**Listing 7.15** Assembly listing for Problem 7.6

```
1        ldr      r3, [fp, #-16];   // load-register
2        mov      r2, r3, lsr #1;   // lsr = shift-right
3        ldr      r3, [fp, #-16];
4        and      r3, r3, #1;
5        rsb      r3, r3, #0;
6        and      r3, r3, #-805306367;
7        eor      r3, r2, r3;
8        str      r3, [fp, #-16];   // store-register
```

**Problem 7.7.** The C in Listing 7.16 was compiled for StrongARM using the following command:

```
/usr/local/arm/bin/arm-linux-gcc -O -S -static loop.c -o loop.s
```

The resulting assembly code is shown in listing 7.17.

(a) Draw a dataflow diagram of the assembly code.

(b) Identify all instructions in this listing that are directly involved in the address calculation of a data memory read.

**Listing 7.16** C program for Problem 7.7

```
1    int a[100]
2    int b[100];
3    int i;
4
5    for (i=0; i<100; ++i) {
6       a[b[i]] = i + 2;
7    }
8    return 0;
```

**Listing 7.17** Assembly listing for Problem 7.7

```
1        mov      r1, #0
2   .L6:
3        add      r0, sp, #800
4        add      r3, r0, r1, asl #2
5        ldr      r3, [r3, #-800]
```

```
6        add      r2, r0, r3, asl #2
7        add      r3, r1, #2
8        str      r3, [r2, #-400]
9        add      r1, r1, #1
10       cmp      r1, #99
11       ble      .L6
```

**Problem 7.8.** Listing 7.18 shows a routine to evaluate the CORDIC transformation (Coordinate Digital Transformation). CORDIC procedures are used to approximate trigonometric operations using simple integer arithmetic. In this case, we are interested in the inner loop of the program, on lines 16–28. This program will be compiled with a single level of optimization as follows:

```
arm-linux-gcc -O1 -g -c cordic.c
```

Next, the assembly code of the program is created using the objdump utility. The command line flags are chosen to generate the assembly code, interleaved with the C code. This is possible if the object code was generated using *debug information* ( -g flag above). The resulting file is shown in Listing 7.19.

```
arm-linux-objdump -S -d cordic.o
```

(a) Study the Listing 7.19 and explain the difference between an addgt and an addle instruction on the ARM processor.
(b) Use objdump find the size of the text segment and the data segment.
(c) Study the Listing 7.19 and point out what are the *callee-saved* registers in this routine.
(d) Estimate the execution time for the cordic routine, ignoring the cache misses.

**Listing 7.18** C listing for Problem 7.8

```
1   #define AG_CONST 163008218
2
3   static const int angles[] = {
4      210828714,  124459457,   65760959,   33381289,
5       16755421,    8385878,    4193962,    2097109,
6        1048570,     524287,     262143,     131071,
7          65535,      32767,      16383,       8191,
8           4095,       2047,       1024,        511 };
9
10  void cordic(int target, int *rX, int *rY) {
11    int X, Y, T, current;
12    unsigned step;
13    X        = AG_CONST;
14    Y        = 0;
15    current = 0;
16    for(step=0; step < 20; step++) {
17      if (target > current) {
18        T         = X - (Y >> step);
```

```
19          Y            = (X >> step) + Y;
20          X            = T;
21          current     += angles[step];
22      } else {
23          T            = X + (Y >> step);
24          Y            = -(X >> step) + Y;
25          X            = T;
26          current     -= angles[step];
27      }
28      }
29      *rX = X;
30      *rY = Y;
31  }
```

**Listing 7.19**  Mixed C-assembly listing for Problem 7.8

```
 1  void cordic(int target, int *rX, int *rY) {
 2      0:   e92d40f0    stmdb   sp!, {r4, r5, r6, r7, lr}
 3      4:   e1a06001    mov     r6, r1
 4      8:   e1a07002    mov     r7, r2
 5  int X, Y, T, current;
 6  unsigned step;
 7  X        = AG_CONST;
 8      c:   e59fe054    ldr     lr, [pc, #84]
 9  Y        = 0;
10     10:   e3a02000    mov     r2, #0  ; 0x0
11  current = 0;
12     14:   e1a01002    mov     r1, r2
13  for(step=0; step < 20; step++) {
14     18:   e1a0c002    mov     ip, r2
15     1c:   e59f5048    ldr     r5, [pc, #72]
16     20:   e1a04005    mov     r4, r5
17      if (target > current) {
18     24:   e1500001    cmp     r0, r1
19          T            = X - (Y >> step);
20     28:   c04e3c52    subgt   r3, lr, r2, asr ip
21          Y            = (X >> step) + Y;
22     2c:   c0822c5e    addgt   r2, r2, lr, asr ip
23          X            = T;
24     30:   c1a0e003    movgt   lr, r3
25          current     += angles[step];
26     34:   c795310c    ldrgt   r3, [r5, ip, lsl #2]
27     38:   c0811003    addgt   r1, r1, r3
28      } else {
29          T            = X + (Y >> step);
30     3c:   d08e3c52    addle   r3, lr, r2, asr ip
31          Y            = -(X >> step) + Y;
32     40:   d0422c5e    suble   r2, r2, lr, asr ip
33          X            = T;
34     44:   d1a0e003    movle   lr, r3
35          current     -= angles[step];
36     48:   d794310c    ldrle   r3, [r4, ip, lsl #2]
37     4c:   d0631001    rsble   r1, r3, r1
```

```
38    50:    e28cc001    add ip, ip, #1
39    54:    e35c0013    cmp ip, #19
40    58:    8586e000    strhi   lr, [r6]
41    }
42  }
43    *rX = X;
44    *rY = Y;
45    5c:    85872000    strhi   r2, [r7]
46  }
47    60:    88bd80f0    ldmhiia sp!, {r4, r5, r6, r7, pc}
48    64:    ea000007    b    24 <cordic+0x24>
49    68:    09b74eda    ldmeqib r7!, {r1, r3, r4, r6, r7, r9, sl,
                                      fp, lr}
50    6c:    00000000    andeq   r0, r0, r0
```

**Problem 7.9.** We discussed the modulo-179 program in Listing 7.7, Sect. 7.4. We compiled the C program for two different processors, ARM and Microblaze, at the highest optimization level. The ARM assembly is shown in Listing 7.20, and the Microblaze assembly is shown in Listing 7.21. Study the assembly listings and answer the questions below.

(a) One of the assembly programs uses delayed branch instructions. Which processor is it?
(b) One of the assembly programs uses predicated instructions, which are instructions that execute conditionally based on the value of a processor flag. Which processor is it?
(c) Estimate the number of clock cycles required on each processor to compute the result of modulo(54).

**Listing 7.20** ARM assembly program for Problem 7.9

```
1  modk:
2        mov     ip, #1
3        mov     r2, ip, asl r1
4        sub     r1, r2, #1
5        and     r0, r0, r1
6        mov     pc, lr
7  divk:
8        mov     r0, r0, lsr r1
9        mov     pc, lr
10 modulo:
11       mov     r2, r0, lsr #8
12       mov     r1, #77
13       and     r0, r0, #255
14 .L9:
15       mul     r3, r1, r2
16       and     ip, r3, #255
17       movs    r2, r3, lsr #8
18       add     r0, r0, ip
19       bne     .L9
20       mov     r2, r0, lsr #8
```

```
21      cmp         r2, #0
22      and         r0, r0, #255
23      bne         .L9
24      cmp         r0, #179
25      subcs       r0, r0, #179
26      mov         pc, lr
```

**Listing 7.21**  Microblaze assembly program for Problem 8.9

```
1  modk:
2          addik      r3,r0,1 # 0x1
3          bsll       r3,r3,r6
4          addik      r3,r3,-1
5          rtsd       r15,8
6          and        r3,r3,r5
7  divk:
8          rtsd       r15,8
9          bsrl       r3,r5,r6
10 modulo:
11         bsrli      r4,r5,8
12         andi       r5,r5,255   #and1
13 $L18:
14         muli       r3,r4,77
15         bsrli      r4,r3,8
16         andi       r3,r3,255 #and1
17         bneid      r4,$L18
18         addk       r5,r5,r3
19         bsrli      r4,r5,8
20         bneid      r4,$L18
21         andi       r5,r5,255 #and1
22         addi       r18,r0,178
23         cmpu       r18,r5,r18
24         bgeid      r18,$L20
25         addk       r3,r5,r0
26         addik      r3,r5,-179
27 $L20:
28         rtsd       r15, 8
29         nop
```

# Chapter 8
# System on Chip

## 8.1 The System-on-Chip Concept

Figure 8.1 illustrates a typical System-on-chip. It combines several components on a bus system. One of these components is a microprocessor (typically a RISC) which acts as a central controller in the SoC. Other components include on-chip memory, off-chip-memory interfaces, dedicated peripherals, hardware coprocessors, and component-to-component communication links.

The application domain greatly affects the type of hardware peripherals, the size of memories and the nature of on-chip communications. A particular configuration of these elements is called a *platform*. Just like a personal computer is a platform for general-purpose computing, a system-on-chip is a platform for domain-specialized computing, i.c. for an ensemble of applications that are typical for a given application domain. Examples of application domains are mobile telephony, video processing, or high-speed networking. The set of applications in the video-processing domain for example could include image transcoding, image compression and decompression, image color transformations, and so forth. Domain specialization in a System-on-Chip is advantageous for several reasons.

- The *specialization* of the platform ensures that its processing efficiency is higher compared to that of general-purpose solutions. Increased processing efficiency means lower power consumption (longer battery lifetime) or higher absolute performance.
- The *flexibility* of the platform ensures that it is a reusable solution that works over multiple applications. As a result, the design cost per-application decreases, applications can be developed faster, and the SoC itself becomes cheaper because it can be manufactured for a larger market.

P.R. Schaumont, *A Practical Introduction to Hardware/Software Codesign*, DOI 10.1007/978-1-4614-3737-6_8, © Springer Science+Business Media New York 2013

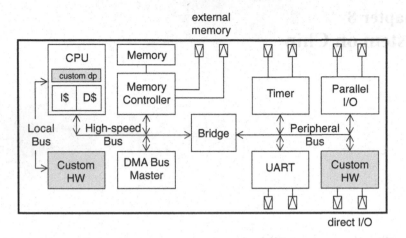

**Fig. 8.1** Generic template for a system-on-chip

## 8.1.1   The Cast of Players

An architecture such as in Fig. 8.1 can be analyzed along four orthogonal dimensions: control, communication, computation, and storage. The role of central controller is given to the microprocessor, who is responsible of issuing control signals to, and collecting status signals from, the various components in the system. The microprocessor may or may not have a local instruction memory. In case it does not have a local instruction memory, caches may be utilized to improve instruction-memory bandwidth. The I$ and D$ symbols in Fig. 8.1 represent the instruction- and data-caches in the microprocessor. In the context of the SoC architecture, these caches will only be of use to the software executing on the microprocessor. This obvious fact has an often-ignored consequence: whenever the microprocessor needs to interact with a peripheral, the data moves outside of the cache. The memory-hierarchy, which makes the CPU so fast and powerful, does not work for data-movement. A microprocessor in an SoC is therefore useful as central controller, but it is not very efficient to carry data around in the system. We will come back to this point in detail further in this Chapter.

The SoC implements communication using system-wide buses. Each bus is a bundle of signals including address, data, control, and synchronization signals. The data transfers on a bus are expressed as read- and write-operations with a particular memory address. The bus control lines indicate the nature of the transfer (read/write, size, source, destination), while the synchronization signals ensure that the sender and receiver on the bus are aligned in time during a data transfer. Each component connected to a bus will respond to a particular range of memory addresses. The ensemble of components can thus be represented in an *address map*, a list of all relevant system-bus addresses.

It is common to split SoC buses into segments. Each segment connects a limited number of components, grouped according to their communication needs.

In the example, a high-speed communication bus is used to interconnect the microprocessor, a high-speed memory interface, and a Direct Memory Access (DMA) controller. A DMA is a device specialized in performing block-transfers on the bus, for example to copy one memory region to another. Next to a high-speed communication bus, you may also find a peripheral bus, intended for lower-speed components such as a timer and input-output peripherals. Segmented buses are interconnected with a *bus bridge*, a component that translates bus transfers from one segment to another segment. A bus bridge will only selectively translate transfers from one bus segment to the other. This selection is done based on the address map.

The bus control lines of each bus segment are under command of the *bus master*, the component that decides the nature of a given bus transfer. The *bus slaves* will follow the directions of the bus master. Each bus segment can contain one or more bus masters. In case there are multiple masters, the identity of the bus master can be multiplexed among bus-master components at run time. In that case, a *bus arbiter* will be needed to decide which component can become a bus master for a given bus transfer. A bus bridge can be either a master or a slave, depending on the direction of the transfers. For example, when going from the high-speed bus to the peripheral bus, the bus bridge will act as a bus slave on the high-speed bus and as a bus master on the peripheral bus. Each of the transfers on the high-speed bus and on the peripheral bus will be handled independently. Therefore the segmentation of buses using bus bridges leads to a dual advantage. First, bus segments can group components with matching read- and write-speed together, thus providing optimal usage of the available bus bandwidth. Second, the bus segments enable communication parallelism.

## 8.1.2 SoC Interfaces for Custom Hardware

Let's consider the opportunities to attach custom hardware modules in the context of an SoC architecture. In the context of this chapter, a 'custom hardware module' means a dedicated digital machine in the form of an FSMD or a micro-programmed machine. Eventually, all custom hardware will be under control of the central processor in the SoC. The SoC architecture offers several possible hardware-software interfaces to attach custom hardware modules. Three approaches can be distinguished in Fig. 8.1 as shaded blocks.

- The most general approach is to integrate a custom hardware module as a standard peripheral on a system bus. The microprocessor communicates with the custom hardware module by means of read/write memory accesses. Of course, the memory addresses occupied by the custom hardware module cannot be used for other purposes (i.e. as addressable memory). For the memory addresses occupied by the custom hardware module, the microprocessors' cache has no meaning, and the caching effect is unwanted. Microcontroller chips with many different peripherals typically use this memory-mapped strategy to attach peripherals. The strong point of this approach is that a universal communication mechanism (memory read/write operations) can be used for a wide range of custom hardware modules. The corresponding disadvantage, of course, is that such a bus-based approach to integrate hardware is not very scalable in terms of performance: the system bus quickly becomes a bottleneck when intense communication between a microprocessor and the attached hardware modules is needed.

- A second mechanism is to attach custom hardware through a local bus system or coprocessor interface provided by the microprocessor. In this case, the communication between the hardware module and the microprocessor will follow a dedicated protocol, defined by the local bus system or coprocessor interface. In comparison to system-bus interfaces, coprocessor interfaces have a high-bandwidth and a low latency. The microprocessor may also provide a dedicated set of instructions to communicate over this interface. Typical coprocessor interfaces do not involve a memory addresses. This type of coprocessor obviously requires a microprocessor with a coprocessor- or local-bus interface.

- Microprocessors may also provide a means to integrate a custom-hardware datapath inside of the microprocessor. The instruction set of the microprocessor is then extended with additional, new instructions to drive this custom hardware. The communication channel between the custom data-path and the processor is typically through the processor register file, resulting in a very high communication bandwidth. However, the very tight integration of custom hardware with a microprocessor also means that the traditional bottlenecks of the microprocessor are a bottleneck for the custom-hardware modules as well. If the microprocessor is stalled because of external events (such as memory-access bandwidth), the custom data-datapath is stalled also.

These observations show that, in the context of SoC, there is no single *best* way to integrate hardware and software. There are many possible solutions, each with their advantages and disadvantages. Selecting the right approach involves trading-off many factors, including the required communication bandwidth, the design complexity of the custom hardware interface, the software, the available design time, and the overall cost budget. The following chapters will cover some of the design aspects of hardware-software interfaces. In the end, however, it is the hardware-software codesigner who must identify the integration opportunities of a given System-on-Chip architecture, and who must realize their potential.

## 8.2   Four Design Principles in SoC Architecture

A SoC is specific to a given application domain. Are there any guiding design principles that are relevant to the design of *any* SoC? This section addresses this question in more detail. The objective is to clarify four design principles that govern the majority of modern SoC architectures. These four principles include heterogeneous and distributed communications, heterogeneous and distributed data processing, heterogeneous and distributed storage, and hierarchical control. We will review each of these four points in detail. This will demonstrate the huge potential of the SoC, and in particular their technological advantage over non-integrated circuits.

## 8.2.1 Heterogeneous and Distributed Data Processing

A first prominent characteristic of an SoC architecture is heterogeneous and distributed data processing. An SoC may contain multiple independent (distributed) computational units. Moreover, these units can be heterogenous. They can include FSMD, micro-programmed engines, or microprocessors.

One can distinguish three forms of data-processing parallelism. The first is word-level parallelism, which enables the parallel processing of multiple bits in a word. The second is operation-level parallelism, which allows multiple instructions to be executed simultaneously. The third is task-level parallelism, which allows multiple independent threads of control to be executing independently. Word-level parallelism and operation-level parallelism are available on all of the machine architectures we discussed so far: FSMD, Micro-programmed machines, RISC, and also SoC. However, only an SoC supports true task-level parallelism. Note that multi-threading in a RISC is not task-level parallelism; it is task-level concurrency on top of a sequential machine (See Sect. 1.7).

Each of the computational units in an SoC can be specialized to a particular function. The overall SoC therefore includes a collection of *heterogeneous* computational units. For example, a digital signal processing chip in a camera may contain specialized units to perform image-processing. Computational specialization is a key ingredient to high-performance. In addition, the presence of all forms of parallelism (word-level, operation-level, task-level) ensures that an SoC can fully exploit the technology.

In fact, integrated circuit technology is *extremely* effective to provide computational parallelism. Consider the following numerical example. A 1-bit full-adder cell can be implemented in about 28 transistors. The Intel Core 2 processor contains 291 million transistors in 65 nm CMOS technology. This is sufficient to implement 325,000 32-bit adders. Assuming a core clock frequency of 2.1 GHz, we thus find that the silicon used to create a Core 2 can theoretically implement 682,000 GOPS. We call this number the *intrinsic computational efficiency* of silicon. Of course, we don't know how to build a machine that would have this efficiency, let alone that such a machine would be able to cope with the resulting power density. The intrinsic computational efficiency merely represents an upperbound for the most efficient use of silicon real-estate.

$$Eff_{intrinsic} = \frac{291.10^6}{28.32}.2.1 \approx 682,000 \; Gops \tag{8.1}$$

The actual Core2 architecture handles around 9.24 instructions per clock cycle, in a single core and in the most optimal case. The actual efficiency of the 2.1 GHz Core 2 therefore is 19.4 GOPS. We make the (strong) approximation that these 9.24 instructions each correspond to a 32-bit addition, and call the resulting throughput the actual Core2 efficiency. The ratio of the intrinsic Core2 efficiency over the actual Core2 efficiency illustrates the efficiency of silicon technology compared to the efficiency of a processor core architectures.

$$Efficiency = \frac{Eff_{intrinsic}}{Eff_{actual}} \approx \frac{682,000}{19.4} = 35,150 \tag{8.2}$$

Therefore, bare silicon can implement computations 35,000 times more efficient than a Core2! While this is a very simple and crude approximation, it demonstrates why specialization of silicon using multiple, independent computational units is so attractive.

## 8.2.2   Heterogeneous and Distributed Communications

The central bus in a system-on-chip is a critical resource. It is shared by many components in an SoC. One approach to prevent this resource from becoming a bottleneck is to split the bus into multiple bus segments using bus bridges. The bus bridge is a mechanism to create distributed on-chip communications. The on-chip communication requirements typically show large variations over an SoC. Therefore, the SoC interconnection mechanisms should be heterogeneous as well. There may be shared busses, point-to-point connections, serial connections and parallel connections.

Heterogeneous and distributed SoC communications enable a designer to exploit the on-chip communication bandwidth. In modern technology, this bandwidth is extremely high. An illustrative example by Chris Rowen mentions the following numbers. In a 90 nm six-layer metal processor, we can reasonably assume that metal layers can be used as follows (Fig. 8.2): two metal layers are used for power and ground, two metal layers are used to route wires in the X direction, and two metal layers are used to route wires in the Y direction. The density of wires in a 90 nm process is 4 wires per micron (a micron is one thousandth of a millimeter), and the bit frequency is at 500 MHz. Consequently, in a chip of 10 mm on the side, we will have 40,000 wires per layer on a side. Such a chip can shuffle and transport 80,000 bits at a frequency of 500 MHz. This corresponds to 40 Tbits per second! Consequently, on-chip communications have a high bandwidth – the real challenge is how to organize it efficiently.

The same efficiency is not available for *off-chip* communication bandwidth. In fact, off-chip bandwidth is very expensive compared to on-chip bandwidth. For example, consider the latest Hypertransport 3.1 standard, a serial link developed for high-speed processor interconnect. Usually, a (high-end) processor will have one to

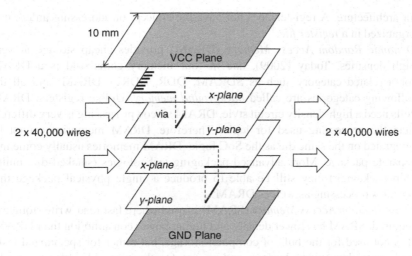

**Fig. 8.2** Demonstration of the routing density in a six-layer metal 90 nm CMOS chip

four of such ports. The maximum aggregate data bandwidth for such a port is around 51.2 Gb per second. Thus we will find less than 204.8 Gb per second input/output bandwidth on a state-of-the-art processor today. That is still 195 times less than the 40 Tb/s on-chip bandwidth in an (older) 90 nm CMOS process! On-chip integration therefore allows very cheap, high-speed interconnect between components. Of course, a similar remark as with the intrinsic computational efficiency applies: the 40 Tb/s number is an ideal estimate, an upperbound. We have no means to build an operational chip with this bandwidth: signal crosstalk, and power density (each wire needs a driver) will prevent this.

### 8.2.3 Heterogeneous and Distributed Storage

A third characteristic of System-on-Chip architectures is a distributed and hetero-geneous storage architecture. Instead of a single, central memory, an SoC will use a collection of dedicated memories. Processors and micro-coded engines may contain local instruction memories. Processors may also use cache memories to maintain local copies of data and instructions. Coprocessors and other active components will use local register files. Specialized accelerators can use dedicated memories for specific applications such as for video frame buffering, or as local scratchpad.

This storage is implemented with a collection of different memory technologies. There are five broad categories of silicon-based storage available today.

- *Registers* are the fastest type of memory available. Registers are also called *foreground memory*. They reside the closest to the computation elements of an architecture. A register does not have the concept of addressing unless it is organized in a *register file*.

*   *Dynamic Random Access Memory* (DRAM) provides cheap storage at very high densities. Today (2009), one in four memory chips sold is a DRAM (or a related category such as SDRAM, DDR, DDR2). DRAM, and all the following categories are called *background memory*. Unlike registers, DRAM cells need a high-density circuit style. DRAM silicon processing is very different than the processing used for logic. Therefore, DRAM memories cannot be integrated on the same die as the SoC logic. DRAM memories usually come in a separate package. More advanced packaging technologies (stacked-die, multi-chip packaging) may still be able to produce a single physical package that contains processing as well as DRAM.

*   *Static Random Access Memory* (SRAM) is used where fast read-write storage is required. SRAM has lower density and higher power consumption than DRAM. It is not used for the bulk of computer storage, but rather for specialized tasks such as caches, video buffers, and so on. On the plus side, SRAM can be implemented with the same process technology as normal logic gates. It is therefore easy to mix SRAM and computational elements in an SoC.

*   *Non-volatile Read-Only Memory* (NVROM) is used for applications that only require read access on a memory, such as for example to store the instructions of a program. Non-volatile memories have a higher density than SRAM. There is a range of technologies that can be used to implement a NVROM (maskprogrammed ROM, PROM, EPROM, EEPROM).

*   *Non-volatile Random Access Memory* (NVRAM) is used for applications that need read-write memories that do not loose data when power is removed. The read- and write-speed in a NVRAM can be asymmetrical (write being slower), so that in the limit the distinction between NVROM and NVRAM is not sharp.

Table 8.1 summarizes the key characteristics of these different types of memory. The entries in the table have the following meaning.

*   The *cell size* is the silicon area required to store a single bit. Storage cells are only one aspect of a complete memory – additional hardware is needed for address decoding, multiplexing bits from the data bus, and so on. High-density storage technologies use only a single transistor per bit, and make use of low-level physical properties of that transistor (parasitic capacitance, floating gate, etc.) to hold the bit.

*   The *retention time* expresses how long a bit can be held in a non-powered memory.

*   The *addressing* mechanism shows how bits are retrieved from memory. In multiplexed addressing, such as used by DRAM, the address is cut in two parts which are provided sequentially to the memory.

*   The *access time* is the time required to fetch a data element from memory. Note that the access time is a coarse number, as it does not capture the detailed behavior of a memory. For example, in NVRAM technologies, the read and write access time is asymmetrical: write takes longer than read. Modern DRAM

**Table 8.1**  Types of memories

| Type | Register Register file | DRAM | SRAM | NVROM (ROM, PROM, EPROM) | NVRAM (Flash, EEPROM) |
|---|---|---|---|---|---|
| Cell size (bit) | 10 transistors | 1 transistor | 4 transistors | 1 transistor | 1 transistor |
| Retention | 0 | Tens of ms | 0 | ∞ | 10 years |
| Addressing | Implicit | Multiplexed | Non-muxed | Non-muxed | Non-muxed |
| Access time | Less then 1ns | Less then 20 ns | Less then 10 ns | 20 ns | 20 ns (read) 100 µs (write) |
| Power consumption | High | Low | High | Very low | Very low |
| Write durability | ∞ | ∞ | ∞ | 1-time | One million times |

memories are very efficient in accessing consecutive memory locations (burst access), but individual random locations take longer. Finally, modern memories can be internally pipelined, such that they can process more than one read or write command at the same time.

- The *power consumption* is a qualitative appreciation for the power consumption of a memory, as measured per access and per bit. Fast read/write storage is much more power-hungry than slow read-only storage.

The presence of distributed storage significantly complicates the concept of a centralized memory address space, which is so useful in SoC. As long as the data within these distributed memories is local to a single component, this does not cause any problem. However, it becomes troublesome when data needs to be shared among components. First, when multiple copies of a single data item exist in different memories, all these copies need to be kept consistent. Second, updating of a shared data item needs to be implemented in a way that will not violate data dependencies among the components that share the data item. It is easy to find a few examples where either of these two requirements will be violated (see Problem 8.1).

In 1994, Wulf and McKee wrote a paper entitled *Hitting the Memory Wall: Implications of the Obvious'*. The authors used the term *memory wall* to indicate the point at which the performance of a (computer) system is determined by the speed of memory, and is no longer dependent on processor speed. While the authors' conclusions were made for general-purpose computing architectures, their insights are also valid for mixed hardware/software systems such as those found in System-on-Chip. Wulf and McKee observed that the performance improvement of processors over time was higher than the performance improvement of memories. They assumed 60% performance improvement per year for processors, and 10% for memories – valid numbers for systems around the turn of the century. The memory wall describes the specific moment when the performance of a computer system becomes performance-constrained by the memory subsystem. It is derived as follows.

In general-purpose computer architectures with a cache, the memory access time of a processor with cycle time $t_c$, cache hit-rate $p$, and memory access time $t_m$, is given by

$$t_{avg} = p \times t_c + (1 - p) \times t_m \tag{8.3}$$

Assume that on the average, one in five processor instructions will require a memory reference. Under this assumption, the system becomes memory-access constrained when $t_{avg}$ reaches five times the cycle time $t_c$. Indeed, no matter how good the processor is, it will spend more time waiting for memory than executing instructions. This point is called the memory wall.

How likely is it for a computer to hit the memory wall? To answer this question, we should observe that the cache hit rate $p$ cannot be 100%. Data-elements stored in memory have to be fetched at least once from memory before they can be stored in a cache. Let us assume a factor of $p = 0.99$ (rather pessimistic), a cycle time $t_c$ of 1, and a cycle time $t_m$ of 10.

Under this assumption,

$$(t_{avg})_{now} = 0.99 \times 1 + 0.01 \times 10 = 1.09 \tag{8.4}$$

One year from now, memory is 1.1 times faster and the processor is 1.6 times faster. Thus, 1 year from now, $t_{avg}$ will change to

$$(t_{avg})_{now+1} = 0.99 \times 1 + 0.01 \times 10 \times \frac{1.6}{.1} = 1.135 \tag{8.5}$$

And after $N$ years it will be

$$(t_{avg})_{now+N} = 0.99 \times 1 + 0.01 \times 10 \times \frac{1.6^N}{1.1} \tag{8.6}$$

The memory wall will be hit when $t_{avg}$ equals 5, which can be solved according to the above equation to be $N = 9.8$ years. Now, more than a decade after this 1994 paper, it is unclear if the doomsday scenario has really materialized. Many other factors have changed in the meantime, making the formula an unreliable predictor. For example, current processor workloads are very different than those from 10 years ago. Multimedia, gaming and internet have become a major factor. In addition, current processor scaling is no longer done by increasing clock frequency but by more drastic changes at the architecture-level (multiprocessors). Finally, new limiting factors, such as power consumption density and technological variability, have changed the quest for performance into one for efficiency.

Despite this, memory remains a crucial element in SoC design, and it still has a major impact on system performance. In many applications, the selection of memory elements, their configuration and layout, and their programming is a crucial design task.

## 8.2.4  Hierarchical Control

The final concept in the architecture of an SoC is the hierarchy of control among components. A hierarchy of control means that the entire SoC operates as a single logical entity. This implies that the components in an SoC receive commands from a central point of control. They can operate loosely, almost independently, from one another, but at some point they need to synchronize and report back to the central point of control. For example, consider a C program (on a RISC) that uses a co-processor implemented as a peripheral. The C program needs to send arguments to the coprocessor, wait for the co-processor to finish execution, and finally retrieve the result from the co-processor. From the perspective of the co-processor, the custom hardware will first wait for operands from the peripheral bus, next it will process them, and finally it will signal completion of the operation (for example, by setting a status flag). The software on the RISC processor and the activities on the co-processor therefore are not independent. The local controller in the co-processor can be developed using an FSM or a micro-programming technique. The RISC processor will maintain overall control in the system, and distribute commands to custom hardware.

The design of a good control hierarchy is a challenging problem. On the one hand, it should exploit the distributed nature of the SoC as good as possible – this implies doing many things in parallel. On the other hand, it should minimize the number of conflicts that arise as a result of running things in parallel. Such conflicts can be the result of overloading the available bus-system or memory bandwidth, or of over-scheduling a coprocessor. Due to the control hierarchy, all components of an SoC are logically connected to each other, and each of them may cause a system bottleneck. The challenge for the SoC designer (or platform programmer) is to be aware of the location of such system bottlenecks, and to control them.

## 8.3  Example: Portable Multimedia System

In this section we illustrate the four key characteristics discussed earlier (distributed and heterogeneous memory, interconnect, computing, and hierarchical control) by means of an example. Figure 8.3 shows the block diagram of a digital media processor by Texas Instruments. The chip is used for the processing of still images, video, and audio in portable, battery-operated devices. It is manufactured in 130 nm CMOS, and the entire chip consumes no more than 250 mW in default-preview mode, and 400 mW when video encoding and decoding is operational.

The chip supports a number of device modes. Each mode corresponds to typical user activity. The modes include the following.

- Live preview of images (coming from the CMOS imager) on the video display.
- Live-video conversion to a compressed format (MPEG, MJPEG) and streaming of the result into an external memory.

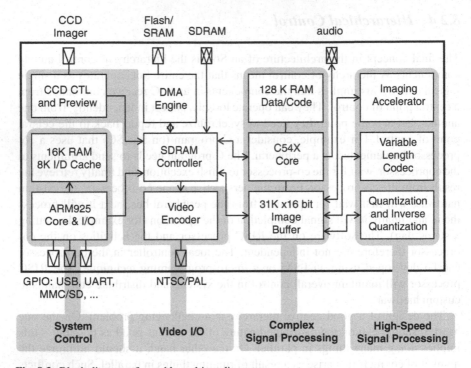

**Fig. 8.3** Block diagram of portable multi-media system

- Still-image capturing of a high-resolution image and conversion to JPEG.
- Live audio capturing and audio compression to MP3, WMA, or AAC.
- Video decode and playback of a recorded stream onto the video display.
- Still image decode and playback of a stored image onto the video display.
- Audio decode and playback.
- Photo printing of a stored image into a format suitable for a photo printer.

The central component of the block diagram in Fig. 8.3 is the SDRAM memory controller. During operation, image data is stored in off-chip SDRAM memory. The SDRAM controller organizes memory traffic to this large off-chip memory. Around the memory controller, four different subsystems are organized. They deal with video input/output, complex signal processing, high-speed signal processing, and system control, respectively.

The video input/output subsystem includes the CCD sensor interface, and a video encoder. The CCD interface is capable of sampling up to 40 MHz at 12 bits per pixel, and it needs to provide high-resolution still images (2–5 Mpixels) as well as moving images (up to 30 frames/s at 640 × 480 pixels). Most CCD sensors record only a single color per pixel. Typically there are 25% red pixels, 25% blue pixels and 50% green pixels, which are arranged in a so-called Bayer pattern. Before images can be processed, the missing pixels need to be filled in (interpolated). This task is a typical example of streaming and dedicated processing. The video subsystem also contains a video encoder, capable of merging two video streams on screen, and capable of providing picture-in-picture functionality. The video coder also includes on-screen menu subsystem functionality. The output of the video coder goes to an attached LCD or a TV. The video coder computes around 100 operations per pixel, while the power budget of the entire video subsystem is less than 100 mW. Hence, we require an energy efficiency on the order of 40 GOPS per W (100 operations per pixel at 40 MHz pixel rate for 100 mW). This energy efficiency is out of range for embedded software on contemporary processors.

The complex signal-processing subsystem is created on top of a C54x digital signal processor (DSP) with 128 KB of RAM and operating at 72 MHz. The DSP processor performs the main processing and control logic for the wide range of signal processing algorithms that the device has to perform (MPEG-1, MPEG-2, MPEG-4, WMV, H.263, H.264, JPEG, JPEG2K, M-JPEG, MP3, AAC, WMA).

The third subsystem is the high-speed signal processing subsystem, needed for encoding and decoding of moving images. Three coprocessors deliver additional computing muscle for the cases where the DSP falls short. There is a DMA engine that helps moving data back and forth between the memory attached to the DSP and the coprocessors. The three coprocessors implement the following functions. The first one is a SIMD-type of coprocessor to provide vector-processing for image processing algorithms. The second is a quantization coprocessor to perform quantization in image encoding algorithms. The third coprocessor performs Huffman encoding for those image encoding standards. The coprocessor subsystem increases to overall processing parallelism of the chip, as they can work concurrently with the DSP processor. This allows the system clock to be decreased.

Finally, the system ARM subsystem is the overall system manager. It synchronizes and controls the different subcomponents of the system. It also provides interfaces for data input/output, and user interface support.

Each of the four properties discussed in the previous section can be identified in this chip.

- The SoC contains *heterogeneous and distributed processing*. There is hardwired processing (video subsystem), signal processing (DSP), and general-purpose processing on an ARM processor. All of this processing can have overlapped activity.
- The SoC contains *heterogeneous and distributed interconnect*. Instead of a single central bus, there is a central 'switchbox' that multiplexes accesses to the off-chip memory. Where needed, additional dedicated interconnections are implemented. Some examples of dedicated interconnections include the bus between the DSP and its instruction memory, the bus between the ARM and its instruction memory, and the bus between the coprocessors and their image buffers.
- The SoC contains *heterogeneous and distributed storage*. The bulk of the memory is contained within an off-chip SDRAM module, but there are also dedicated instruction memories attached to the TI DSP and the ARM, and there are dedicated data memories acting as small dedicated buffers.

• Finally, there is a hierarchy of control to ensures that the overall parallelism in the architecture is optimal. The ARM will start/stop components and control data streams depending on the mode of the device.

The DM310 chip is an excellent example of the effort it takes to support real-time video and audio in a portable device. The architects (hardware and software people) of this chip have worked closely together to come up with the right balance between flexibility and energy-efficiency.

## 8.4   SoC Modeling in GEZEL

In the last section of this chapter, we consider how a System-on-Chip can be modeled in GEZEL, building on our previous experience with FSMD design, micro-programmed design, and general-purpose processors. GEZEL models and simulates microprocessors as well as SoC hardware. A typical example configuration is shown in Fig. 8.4. It includes several components. Custom hardware modules are captured as FSMD models. The microprocessor cores are captured as custom library modules, called `ipblock`. Each microprocessor core offers different types of interfaces. Each of these interfaces is captured using a different `ipblock`. The software executed by the microprocessor core is developed in C or assembly, and converted to binary format using a cross-compiler or cross-assembler. The binary is used to initialize the instruction-set simulator, embedded in an `ipblock`. The entire system simulation is executed by the GEZEL platform simulator `gplatform`.

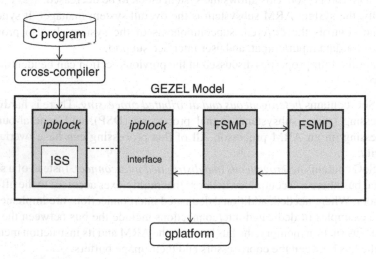

**Fig. 8.4**  A GEZEL system-on-chip model

Listing 8.1  A GEZEL top-level module with a single ARM core

```
1  ipblock myarm {
2    iptype "armsystem";
3    ipparm "exec = hello";
4  }
5
6  system S {
7    myarm;
8  }
```

### 8.4.1  An SoC with a StrongARM Core

We describe some of the features of SoC modeling using the cores included in the gplatform simulator. The first core is a StrongARM core, modeled with the Simit-ARM v2.1 simulator developed by W. Qin at Boston University. Listing 8.1 shows a simple, standalone ARM core. Line 2 of this listing tells that this module is an ARM core with attached instruction memory (armsystem). Line 3 names the ELF executable that must be loaded into the ARM simulator when the simulation starts. The syntax of an ipblock is generic and is used for many different types of cosimulation entities.

The model shown in Listing 8.1 is not very exciting since it does not show any interaction between hardware and software. We will extended this model with a *memory-mapped* interface, as shown in Listing 8.2. Figure 8.5 illustrates how this model corresponds to a System-on-Chip architecture. In Listing 8.2, a hardware-to-software interface is defined on lines 6–10. This particular example shows a memory-mapped interface. The interface has a single output port data. The port can be thought of as a register that is written by the software. The software can update the value of the register by writing to memory address 0x80000000. After each update, the output port  data will hold this value until the software writes to the register again. Note that the association between the memory-mapped interface and the ARM core is established using the name of the core (line 8 in Listing 8.2). Lines 12–28 show a custom hardware module, modeled as an FSMD, which is attached to this memory-mapped interface. The FSM uses the least-significant bit from the memory-mapped register as a state transition condition. Whenever this bit changes from 0 to 1, the FSMD will print the value of the memory-mapped register.

To cosimulate this model, we proceed as follows. First, we cross-compile a C program to run on the ARM. Next, we execute the cosimulation. The following is an example C program that we will run on top of this system architecture.

```
#include <stdio.h>

int main() {
  int y;
  volatile int * a = (int *) 0x80000000;

  *a = 25;
```

**Listing 8.2**  A GEZEL module with an ARM core and a memory-mapped interface on the ARM

```
 1  ipblock myarm {
 2     iptype "armsystem";
 3     ipparm "exec=hello";
 4  }
 5
 6  ipblock port1(out data     : ns(32)) {
 7     iptype "armsystemsource";
 8     ipparm "core=myarm";
 9     ipparm "address = 0x80000000";
10  }
11
12  dp portreader {
13     sig data : ns(32);
14     use myarm;
15     use port1(data);
16     reg changed : ns(1);
17     always    { changed = data[0]; }
18     sfg show { $display($cycle,": The MM interface is now ",
                                  $dec, data); }
19     sfg nil   { }
20  }
21  fsm f_portreader(portreader) {
22     initial s0;
23     state   s1;
24     @s0 if (~changed) then (nil)   -> s0;
25                       else (show)  -> s1;
26     @s1 if (changed)  then (nil)   -> s1;
27                       else (nil)   -> s0;
28  }
29
30  system S {
31     portreader;
32  }
```

```
  *a = 0;
  *a = 39;
  *a = 0;

  return 0;
}
```

This program creates a pointer to the absolute memory address 0x80000000, which corresponds to the memory-mapped port of the custom hardware module in Listing 8.2. The C program then writes a sequence of values to this address. The nonzero values will trigger the $display statement shown on line 18 of Listing 8.2. Compilation of this program, and execution of the cosimulation, is done through the following commands.

```
> arm-linux-gcc -static  hello.c -o hello
```

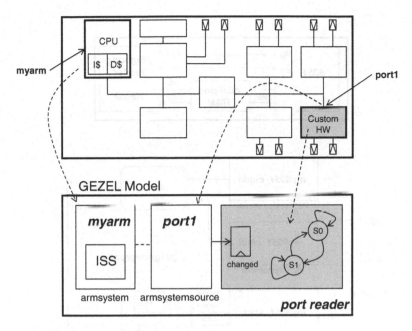

**Fig. 8.5** Correspondence of Listing 8.2 to SoC architecture

```
> gplatform armex.fdl
core myarm
armsystem: loading executable [hello]
7063: The MM interface is now 25
7069: The MM interface is now 39
Total Cycles: 7595
```

The cosimulation verifies that data is passed correctly from software to hardware. The first print statement executes at cycle 7063. This startup delay is used to set up the C runtime environment on the ARM; changing the C runtime environment to a faster, leaner library may reduce this delay significantly.

The relation between the GEZEL model and the System-on-Chip architecture, as illustrated in Fig. 8.5, shows that the FSMD captures the internals of a shaded 'custom hardware' module in a System-on-Chip architecture. The memory-mapped register captured by port1 is located at the input of this custom hardware module. Thus, the GEZEL model in Listing 8.2 does not capture the bus infrastructure (the peripheral bus, the bus bridge, the high-speed bus) of the SoC. This has an advantage as well as a disadvantage. On the plus side, the resulting simulation model is easy to build, and will have a high simulation speed. On the down side, the resulting simulation does not capture the bus conflicts that occur in the real SoC architecture, and therefore the simulation results may show a difference with the real chip. Ultimately, the choice of modeling accuracy is with the designer. A more detailed GEZEL model could capture the transactions on an SoC bus as well, but this would cost an additional effort, and the resulting model may simulate at a lower

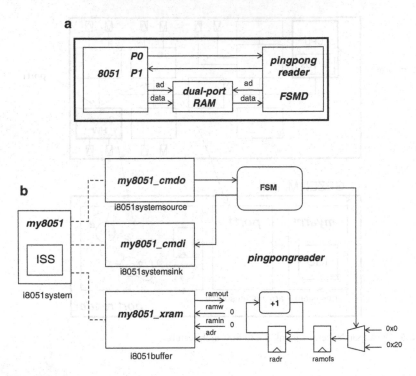

**Fig. 8.6** (**a**) 8051 microcontroller with a coprocessor; (**b**) Corresponding GEZEL model structure

speed. For a cosimulation that focuses on verifying the hardware/software modules, a model such as shown on Listing 8.2 is adequate.

### 8.4.2   Ping-Pong Buffer with an 8051

As a second example, we show how an 8051 microcontroller core can be cosimulated in a GEZEL system model. Figure 8.6a shows a system with an 8-bit 8051 micro-controller, a dual-port RAM with 64 locations, and a hardware module. The micro-controller, as well as the hardware module, can access the RAM. The 8051 microcontroller has several 8-bit I/O ports, and two of them are used in this design. Port P0 is used to send a data byte to the hardware, while port P1 is used to retrieve a data byte from the hardware.

The idea of this design is a *ping-pong buffer*. The RAM is split up in two sections of 32 locations each. When the 8051 controller is writing into the lower section of the RAM, the hardware will read out the upper section of the RAM. Next, the 8051 will switch to the higher section of the RAM, while the hardware module will scan out the lower section of the RAM. This double-buffering technique is frequently used to emulate a dual-port shared RAM with single-port RAM modules. Switching between the two operational modes of the system is implemented using a two-way handshake between the 8051 controller and the hardware. The two ports on the 8051 are used for this purpose.

Figure 8.6b and Listing 8.3 show the GEZEL design that implements this model. The 8051 microcontroller is captured with three different ipblock: one for the microcontroller (lines 1–6), a second one for port P0 configured as input port (lines 8–12), and a third one for port P1 configured as output port (lines 14–18). Similar to the StrongARM simulation model, the 8051 microcontroller is captured with an instruction-set simulator, in this case the Dalton ISS from the University of California at Riverside. The shared buffer is captured in an ipblock as well, starting on line 20. The shared buffer is specific to the 8051 microcontroller, and it is attached to the 8051's xbus (expansion bus). The buffer provides one read/write port for the hardware, while the other port is only accessible from within the 8051 software. The hardware module that accesses the ping-pong buffer is listed starting at line 30. The FSMD will first read locations 0 through $0 \times 1F$, and next locations $0 \times 20$ through $0 \times 3F$. The handshake protocol is implemented through the 8051's P0 and P1 port.

**Listing 8.3** GEZEL model of a ping-pong buffer between an 8051 microcontroller and an FSMD

```
 1  ipblock my8051 {
 2      iptype "i8051system";
 3      ipparm "exec=ramrw.ihx";
 4      ipparm "verbose=1";
 5      ipparm "period=1";
 6  }
 7
 8  ipblock my8051_cmdo(out data : ns(8)) {
 9      iptype "i8051systemsource";
10      ipparm "core=my8051";
11      ipparm "port=P0";
12  }
13
14  ipblock my8051_cmdi(in data : ns(8)) {
15      iptype "i8051systemsink";
16      ipparm "core=my8051";
17      ipparm "port=P1";
18  }
19
20  ipblock my8051_xram(in  idata   : ns(8);
21                      out odata   : ns(8);
22                      in  address : ns(6);
23                      in  wr      : ns(1)) {
24      iptype "i8051buffer";
25      ipparm "core=my8051";
26      ipparm "xbus=0x4000";
27      ipparm "xrange=0x40"; // 64 locations at address 0x4000
28  }
29
30  dp pingpongreader {
```

```
31      reg rreq, rack, rid : ns(1);
32      reg radr              : ns(6);
33      reg ramofs            : ns(6);
34      sig adr               : ns(6);
35      sig ramin, ramout     : ns(8);
36      sig ramw              : ns(1);
37      sig P0o, P0i          : ns(8);
38      use my8051;
39      use my8051_cmdo(P0o);
40      use my8051_cmdi(P0i);
41      use my8051_xram(ramin, ramout, adr, ramw);
42      always        { rreq   = P0o[0];
43                      adr    = radr;
44                      ramw   = 0;
45                      ramin  = 0;            }
46      sfg noack     { P0i    = 0;            }
47      sfg doack     { P0i    = 1;            }
48      sfg getramofs0 { ramofs = 0x0;         }
49      sfg getramofs2 { ramofs = 0x20;        }
50      sfg readram0   { radr   = ramofs;      }
51      sfg readram1   { radr   = radr + 1;
52                       $display($cycle, " ram radr ", radr, "
                             data ", ramout);
53                     }
54  }
55
56  fsm fpingpongreader(pingpongreader) {
57    initial s0;
58    state s1, s2, s3, s4, s5, s6;
59    @s0 if (~rreq) then (noack)                   -> s1;
60                   else (noack)                   -> s0;
61
62    @s1 if (rreq) then (doack, getramofs0)        -> s2;
63                  else (noack)                    -> s1;
64
65    @s2 (readram0, doack)                         -> s3;
66    @s3 if (radr == 0x5) then (doack)             -> s4;
67                         else (readram1, doack)   -> s3;
68
69    @s4 if (~rreq) then (noack, getramofs2)       -> s5;
70                   else (doack)                   -> s4;
71
72    @s5 (readram0, noack)                         -> s6;
73    @s6 if (radr == 0x25) then (doack)            -> s1;
74                          else (readram1, doack)  -> s6;
75  }
76
77  system S {
78    pingpongreader;
79  }
```

Listing 8.4 shows the driver software for the 8051 microcontroller. This software
was written for the Small Devices C Compiler (sdcc), a C compiler that supports a

**Listing 8.4** 8051 software driver for the ping-point buffer

```
1   #include <8051.h>
2
3   void main() {
4     int i;
5
6     volatile xdata unsigned char *shared =
7                   (volatile xdata unsigned char *) 0x4000;
8
9     for (i=0; i<64; i++) {
10      shared[i] = 64 - i;
11    }
12
13    P0 = 0x0;
14    while (1) {
15
16      P0 = 0x1;
17      while (P1 != 0x1) ;
18
19      // hw is accessing section 0 here.
20      // we can access section 1
21      for (i = 0x20; i < 0x3F; i++)
22        shared[i] = 0xff - i;
23
24      P0 = 0x0;
25      while ((P1 & 0x1)) ;
26
27      // hw is accessing section 1 here
28      // we can access section 0
29      for (i = 0x00; i < 0x1F; i++)
30        shared[i] = 0x80 - i;
31    }
32  }
```

broad range of microcontrollers. This compiler directly supports 8051 port access through symbolic names (P0, P1, and so on). In addition, the shared memory accesses can be modeled through an initialized pointer.

To cosimulate the 8051 and the hardware, we start by cross-compiling the 8051 C code to binary format. Next, we use the gplatform cosimulator to run the simulation. Because the microcontroller will execute an infinite program, the cosimulation is terminated after 60,000 clock cycles. The program output shows that the GEZEL model scans out the lower part of the ping-pong buffer starting at cycle 36952, and the upper part starting at cycle 50152. The cycle count is relatively high because the instruction length of a traditional 8051 microcontroller is high: each instruction takes 12 clock cycles to execute.

```
> sdcc --model-large ram.c
> gplatfrom -c 60000 block8051.fdl
```

```
i8051system: loading executable [ramrw.ihx]
0x00      0x00      0xFF      0xFF
0x01      0x00      0xFF      0xFF
36952 ram radr 0/1 data 40
36953 ram radr 1/2 data 3f
36954 ram radr 2/3 data 3e
36955 ram radr 3/4 data 3d
36956 ram radr 4/5 data 3c
0x00      0x01      0xFF      0xFF
50152 ram radr 20/21 data df
50153 ram radr 21/22 data de
50154 ram radr 22/23 data dd
50155 ram radr 23/24 data dc
50156 ram radr 24/25 data db
Total Cycles: 60000
```

### 8.4.3   UART on the AVR ATMega128

A third example shows how GEZEL can be used to report low-level behavior
in hardware and how that can be correlated to high-level (embedded) software
behavior. This application makes use of an ATmega128 AVR microcontroller,
integrated in GEZEL through the SimulAVR ISS. The AVR application software
will send characters through a serial communications link, and the GEZEL model
will reveal the physical communications format of bits flowing over the serial link.
The ATmega128 core contains several dedicated peripherals, including a Universal
Asynchronous Receiver/Transmitter (UART). The use of the UART peripheral
simplifies development of communications software on the ATmega128 microcontroller. Specifically, the UART peripheral supports serial-to-parallel conversion
(for reception) and the parallel-to-serial conversion (for transmission) directly in
hardware. In the ATmega128 software application, communication using a UART
is implemented by means of reading from, and writing to, special-special purpose
registers that control the UART hardware.

Listing 8.5 shows the system architecture, which instantiates an ATmega128 core
(lines 1–5), and an input/output port (lines 7–13). The UART is included within
the ATmega128 core; the input/output port only serves to provide access to the
input/output pins of the UART peripheral. The ATmega128 has six general-purpose
input/output ports, named A through F. Special peripherals, including the UART,
are multiplexed on those ports. For UART device 0, used in this example, the least
significant bits of port E are used for input (rxd) and output (txd) respectively.
The parameter port=E on line 10 of Listing 8.5 configures the interface as port E of
the Atmega128 core. The parameter pindir=xxxxxxx10 on line 11 configures
the direction of the port bits. Bit 0 will be used for serial-input, while bit 1 will be
used as serial-output. The pins are thus configured as input and output respectively.

The hardware attached to the transmit and receive pins of the UART is contained
within lines 15–32 of Listing 8.5. The baudmeter_ctl module implements

**Listing 8.5** GEZEL UART monitor

```
1   ipblock avrcore {
2     iptype "atm128core";
3     ipparm "exec=avruart.elf";
4     ipparm "fcpuMhz=8";
5   }
6
7   ipblock avr_E_port (in   rxd : ns(1);
8                       out  txd : ns(1)) {
9     iptype "atm128port";
10    ipparm "core=avrcore";
11    ipparm "port=E";
12    ipparm "pindir=xxxxxx10";
13  }
14
15  dp baudmeter(out rxd : ns(1);
16               in  txd : ns(1)) {
17    reg b : ns(1);
18    always    { b = txd;
19                rxd = b;
20              }
21    sfg bit10 { $display("@", $cycle, ": ->", txd); }
22    sfg idle  { }
23  }
24  fsm baudmeter_ctl(baudmeter) {
25    initial s0;
26    state s1;
27    @s0 if ( b) then (bit10) -> s1;
28              else (idle)  -> s0;
29    @s1 if (~b) then (bit10) -> s0;
30              else (idle)  -> s1;
31  }
32
33  dp top {
34    sig txd, rxd : ns(1);
35    use avrcore;
36    use avr_E_port(rxd, txd);
37    use baudmeter(rxd, txd);
38  }
39
40  system S {
41    top;
42  }
```

**Listing 8.6**  UART driver program

```
1  #include <avr/io.h>
2  #include <stdio.h>
3
4  #define F_CPU 8000000UL
5  #define UART_BAUD 38400
6
7  void uart_init(void) {
8    UBRR0L = (F_CPU / (16UL * UART_BAUD)) - 1;
9    UCSR0B = _BV(TXEN0) | _BV(RXEN0); /* tx/rx enable */
10  }
11
12  int uart_putchar(char c, FILE *stream) {
13    loop_until_bit_is_set(UCSR0A, UDRE0);
14    UDR0 = c;
15    return 0;
16  }
17
18  int uart_getchar(FILE *stream) {
19    uint8_t c;
20    char *cp, *cp2;
21    static char *rxp;
22    loop_until_bit_is_set(UCSR0A, RXC0);
23    if (UCSR0A & _BV(FE0))
24      return _FDEV_EOF;
25    if (UCSR0A & _BV(DOR0))
26      return _FDEV_ERR;
27    c = UDR0;
28    return c;
29  }
30
31  FILE uart_stream = FDEV_SETUP_STREAM(uart_putchar,
32                                       uart_getchar,
33                                       _FDEV_SETUP_RW);
34
35  int main() {
36    uint8_t c;
37    uart_init();
38    stdout = stdin = &uart_stream;
39    putchar('/');  // ascii 0x2F
40    c = getchar();
41    putchar(c);
42  }
```

a loopback: it connects the transmit pin to the receive pin, so that every byte transmitted by the ATmega128 application will also be received by it. The module also detects every level transition on the serial line, and prints the direction of the transition as well as the clock cycle. By analyzing this debug output, we will reconstruct the waveform generated on the serial line.

The driver software for the ATmega128 is shown in Listing 8.6. This software is written for the `avr-gcc` compiler and uses the definitions of `avr libc`. Three UART functions (`uart_init`, `uart_putchar`, and `uart_getchar`) support initialization, character transmission, and character reception, respectively. The selected baudrate is 38,400 for a CPU clock rate of 8 MHz. Therefore, every UART symbol (a single bit) takes 208 CPU clock cycles. The `uart_putchar` function transmits a character. The function first busy-waits until the UART hardware is able to accept a character, and next initiates transmission by writing the character ASCII code in the transmitter data register. The `uart_getchar` function receives a character. The function busy-waits until the UART signals that a character has been received, then checks the UART error flags, and finally returns the received character. The `FDEV_SETUP_STREAM` macro is specific to the AVR C library. This macro creates a streaming input/output structure for streaming data to/from the UART. The `main` function of the test program transmits a single character, then receives one, and transmits it again. If the loopback is operating correctly, we can expect that every character will be received twice.

To simulate this design, we compile the AVR application with the AVR cross compiler. Next, we use `gplatform` to simulate it. We selected a run of 8,000 clock cycles.

```
> avr-gcc -mmcu=atmega128  avruart.c -o avruart.elf
> gplatform -c 8000 avruart.fdl
atm128core: Load program avruart.elf
atm128core: Set clock frequency 8 MHz
@238:  ->1
@337:  ->0
@545:  ->1
@1377: ->0
@1585: ->1
@1793: ->0
@2209: ->1
@2625: ->0
@2833: ->1
@3665: ->0
@3873: ->1
@4081: ->0
@4497: ->1
Total Cycles: 8000
```

The transitions can be explained using Fig. 8.7. The application software transmits the slash character, which corresponds to ASCII code 0x2F, or 00101111 in binary. A UART serial transmission line sends data bytes starting with the least-significant-bit, so that we expect to find the pattern 11110100 as part of the serial transmission. The UART also transmits a start bit (always 0) and a stop bit (always 1) to mark the beginning and end of the transmission. The waveform for a

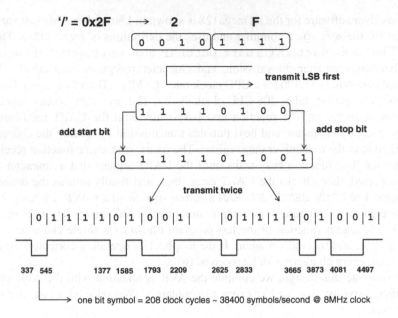

**Fig. 8.7** Serial transmission of '//' as observed by GEZEL simulation

single slash character therefore corresponds to the pattern 0111101001, and the waveform for the application will be a repetition of two times this character.

While the example discussed here is very basic, it illustrates a situation which occurs often in hardware/software codesign: activities in hardware need to be interpreted in terms of activities in software, and vice versa. This requires a designer to think across multiple, and possibly very different, abstraction levels. Just compare Listing 8.6 and Fig. 8.7, for example.

## 8.5 Summary

System-on-chip architecture is a balance between flexibility and specialization. The RISC core, the champion of flexibility in embedded designs, takes care of general-purpose processing, and acts as a central controller in SoC. Multiple additional specialized components, including memories, peripherals, and coprocessors, assist the RISC processor to handle specialized tasks. The interconnect infrastructure, consisting of on-chip bus segments, bus bridges, and specialized connections, help integrating everything together.

All of this makes the SoC a wide-spread paradigm that will be around for some years to come. It is a pragmatic solution that addresses several problems of modern electronic design at the same time. First, an SoC maintains flexibility and is applicable as a platform for several applications within an application domain. This reusability makes the SoC economically advantageous. Compared to a dedicated hardware design, the SoC chip is more general, and a given application can be created quicker. Second, an SoC contains specialized processing capabilities where needed, and this allows it to be energy-efficient. This greatly expands to potential applications of SoC.

In this chapter, we have reached the end of the second part in this book. The key objective of our journey was to investigate how dedicated hardware becomes flexible and programmable. We started from custom-hardware models coded as FSMD models. Next, we replaced the fixed finite state machine of an FSMD with a flexible micro-coded controller, and obtained a micro-programmed architecture. Third, we turned to RISC processors, which are greatly improved micro-programmed architectures that shield of software from hardware. Finally, we used the RISC as a central controller in the System-on-Chip architecture.

## 8.6 Further Reading

System-on-chip is a broad concept with many different dimensions. One of these dimensions is easier and faster design through reuse (Saleh et al. 2006). Another is that SoC technology is critical for modern consumer applications because of the optimal balance between energy-efficiency and flexibility (Claasen 2006). In recent years, alternative visions on SoC architectures have been given, and an interesting one is given in the book of Chris Rowen (2004). The example on the efficiency of on-chip interconnect comes from the same book.

The definition of *intrinsic computational power* of silicon is elaborated in the ISSCC99 article by Claasen (1999). The paper by Wulf and McKee on the Memory Wall can be found online (Wulf and McKee 1995). In 2004, one of the authors provided an interesting retrospective (McKee 2004).

The digital media processor discussed in this chapter is described in more detail by Talla and colleagues in (2004).

The instruction simulators integrated in GEZEL include a StrongARM core (based on Simit-ARM), an 8051 core (based on Dalton (Vahid 2009)), an AVR core (based on simulavr), and a Picoblaze core (based on kpicosim). Appendix A provides a summary on the use of each of those, including a few guidelines that show how to integrate your own instruction-set simulator.

## 8.7 Problems

**Problem 8.1.** Consider Fig. 8.1 again.

(a) Explain why the memory area occupied by the UART peripheral cannot be cached by the RISC processor.

**Fig. 8.8** System-on-chip
model for Problem 8.2

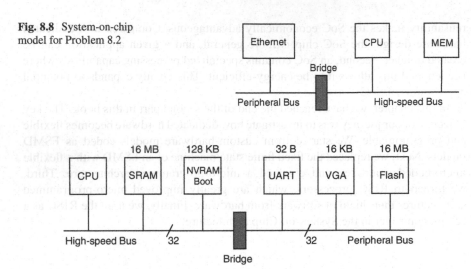

**Fig. 8.9** System-on-chip model for Problem 8.3

(b) Assume that the high-speed bus would include a second RISC core, which also has an instruction-cache and a data-cache. Explain why, without special precautions, caching can cause problems with the stable operation of the system.

(c) A quick fix for the problem described in (b) could be obtained by dropping one of the caches in each processor. Which cache must be dropped: the instruction-cache or the data-cache?

**Problem 8.2.** Consider the simple SoC model in Fig. 8.8. Assume that the high-speed bus can carry 200 MWord/s, and the peripheral bus can carry 30 MWord/s. The CPU has no cache and requests the following data streams from the system: 80 MWord/s of read-only bandwidth for instructions, 40 Mword/s of read/write bandwidth for data, and 2 MWord/s for Ethernet packet input/output.

(a) What is the data bandwidth through the bus bridge?

(b) Assume you have to convert this architecture into a dual-core architecture, where the second core has the same data stream requirements as the first core. Discuss how you will modify the SoC. Keep in mind that you can add components and busses, but that you cannot change their specifications. Don't forget to add bus arbitration units, if you need them.

**Problem 8.3.** You have to design a memory map for the SoC shown in Fig. 8.9. The system contains a high-speed bus and a peripheral bus, both of them with a 32-bit address space and both of them carrying words (32 bit). The components of the system include a RISC, a 16 MB RAM memory, 128 KB of non-volatile program memory, a 16 MB Flash memory. In addition, there is a VGA peripheral and a UART peripheral. The VGA has a 16 KB video buffer memory, and the UART contains 32 bytes of transmit/receive registers.

(a) Draw a possible memory map for the processor. Keep in mind that the Bus Bridge can only convert bus transfers within a single, continuous address space.
(b) Define what address range can be cached by the processor. A 'cached address range' means that a memory-read to an address in that range will result in a backup copy stored in the cache.

**Problem 8.4.** Consider Listings 8.3 and 8.4 again. Modify the GEZEL program and the C program so that the FSMD *writes* into the shared memory, and the C program *reads* from the shared memory. Co-simulate the result to verify the solution is correct.

# Part III
# Hardware/Software Interfaces

The third part of this book makes a walkthrough of all the elements involved in connecting software and custom-hardware. We will discuss the general design principles of a hardware/software interface, and describe all the building blocks needed to implement it. This includes on-chip busses, microprocessor interfaces, and hardware interfaces.

# Chapter 9
# Principles of Hardware/Software Communication

## 9.1  Connecting Hardware and Software

Over the next few chapters, we will discuss various forms of interconnection
between hardware components and software drivers. Figure 9.1 presents a synopsis
of the elements in a hardware/software interface. The objective of the hardware/-
software interface is to connect the software application to the custom-hardware
module. There are five elements involved.

1. The microprocessor and the coprocessor are both connected to an on-chip
   communication mechanism, such as an on-chip bus. The *on-chip bus* transports
   data from the microprocessor module to the custom-hardware module. While
   typical on-chip buses are shared among several masters and slaves, they can
   also be implemented as dedicated point-to-point connections. For example,
   coprocessors are often attached to a dedicated link.
2. Both the microprocessor and the coprocessor need an interface to the on-chip
   communication bus. The *microprocessor interface* includes the hardware and
   low-level firmware to allow a software program to get 'out' of the microprocessor. A microprocessor can use several different mechanisms for this, such
   as coprocessor instructions, or memory load/store instructions.
3. The *hardware interface* includes the hardware needed to attach the coprocessor
   to the on-chip communication subsystem. For example, in the case of an on-chip
   bus, the hardware interface handles data coming from, and going to, the on-chip
   bus. The hardware interface will decode the on-chip bus protocol, and make the
   data available to the custom-hardware module through a register or a dedicated
   memory.
4. The software application is connected to the microprocessor interface through
   a *software driver*, a small module that wraps transactions between hardware
   and software into software function calls. This driver converts software-centric
   paradigms (pointers, arrays, variable-length data structures) into structures that
   are suited for communication with hardware. To achieve this goal, the software
   driver may require the introduction of additional data structures and commands.

P.R. Schaumont, *A Practical Introduction to Hardware/Software Codesign*,
DOI 10.1007/978-1-4614-3737-6_9, © Springer Science+Business Media New York 2013

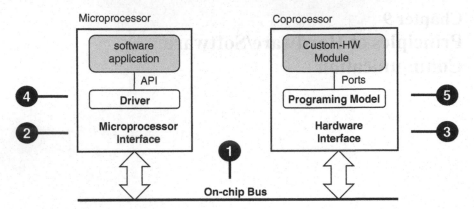

**Fig. 9.1** The hardware/software interface

5. The custom-hardware module is connected to the custom-hardware interface through a *programming model*, a structure that presents an abstraction of the hardware to the software application. The hardware interface encapsulates the custom-hardware module, which may have an arbitrary number of ports, parameters. The programming model, on the other hand, should be easy to handle by the microprocessor: it uses memory locations, coprocessor instructions, and so on. To implement this conversion, the hardware interface may require the introduction of additional storage and controls.

Clearly, the link from hardware to software represents a rich and complex design space. In any given codesign problem, you'll find that there are many different ways to implement a link from hardware to software. However, there may be just a few of them that have acceptable overhead in the application.

In this chapter, we discuss the principles of hardware/software interface design. In the chapter after that, we will describe on-chip busses. Further, there will be one chapter on the microprocessor interface (with driver), and one on the custom-hardware interface (with hardware interface).

## 9.2   Synchronization Schemes

How can we guarantee that the software application and the custom-hardware module will remain synchronized, given that they are independently executing entities? How does a hardware module know that a software program wishes to communicate with it? Answering these questions requires us to select a synchronization scheme.

**Fig. 9.2** Synchronization point

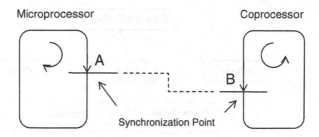

Microprocessor                                                                        Coprocessor

Synchronization Point

## 9.2.1 Synchronization Concepts

We define *synchronization* as the structured interaction of two otherwise independent and parallel entities. Figure 9.2 illustrates the key idea of synchronization. Two entities, in this case a micro-processor and a coprocessor, each have an independent thread of execution. Through synchronization, one point in the execution thread of the microprocessor is tied to one point in the control flow of the coprocessor. This is the synchronization point. Synchronization must guarantee that, when the microprocessor is at point A, then the coprocessor will be at point B.

Synchronization is needed to support communication between parallel subsystems: every *talker* needs to have a *listener* to be heard. Obviously, if parallel components never interact, there's no point in keeping them synchronized. We discussed communication within parallel systems before: recall our discussion on the implementation of data-flow (Chap. 2). In data-flow, different actors communicate with one another through the exchange of tokens. Assume that one actor is implemented in software and another one is implemented as a custom-hardware module. Also, assume that the software actor sends tokens to the hardware actor. According to the rules of data-flow, each token produced must eventually be consumed, and this implies that the hardware actor must know when the software actor is sending that token. In other words: the hardware and software actors will need to synchronize when communicating a token. Of course, there are many different ways to realize a data-flow communication, depending on how we realize the data-flow edge. But, regardless of the realization, the requirement to synchronize does not go away. For example, one may argue that a FIFO memory could be used to buffer the tokens going from software to hardware, thereby allowing hardware and software actors to run 'more independently'. Well, FIFO memories do not remove the requirement to synchronize. When the FIFO is empty, the hardware actor will need to wait until a token appears, and when the FIFO is full, the software actor will need to wait until a free space appears.

Synchronization is an interesting problem because it has several dimensions, each with several levels of abstraction. Figure 9.3 shows the three dimensions of interest: time, data, and control. In this section, we explain the meaning of these dimensions. In further sections, we discuss several examples of synchronization mechanisms.

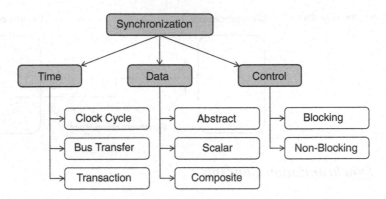

**Fig. 9.3** Dimensions of the synchronization problem

The *dimension of time* expresses the granularity at which two parallel entities synchronize. Clock-cycle accuracy is needed when we interface two hardware components with each other. Bus-transfer accuracy is needed when the granularity of synchronization is expressed in terms of a specific bus protocol, such as a data transfer from a master to a slave. Finally, transaction accuracy is needed when the granularity of synchronization is a logical transaction from one entity to the next. Note that the meaning of *time* varies with the abstraction level, and it does not always have to correspond to wall-clock time. Instead, time synchronization may also refer to clock cycles, bus transfers, and logical transfers. Time synchronization may be even limited to a partial ordering, as in *A does not happen before B*, for example.

In practice, a synchronization scheme between hardware and software covers all abstraction levels in time: A high-level, logical synchronization implemented with bus transfers will only work if the bus transfers themselves are synchronized. In turn, each bus transfer may require a protocol that takes several clock cycles, and the communicating subsystems will synchronize at each clock edge. However, for a hardware-software codesigner, being able to think about synchronization problems at higher levels of abstraction is fundamental, especially in the earlier phases of a design.

The *data dimension of synchronization* determines the size of the container involved in the synchronization. When no data is involved at all, the synchronization between two entities is abstract. Abstract synchronization is useful to handle access to a shared resource. On the other hand, when two parallel entities communicate data values, they will need a shared data container. The communication scheme works as follows: one entity dumps information in the data container, and synchronizes with the second entity. Next, the second entity retrieves information of the data container. The first entity should not overwrite the value in the data container until it can be sure that the second entity has synchronized. In this scheme, synchronization is used to indicate when there is something of interest in the shared data container. Under scalar data synchronization, the two entities will synchronize for every single data item transferred. Under composite data synchronization, the two entities will synchronize over a group of data items.

The *control dimension of synchronization* indicates how the local behavior in each entity will implement synchronization. In a blocking scheme, the synchronization can stall the local thread of control. In a non-blocking scheme, the local behavior will not be stalled, but instead a status signal is issued to indicated that the synchronization primitive did not succeed.

A hardware-software co-designer is able to make decisions along each of these three dimensions separately. In the following sections, several examples of synchronization will be described.

### 9.2.2 *Semaphore*

A *semaphore* is a synchronization primitive which does not involve the transfer of data, but instead controls access over an abstract, shared resource. A semaphore $S$ is a shared resource that supports two operations: grabbing the semaphore ($P(S)$) and releasing the semaphore ($V(S)$). These operations can be executed by several concurrent entities. In this case, we will assume there are two entities competing for the semaphore. The $P$ and $V$ are the first letters of the Dutch verbs 'proberen' and 'verhogen', chosen by the scientist who proposed using semaphores in system software, Edgser Dijkstra.

The meaning of $P(S)$ and $V(S)$ is as follows. $P(S)$ and $V(S)$ are indivisible operations that manipulate the value of a semaphore. Initially, the value of the semaphore is 1. The operation $P(S)$ will decrement the semaphore by one. If an entity tries to $P(S)$ the semaphore while it is zero, then $P(S)$ will stall further execution of that entity until the semaphore is nonzero. Meanwhile, another entity can increment the semaphore by calling $V(S)$. When the value of the semaphore is non-zero, any entity which was stalled on a $P(S)$ operation will decrement the semaphore and proceed. In case multiple entities are blocked on a semaphore, one of them, chosen at random, will be able to proceed. The maximum value of the basic binary semaphore is 1. Calling $V(S)$ several times will not increase the semaphore above 1, but it will not stall either. There are more elaborate semaphore implementations as well, such as counting semaphores. For our purpose however, binary semaphores are sufficient.

Using semaphore operations, it is possible to describe the synchronization of two concurrent entities. The pseudocode in Listing 9.1 is an example using a single semaphore. The first of two concurrent entities needs to send data to the second entity through a shared variable shared_data. When the first entity starts, it immediately decrements the semaphore. Entity two, on the other hand, waits for a short while, and then will stall on the semaphore. Meanwhile, entity one will write into the shared variable, and increment the semaphore. This will unlock the second entity, which can now read the shared variable. The moment when entity

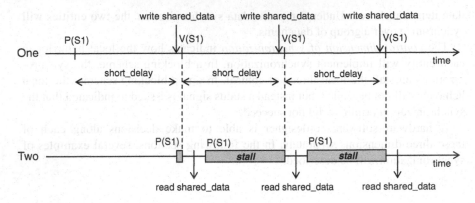

**Fig. 9.4** Synchronization with a single semaphore

**Listing 9.1** One-way synchronization with a semaphore

```
int shared_data;
semaphore S1;

entity one {
  P(S1);
  while (1) {
    short_delay();
    shared_data = ...;
    V(S1);                    // synchronization point
  }
}

entity two {
  short_delay();
  while (1) {
    P(S1);                    // synchronization point
    received_data = shared_data;
  }
}
```

one calls $V(S1)$ and entity two is stalled on $P(S1)$ is of particular interest: it is the synchronization point between entities one and two.

Figure 9.4 illustrates the interaction between entities one and two. The dashed lines indicate the synchronization points. Because entity two keeps on decrementing the semaphore faster than entity one can increment it, entity two will always stall. As a result, each write of shared_data by entity one is followed by a matching read in entity two.

Yet, this synchronization scheme is not perfect, because it assumes that entity two will always arrive first at the synchronization point. Now assume that the slowest entity would be entity two instead of entity one. Under this assumption, it is possible

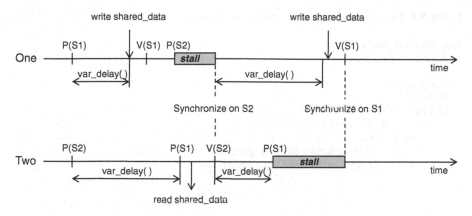

**Fig. 9.5** Synchronization with two semaphores

that entity one will write `shared_data` several times before entity two can read a single item. Indeed, $V(S1)$ will not stall even if it is called several times in sequence. Such a scenario is not hard to envisage: just move the *short_delay()* function call from the while-loop in entity one to the while-loop in entity two.

This observation leads to the conclusion that the general synchronization of two concurrent entities needs to work in two directions: one entity needs to be able to wait on the other, and vice versa. In the *producer/consumer* scenario explained above, the producer will need to wait for the consumer if that consumer is slow. Conversely, the consumer will need to wait for the producer if the producer is slow. We can address the situation of unknown delays with a two-semaphore scheme, as shown in Listing 9.2. At the start, each entity decrements a semaphore. $S1$ is used to synchronize entity two, while $S2$ is used to synchronize entity one. Each entity will release its semaphore only after the read-operation (or write-operation) is complete.

Figure 9.5 illustrates the case where two semaphores are used. On the first synchronization, entity one is quicker than entity two, and the synchronization is done using semaphore $S2$. On the second synchronization, entity two is faster, and in this case the synchronization is done using semaphore $S1$.

### 9.2.3 One-Way and Two-Way Handshake

In parallel systems, concurrent entities may be physically distinct, and implementing a centralized semaphore may not be feasible. To handle this situation, we will use a *handshake*: a signaling protocol based on signal levels. The concepts of semaphore-based synchronization still apply: we implement a synchronization point by making one entity wait for another one.

The most simple implementation of a handshake is a *one-way handshake*, which needs only one wire. Figure 9.6 clarifies the implementation of this handshake

**Listing 9.2**  Two-way synchronization with two semaphores

```
int shared_data;
semaphore S1, S2;

entity one {
  P(S1);
  while (1) {
    variable_delay();
    shared_data = ...;
    V(S1);  // synchronization point 1
    P(S2);  // synchronization point 2
  }
}

entity two {
  P(S2);
  while (1) {
    variable_delay();
    P(S1);  // synchronization point 1
    received_data = shared_data;
    V(S2);  // synchronization point 2
  }
}
```

for the case of two hardware modules. When we will discuss hardware/software interfaces, we will also consider handshakes between hardware and software. In this figure, entity one transmits a query signal to entity two. Entity two captures this signal in a register, and uses its value as a state transition condition. The synchronization point is the transition of S0 to S1 in entity one, with the transition of S2 to S3 in entity two. Entity two will wait for entity one until both of them can make these transitions in the same clock cycle. Entity one needs to set of acknowledge signal to high one cycle *before* the actual synchronization point, because the request input in entity two is captured in a register.

The limitation of a one-way handshake is similar to the limitation of a one-semaphore synchronization scheme: it only enables a single entity to stall. To accommodate arbitrary execution orderings, we need a two-way handshake as shown in Fig. 9.7. In this case, two symmetrical handshake activities are implemented. Each time, the query signal is asserted during the transition preceding the synchronization point. Then, the entities wait until they receive a matching response. In the timing diagram of Fig. 9.7, entity one arrives first in state S0 and waits. Two clock cycles later, entity two arrives in state S2. The following clock cycle is the synchronization point: as entity one proceeds from S0 to S1, entity two makes a corresponding transition from S2 to S3. Because the handshake process is bidirectional, the synchronization point is executed correctly regardless of which entity arrives first at that point.

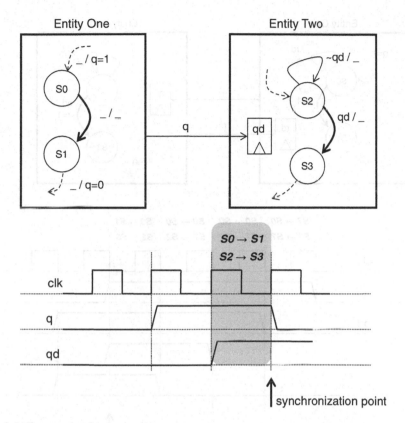

**Fig. 9.6** One-way handshake

There are still some opportunities for optimization. For example, we can de-assert the response signal already during the synchronization point, which will make the complete handshake cycle faster to complete. We can also design the protocol such that it uses level transitions rather than absolute signal levels. Some of these optimizations are explored in the Problems at the end of this Chapter.

### 9.2.4 Blocking and Non-blocking Data-Transfer

Semaphores and handshakes are different ways to implement a synchronization point. A hardware/software interface uses a synchronization point to transfer data. The actual data transfer is implemented using a suitable hardware/software interface, as will be described later in this chapter.

An interesting aspect of the data transfer is how a synchronization point should be implemented in terms of the execution flow of the sender or receiver. If a sender or receiver arrives too early at a synchronization point, should it wait idle until the proper condition comes along, or should it go off and do something else? In

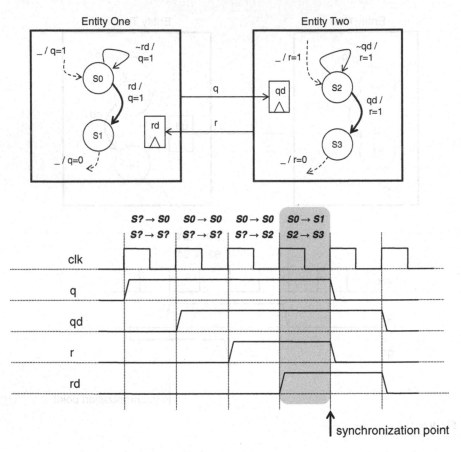

**Fig. 9.7**  Two-way handshake

terms of the send/receive operations in hardware and software, these two cases are distinguished as *blocking* data transfers and *non-blocking* data transfers.

A blocking data transfer will stall the execution flow of the software or hardware until the data-transfer completes. For example, if software has implemented the data transfer using function calls, then a blocking transfer would mean that these functions do not return until the data transfer has completed. From the perspective of the programmer, blocking primitives are the easiest to work with. However, they may stall the rest of the program.

A non-blocking data transfer will not stall the execution flow of software or hardware, but the data transfer may be unsuccessful. So, a software function that implements a non-blocking data transfer will need to introduce an additional status flag that can be tested. Non-blocking data transfers will not stall, but they require additional attention of the programmer to deal with exception cases.

Both of the semaphore and handshake schemes discussed earlier implement a blocking data-transfer. To use these primitives for a non-blocking data transfer, a sender or receiver should be able to test the outcome of the synchronization operation, without actually engaging in it.

## 9.3 Communication-Constrained Versus Computation-Constrained

In the selection of a suitable hardware/software communication interface, the resulting performance of the system is of crucial importance. Very often, the main reason for designing a custom hardware module is that the designer hopes to increase overall system performance with it. The argument of hardware acceleration is very often made in terms of computational performance. Consider the following example.

> A function XYZ executes on a slow software processor and takes 100 ms to complete. By writing a hardware implementation of XYZ, the execution time decreases to 1 ms. Hence, the system can be accelerated by a factor of 100.

There is an important pitfall in this reasoning. The overall application still runs on the slow software processor. Having XYZ execute in fast hardware does not help, if the software application cannot efficiently make use of the hardware module. For example, let's say that, due to an inefficient hardware/software interface, invoking the hardware version of XYZ takes 20 ms. Then, the system speedup is only a factor of 5, not 100!

In practical situations, we may analyze such performance limits. An example is given in Fig. 9.8. This hardware module has three input ports and one output port. Thus, each time we invoke the hardware coprocessor, we need to transfer $128 + 128 + 32 + 32 = 320$ bits. Let's assume that this custom hardware module takes

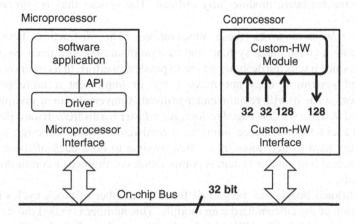

**Fig. 9.8** Communication constraints of a coprocessor

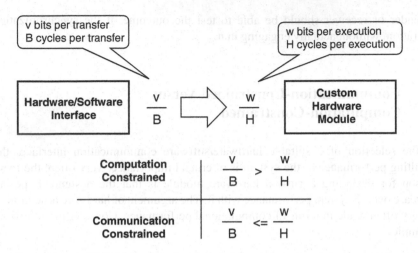

**Fig. 9.9**  Communication-constrained system vs. computation-constrained system

five cycles to compute a result. Hence, when this module is connected to software, and we wish to run the module at full performance, we will need to support a data bandwidth of $320/5 = 64$ bits per cycle. This data bandwidth needs to be delivered through a hardware/software interface. As illustrated in Fig. 9.8, a 32-bit bus is used to deliver data to the coprocessor. Since each bus transfers requires at least one clock cycle, the bus cannot provide more than 32 bits per cycle. Clearly, for full utilization, the coprocessor needs a larger data bandwidth than can be provided through the hardware/software interface. In this case, the system is *communication-constrained*.

Now, let's assume that the hardware coprocessor takes 50 cycles (instead of 5) to complete the operation. In this case, full utilization of the hardware will require $320/50 = 6.4$ bits per cycle. A 32-bit bus may be able to deliver the data that will keep the hardware module fully utilized. The system thus is *computation-constrained*.

Figure 9.9 summarizes these observations. The distinction between a communication-constrained system and a computation-constrained system is important, since it tells the designer where to put design effort. In a communication-constrained system, it does not make sense to implement a more powerful coprocessor, since it will remain under-utilized. Conversely, in a computation-constrained system, we don't need to look for a faster hardware/software interface. Even if the exact performance limits in a hardware-software codesign problem may be very hard to determine, it is often feasible to do a back-of-the-envelope calculation, and find out if a system is computation-constrained or communication-constrained.

An additional insight can be gained from the number of clock cycles needed per execution of the custom hardware module. This number is called the *hardware sharing factor* or HSF. The HSF is defined as the number of clock cycles that are available in between each input/output event. For example, an HSF of 10 would

**Table 9.1** Hardware sharing factor

| Architecture | HSF |
|---|---|
| Systolic array processor | 1 |
| Bit-parallel processor | 1–10 |
| Bit-serial processor | 10–100 |
| Micro-coded Processor | >100 |

mean that a given hardware architecture has a cycle budget of 10 clock cycles between successive input/output events. Thus, if this architecture would contain two multiplier operators, then these 10 clock cycles are adequate to support up to 20 multiplications. The HSF is an indication if a given architecture is powerful enough to sustain a computational requirement. Indeed, there is a strong correlation between the internal architecture of a hardware module and its HSF. This is illustrated in Table 9.1.

- A systolic-array processor is a multi-dimensional arrangement of computation units (datapaths or dataflow-actor-like processors) that operate on one or more parallel streams of data items.
- A bit-parallel processor is a processor with bit-parallel operators such as adders and multipliers that operates under control of a simple engine (such as a FSMD or a micro-programmed controller).
- A bit-serial processor is a processor with bit-serial operators, i.e. operators that compute on a single bit at a time, under control of a simple engine (such as a FSMD or a micro-programmed controller).
- A micro-coded processor is a processor with an instruction-fetch, similar to a general purpose processor.

This means that knowledge of the HSF, at the start of a design, may help a designer to select the right architecture style for the coprocessor.

## 9.4 Tight and Loose Coupling

The third generic concept in hardware/software interfaces is that of *coupling*. Coupling indicates the level of interaction between the execution flow in software and the execution flow in custom-hardware. In a tight coupling scheme, custom-hardware and software synchronize often, and exchange data often, for example at the granularity of a few instructions in software. In a loose coupling scheme, hardware and software synchronize infrequently, for example at the granularity of a function or a task in software. Thus, *coupling* relates the ideas of synchronization with performance.

We don't give a formal description of coupling, but instead describe it in terms of an example. First, let's compare the difference between a hardware module attached to the memory bus of a processor, and a hardware module that is attached directly to a dedicated port on the processor. These two interfaces – which we will

| | Coprocessor | Memory-mapped |
| --- | --- | --- |
| Factor | interface | interface |
| Addressing | Processor-specific | On-chip bus address |
| Connection | Point-to-point | Shared |
| Latency | Fixed | Variable |
| Throughput | Higher | Lower |

**Table 9.2** Comparing a coprocessor interface with a memory-mapped interface

describe in detail in the next Chapters – are called memory-mapped interface and coprocessor interface respectively. Table 9.2 compares the key features of these two interfaces.

These two interfaces each take a different approach to synchronization. In the case of a coprocessor interface, synchronization between a hardware module and software is at the level of a single instruction. Such a coprocessor instruction typically carries both operands (from software driver to hardware coprocessor) as well as result (from hardware coprocessor to software driver). In the case of a memory-mapped interface, synchronization between a hardware module and software is at the level of a single bus transfer. Such a bus transfer is unidirectional (read or write), and either carries operands from the software driver to the hardware coprocessor, or else results from the hardware coprocessor to the software driver. Clearly, the use of a coprocessor interface versus a memory-mapped interface imply a different style of synchronization between hardware and software.

A given application can use either tight-coupling or loose-coupling. Figure 9.10 shows how the choice for loose-coupling of tight-coupling can affect the latencies of the application. The left side of the figure illustrates a tight-coupling scheme. The software will send four separate data items to the custom hardware, each time collecting the result. The figure assumes a single synchronization point which sends the operand and retrieves the result. This is the scheme that could be used by a coprocessor interface. The synchronization point corresponds to the execution of a coprocessor instruction in the software.

The right side of the figure illustrates a loosely coupled scheme. In this case, the software provides a large block of data to the custom hardware, synchronizes with the hardware, and then waits for the custom hardware to complete processing and return the result. This scheme would be used by a memory-mapped interface, for example using a shared-memory.

Loosely-coupled schemes tend to yield slightly more complex hardware designs because the hardware needs to deal more extensively with data movement between hardware and software. On the other hand, tightly-coupled schemes lean more on the software to manage overall system execution. Achieving a high degree of parallelism in the overall design may be easier to achieve with a loosely-coupled scheme than with a tightly-coupled scheme.

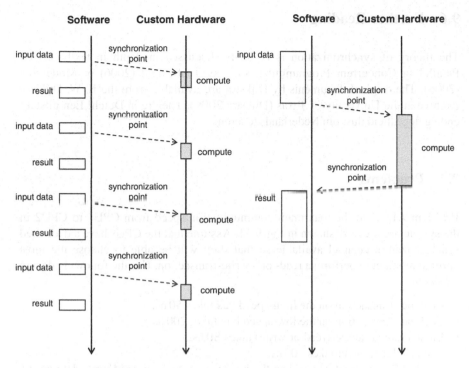

**Fig. 9.10** Tight coupling versus loose coupling

## 9.5   Summary

In this Chapter, we discussed three concepts in hardware/software interface design. The first concept is that of synchronization, the structured interaction of parallel and independent entities. The second concept is the difference between communication-constrained and computation-constrained systems. The final concept is the difference between loose coupling and tight coupling.

All hardware/software interfaces need to deal with synchronization, and often you will need to build your own, application-level synchronization using the low-level primitives provided by a hardware/software interface. For example, the implementation of an internet-protocol component may require to process a full packet of data at a time.

Hardware coprocessors only make sense in computation-constrained systems. When a system becomes communication-constrained, one must carefully examine the overall design to determine if the system architecture (i.e. the application software *and* the hardware architecture) is still adequate for the problem at hand.

Coupling is arguably the most complex concept in this chapter, because it is the most abstract one. Often, by defining the system operation as a set of collaborating tasks, one can also establish the required coupling between components.

## 9.6  Further Reading

The theory of synchronization is typically discussed in depth in textbooks on Parallel or Concurrent Programming, such as Taubenfeld (2006) or Moderchai (2006). The original documents by Dijkstra are available from the E. W. Dijkstra Archive at the University of Texas (Dijkstra 2009). They're in Dutch! Een uitstekende gelegenheid dus, om Nederlands te leren.

## 9.7  Problems

**Problem 9.1.** Find the maximum communication speed from CPU1 to CPU2 in the system architecture shown in Fig. 9.11. Assume that the CPUs have a dedicated synchronization channel available so that they will be able to choose the most optimal moment to perform a read- or a write-transaction. Use the following design constants.

- Each bus transaction on the high-speed bus takes 50 ns.
- Each bus transaction on the low-speed bus takes 200 ns.
- Each memory access (read or write) takes 80 ns.
- Each bridge transfer takes 100 ns.
- The CPU's are much faster than the bus system, and can read/write data on the bus at any chosen data rate.

**Problem 9.2.** Consider the two-way handshake in Fig. 9.12. A sender synchronizes with a receiver and transmits a sequence of data tokens through a data register.

(a) Describe under what conditions register $r1$ can be removed without hurting the integrity of the communication. Assume that, after taking $r1$ away, the req input of the receiver is tied to logic-1.
(b) Describe under what conditions register $r3$ can be removed without hurting the integrity of the communication. Assume that, after taking $r3$ away, the req input of the sender is tied to logic-1.

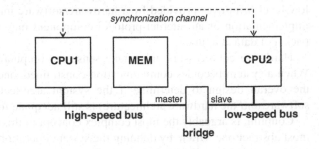

**Fig. 9.11**  System topology
for Problem 9.1

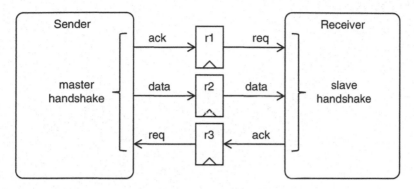

**Fig. 9.12** Two-way handshake for Problems 9.1 and 9.4

(c) Assume that you would substitute register r1 by two registers in series, so that the entire transition from sender-ack to receiver-req now takes two clock cycles instead of one. Describe the effect of this change on the throughput of the communication, and describe the effect of this change on the latency of the communication.

**Problem 9.3.** A C function has ten inputs and ten outputs, all of them integers. The function takes 1,000 cycles to execute in software. You need to evaluate if it makes sense to build a coprocessor for this function. Assume that the function takes K cycles to execute in hardware, and that you need Q cycles to transfer a word between the software and the coprocessor over a system bus. Draw a chart that plots Q in terms of K, and indicate what regions in this chart justify a coprocessor.

**Problem 9.4.** Consider the two-way handshake in Fig. 9.12. Implement this two-way handshake by developing an FSMD for the sender and the receiver. Next, optimize the two-way handshake so that *two* tokens have been transferred each time req and ack have make a complete handshake and returned to the logic-0 state.

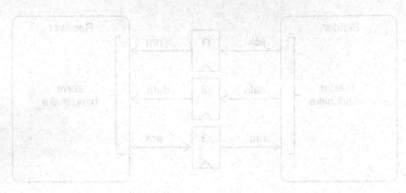

Fig. 9.12 Two-way handshake for Problems 9.x and 9.x.

(c) Assume that you would substitute register $r2$ by two registers in series, so that the entire transition from sender-ack to receiver-req now takes two clock cycles instead of one. Describe the effect of this change on the throughput of the communication, and describe the effect of this change on the latency of the communication.

Problem 9.x. A C function takes an input, and then outputs all of their answers. The function takes 1,000 cycles to execute in software. You need to evaluate if it makes sense to build a coprocessor for this function. Assume that the handshaking takes K cycles to execute in hardware, and that you need Q cycles to transfer a word between the software and the coprocessor over a system bus. Draw a chart that plots Q in terms of K, and indicate what regions in this chart result in a coprocessor.

Problem 4.x. Consider the two-way handshake in Fig. 9.12. Implement this two-way handshake by developing an FSMD for the sender and the receiver. Make it operationally correct, handshake so that two tokens have been transmitted at each time an end after one token is complete handshake and returned to the logical 0 state.

# Chapter 10
# On-Chip Busses

## 10.1 On-Chip Bus Systems

This section describes the basic structure of an on-chip bus, and defines common terminology and notations.

### 10.1.1 A Few Existing On-Chip Bus Standards

The discussions in this chapter are based on four different bus systems: the AMBA bus, the CoreConnect bus, the Avalon bus, and the Wishbone bus. There are many other bus systems, but these are commonly used in contemporary System-on-Chip design.

- **AMBA** (Advanced Microcontroller Bus Architecture) is the bus system used by ARM processors. Originally introduced in 1995, the AMBA bus is now in its fourth generation, and it has evolved into a general on-chip interconnect mechanism. The fourth generation of AMBA provides five variants of interconnect: A general-purpose, low-bandwidth bus called APB (Advanced Peripheral Bus), a high-speed single-frequency bus called AHB (Advanced High-performance Bus), a high-speed multi-frequency bus called AXI (AMBA Advanced Extensible Interface), a system control bus called ASB (Advanced System Bus) and a debug control bus called ATB (Advanced Trace Bus). We will focus on the peripheral bus (APB) and, for selected cases, on the high-speed bus (AHB).

- **CoreConnect** is a bus system proposed by IBM for its PowerPC line of processors. Similar the AMBA bus, the CoreConnect bus comes in several variants. The main components include a general-purpose, low-bandwidth bus called OPB (On-chip Peripheral Bus) and a high-speed single-frequency bus called PLB (Processor Local Bus).

P.R. Schaumont, *A Practical Introduction to Hardware/Software Codesign*,     287
DOI 10.1007/978-1-4614-3737-6_10, © Springer Science+Business Media New York 2013

**Table 10.1** Bus configurations for existing bus standards

| Bus | High-performance shared bus | Peripheral shared bus | Point-to-point bus |
|---|---|---|---|
| AMBA v3 | AHB | APB | |
| AMBA v4 | AXI4 | AXI4-lite | AXI4-stream |
| Coreconnect | PLB | OPB | |
| Wishbone | Crossbar topology | Shared topology | Point to point topology |
| Avalon | Avalon-MM | Avalon-MM | Avalon-ST |

*AHB* AMBA highspeed bus, *APB* AMBA peripheral bus, *AXI* advanced extensible interface, *PLB* processor local bus, *OPB* onchip peripheral bus, *MM* memory-mapped, *ST* streaming

- **Avalon** is a bus system developed by Altera for use in SoC applications of its Nios processor. Avalon is defined in terms of the different types of interfaces it provides to SoC components. The most important of these are Avalon-MM for a memory-mapped interface, and Avalon-ST for a streaming interface. These two interfaces are used on shared busses and point-to-point busses, respectively.
- **Wishbone** is an open-source bus system proposed by SiliCore Corporation. The bus is used by many open-source hardware components, for example those in the OpenCores project (http://www.opencores.org). The Wishbone bus is simpler than the previous standards. The specification defines two interfaces (a master-interface and a slave-interface) from which various bus topologies can be derived.

Rather than describing each bus separately, we will unify them in a generic bus that reflects the common characteristics of all of them. We will occasionally point how each of AMBA, CoreConnect, Avalon and Wishbone implement the features of this generic bus.

We distinguish two bus configurations: the shared bus, and the point-to-point bus. The difference between these two is simply the number of components sharing the bus. Shared busses are most common, and several standards (AMBA, CoreConnect) define multiple versions of them, differing in complexity and performance. Table 10.1 shows the configurations defined for four different bus interface standards. The following two subsections provide a generic definition of a shared bus and of a point-to-point bus.

## 10.1.2   Elements in a Shared Bus

An on-chip bus system implements a bus protocol: a sequence of steps to transfer data in an orderly manner. A typical on-chip bus system will consist of one or more bus segments, as shown in Fig. 10.1. Each bus segment groups one or more bus *masters* with bus *slaves*. Bus bridges are directional components to connect bus segments. A bus-bridge acts as a slave at the input, and as a master at the output. At any particular moment, a bus segment is under control of either a bus master or a bus arbiter. A bus arbiter's role is to decide which bus master is allowed to control

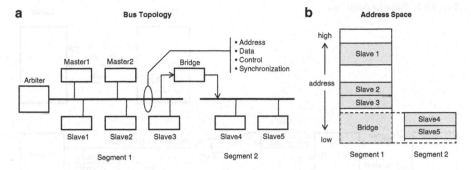

**Fig. 10.1** (a) Example of a multi-master segmented bus system (b) Address space for the same bus

the bus. The arbitration is done for each bus transaction. It should be done in a fair manner such that no bus masters gets permanently locked out of bus access. The bus slaves can never obtain control over a bus segment, but instead have to follow the directions of the bus master that owns the bus.

A bus system uses an *address space* to organize the communication between components. A sample address space is shown on the right of Fig. 10.1. Usually, the smallest addressable entity in an address space is 1 byte. Each data transfer over the bus is associated with a given destination address. The destination address determines what component should pick up the data. Bus bridges are address-transparent: they will merge the slave address spaces from their output, and transform it to a single slave address space at their input.

An on-chip bus physically consists of a bundle of wires, which includes the following four categories: address wires, data wires, command wires, and synchronization wires.

- Data wires transfer data items between components. As discussed in the previous chapter, on-chip wiring is very dense, and data wires do not have to be multiplexed. Masters, slaves and bridges will have separate data inputs and data outputs.
- Address wires carry the address that goes with a given data item. The process of recognizing the destination address is called *address decoding*. One approach to implement address decoding is to implement it inside of the bus slave. Another approach is to perform address decoding centrally, and to distribute the decoded address signals directly to the slaves.
- Command wires describe the nature of the transfer to be performed. Simple commands include *read* and *write*, but larger on-chip bus systems may contain a wide variety of commands, that qualify a given read or write command. Several examples will be discussed later for actual on-chip buses.
- Synchronization wires ensure that bus masters and bus slaves are synchronized during data transfer. Common on-chip bus systems today are synchronous. They use a single clock signal per bus segment: all data, address, and command

**Fig. 10.2** Point-to-point bus

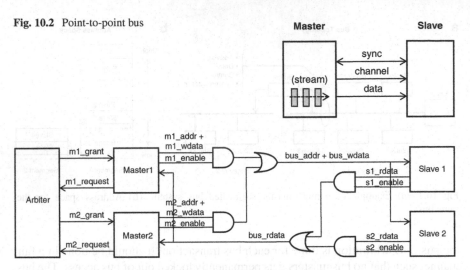

**Fig. 10.3** Physical interconnection of a bus. The *_addr, *_wdata, *_sdata signals are signal vectors. The *_enable, *_grant, *_request signals are single-bit signals

wires are referenced to the edges of the bus clock. Besides the clock signal, additional control signals can be used to synchronize a bus master and bus slave, for example to indicate time-outs and to support request-acknowledge signalling.

### 10.1.3  Elements in a Point-to-Point Bus

For dedicated connections, bus systems may also support a point-to-point communication mode. Figure 10.2 illustrates such a point to point connection. Like a shared bus, a point-to-point bus also identifies a master and a slave. There is no concept of address space. Instead, the data is looked upon as an infinite stream of items. Data items may still be assigned to logical *channels*. This enables multiple streams to be multiplexed over the same physical channel. Each data transfer will then consist of a tuple (datavalue, channel).

The point-to-point bus has synchronization wires to handle the interaction between master and slave. The synchronization mechanism is very similar to the one we described in the previous chapter (Sect. 9.2.3).

### 10.1.4  Physical Implementation of On-Chip Busses

Figure 10.3 shows the physical layout of a typical on-chip bus segment with two masters and two slaves. The AND and OR gates in the center of the diagram serve as multiplexers. Several signals are merged this way into bus-wide address and data signals. For example, the address generated by the bus masters is merged into a single bus address, and this bus address is distributed to the bus slaves. Similarly, the data transfer from the masters to the slaves is merged into a bus-wide write-data signal, and the data transfer from the slaves to the masters is merged into a bus-wide read-data signal. This implementation is typical for on-chip bus systems. Due to the low cost of an on-chip wire, multiplexing is not done for the purpose of saving silicon real-estate, but for the purpose of logical merging of signals. Notice, for example, how the bus system implements a separate bundle of wires for reading data and for writing data. This is common for each of the four bus systems discussed earlier.

The convention that associates the direction of data with *reading* and *writing* the data is as follows. Writing data means: sending it from a master to a slave. Reading data means: sending it from a slave to a master. This convention affects the input/output direction of bus signals on slave components and master components.

In Fig. 10.3, each master generates its own bus-enable signal in order to drive a data item or an address onto the bus. For example, when `Master1` will write data, `m1_enable` will be high while `m2_enable` will be low. If both enable signals would be high, the resulting bus address and write-data will be undefined. Thus, the bus protocol will only work when the components collaborate and follow the rules of the protocol. The bus arbiter has to ensure that only one bus master at a time takes control of the bus. Bus arbitration can be done globally, for an entire bus segment, or it can be done per slave component. We will describe bus arbitration further in the context of Multi-Master Systems (Sect. 10.3).

### 10.1.5 Bus Naming Convention

Since a bus segment can group a potentially large amount of signals, bus systems will follow a *naming convention*. The objective of a naming convention is to infer the functionality and connectivity of a wire based on its name. For example, a naming convention is very helpful to read a timing diagram. A naming convention can also help engineers to visualize the connectivity in a (textual) netlist of a circuit.

A component pin name will reflect the functionality of that pin. Each bus standard has its own favorite names for address signals, data signals, control signals and synchronization signals. For example, the IBM/Coreconnect bus uses `ABUS` for the address bus, AMBA uses `PADDR` or `HADDR`, Avalon uses `address` and Wishbone uses `AD`. Using common pin names allows a designer to easily recognize the interfaces on a hardware module.

Bus signals are created by interconnecting component pins. Bus signals follow a convention, too. The key issue is to avoid confusion between similar signals. For example, in Fig. 10.3, there are two master components, each with a `wdata` signal. To distinguish these signals, the component instance name is included as a prefix in bus signal name, such as `m2_wdata`. Other bus systems use this technique as well. In some cases, both the master instance name and the slave instance name will be included as part of the bus signal name. This results in very long identifiers,

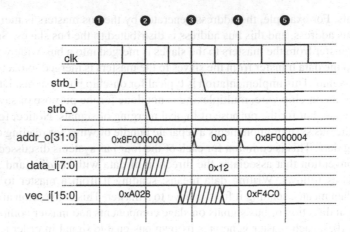

**Fig. 10.4**  Bus timing diagram notation

such as   `cpu1_read_data_valid_sdram0`. While such identifiers are tedious
to type, they precisely reflect the function of the wire.

Bus systems may use additional naming conventions to help a designer. For
example, AMBA prefixes all signals on the high-speed bus (AHB) with the letter
H, and all signals on the peripheral bus (APB) with the letter P. Wishbone suffixes
all input pins with the letter I, and all output pins with the letter O. Obviously, for a
designer, it will greatly help to adopt and follow the bus naming conventions of the
particular bus in use: it avoids coding mistakes and connection errors, it makes code
reusable, and it simplifies debugging and verification.

### 10.1.6  Bus Timing Diagram

Because a bus system reflects a complex, highly parallel entity, timing diagrams
are extensively used to describe the timing relationships of one signal to the other.
Figure 10.4 illustrates a timing diagram of the activities in a generic bus over five
clock cycles. The clock signal is shown on top, and all signals are referenced to the
upgoing clock edge. Dashed vertical lines indicate the timing reference.

Signal buses of several wires can be collapsed into a single trace in the
timing diagram. Examples in Fig. 10.4 are addr_o, data_i, and vec_i. The
label indicates when the bus changes value. For example, addr_o changes from
0x8F000000 to 0x00000000 at the third clock edge, and it changes back to
0x8F000004 one clock cycle later. Various schemes exist to indicate that a signal
or a bus has an unknown or don't care value. The value of data_i at the second
clock edge, and the value of vec_i at the third clock edge are all unknown.

When discussing timing diagrams, one must make a distinction between clock
edges and clock cycles. The difference between them is subtle, but often causes

confusion. The term *clock cycle* is ambiguous, because it does not indicate a singular point in time: a clock cycle has a beginning and an end.

A *clock edge*, on the other hand, is an atomic event that cannot be partitioned further (at least not under a single-clock synchronous paradigm). A consequence of the ambiguous term *clock cycle* is that the meaning of the term changes with the direction of the signals. When discussing the value of input signals, designers usually mean to say that these signals must be stable at the start of the clock cycle, just *before* a clock edge. When discussing the value of output signals, designers usually talk about signals that are stable at the end of the clock cycle, so *after* a clock edge. Consider for example signal strb_o in Fig. 10.4. The signal goes down just after the clock edge labeled 2. As strb_o is an output signal, a designer would say that the signal is low in clock cycle 2; the output should reach a stable value after clock edge 2. In contrast, consider the signal strb_i. This input signal is high at the clock edge labeled 2. Therefore, a designer would say this input is high in clock cycle 2. This means that the signal should reach a stable value before clock edge 2. To avoid this ambiguity, we will discuss timing diagrams in terms of clock edges rather than clock cycles.

Bus timing diagrams are very useful to describe the activities on a bus as a function of time. They are also a central piece of documentation for the design of a hardware-software interface.

## 10.1.7 Definition of the Generic Bus

Even though different bus systems may use a different naming convention for signals, they have a similar functionality. This chapter is not written towards any specific bus system, but instead emphasizes the common concepts between them. We will therefore define a generic bus. Where appropriate, we will indicate the relationship of this generic bus with a specific implementation on CoreConnect, AMBA, Avalon or Wishbone. The signals that make up the generic bus are listed in Table 10.2. The exact meaning of these signals will be explained throughout this chapter.

Table 10.3 makes a comparison between the signal names of the generic bus, and equivalent signals on the CoreConnect/OPB bus, the AMBA/APB bus, the Avalon-MM bus, and the Wishbone bus.

## 10.2 Bus Transfers

In this section, we will discuss several examples of data transfers between a bus master and a bus slave. We will also discuss common strategies used by on-chip busses to improve the overall system performance.

**Table 10.2** Signals on the generic bus

| Signal name | Meaning |
|---|---|
| clk | Clock signal. All other bus signals are references to the upgoing clock edge |
| m_addr | Master address bus |
| m_data | Data bus from master to slave (write operation) |
| s_data | Data bus from slave to master (read operation) |
| m_rnw | Read-not-Write. Control line to distinguish read from write operations |
| m_sel | Master select signal, indicates that this master takes control of the bus |
| s_ack | Slave acknowledge signal, indicates transfer completion |
| m_addr_valid | Used in place of m_sel in split-transfers |
| s_addr_ack | Used for the address in place of s_ack in split-transfers |
| s_wr_ack | Used for the write-data in place of s_ack in split-transfers |
| s_rd_ack | Used for the read-data in place of s_ack in split-transfers |
| m_burst | Indicates the burst type of the current transfer |
| m_lock | Indicates that the bus is locked for the current transfer |
| m_req | Requests bus access to the bus arbiter |
| m_grant | Indicates bus access is granted |

**Table 10.3** Bus signals for simple read/write on Coreconnect/OPB, ARM/APB, Avalon-MM and Wishbone busses

| generic | CoreConnect/OPB | AMBA/APB | Avalon-MM | Wishbone |
|---|---|---|---|---|
| clk | OPB_CLK | PCLK | clk | CLK_I (master/slave) |
| m_addr | Mn_ABUS | PADDR | Mn_address | ADDR_O (master) |
| | | | | ADDR_I (slave) |
| m_rnw | Mn_RNW | PWRITE | Mn_write_n | WE_O (master) |
| m_sel | Mn_Select | PSEL | | STB_O (master) |
| m_data | OPB_DBUS | PWDATA | Mn_writedata | DAT_O (master) |
| | | | | DAT_I (slave) |
| s_data | OPB_DBUS | PRDATA | Mb_readdata | DAT_I (master) |
| | | | | DAT_O (slave) |
| s_ack | Sl_XferAck | PREADY | Sl_waitrequest | ACK_O (slave) |

## 10.2.1   Simple Read and Write Transfers

Figure 10.5 illustrates a write transfer on a generic peripheral bus. A bus master will write the value 0xF000 to address 0x8B800040. We assume that this bus only has a single bus master and that it does not need arbitration. On clock edge 2, the master takes control of the bus by driving the master select line m_sel high. This indicates to the bus slave that a bus transaction has started. Further details on the nature of the bus transaction are reflected in the state of the bus address m_addr and the bus read/write control signal m_rnw. In this case, the transfer is a write, so the *read-not-write* (m_rnw) signal goes low.

Bus requests from the master are acknowledged by the slave. A slave can extend the duration of a transfer in case the slave cannot immediately respond to the request of a master. In Fig. 10.5, the slave issues an acknowledge signal s_ack on clock

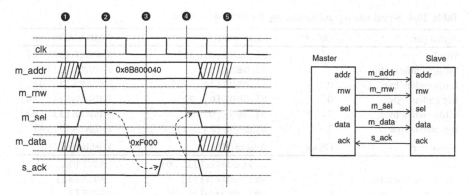

**Fig. 10.5** Write transfer with one wait state on a generic peripheral bus

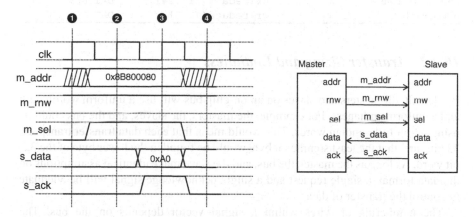

**Fig. 10.6** Read transfer with no wait state on a generic peripheral bus

edge 4. This is one clock cycle later than the earliest possible clock edge 3. Such a cycle of delay is called a *wait state*: the bus transaction is extended for one clock cycle. Wait states enable communication between bus components of very different speed. However, wait states are also a disadvantage. During a wait state, the bus is tied-up and inaccessible to other masters. In a system with many slow slaves, this will significantly affect the overall system performance. A *bus timeout* can be used to avoid that a slave completely takes over a bus. If, after a given amount of clock cycles, no response is obtained from the bus slave, the bus arbiter can declare a timeout condition. The timeout condition will alert the bus master to give up the bus and abort the transfer.

Figure 10.6 shows a read transfer with no wait states. The protocol is almost identical as a write transfer. Only the direction of data is reversed (from slave to master), and the m_rnw control line remains high to indicate a read transfer. The bus protocols for read and write described here are typical for peripheral buses.

**Table 10.4** Signal naming and numbering for a bus slave input

| Signal part | Offset | CoreConnect/OPB | AMBA/APB |
|---|---|---|---|
| Word | | S1_DBUS[0..31] | PWDATA[31..0] |
| Most significant bit | | S1_DBUS[0] | PWDATA[31] |
| Little endian byte | 0 | S1_DBUS[24..31] | PWDATA[7..0] |
| Big endian byte | 0 | S1_DBUS[0..7] | PWDATA[31..24] |
| Little endian byte | 3 | S1_DBUS[0..7] | PWDATA[31..24] |
| Big endian byte | 3 | S1_DBUS[24..31] | PWDATA[7..0] |
| Signal part | Offset | Avalon-MM | Wishbone |
| Word | | writedata[31..0] | DAT_I[31..0] |
| Most significant bit | | writedata[31] | DAT_I[31] |
| Little endian byte | 0 | writedata[7..0] | DAT_I[7..0] |
| Big endian byte | 0 | writedata[31..24] | DAT_I[31..24] |
| Little endian byte | 3 | writedata[31..24] | DAT_I[31..24] |
| Big endian byte | 3 | writedata[7..0] | DAT_I[7..0] |

## 10.2.2   Transfer Sizing and Endianess

By default, all masters and slaves on an on-chip bus will use a uniform wordlength and a uniform endianess. For example, the masters, the slaves, and the bus could be using 32-bit little-endian words. This would mean that each data transfer transports 32 bits, and that the least significant byte would be found in the lower byte of the 32-bit word. As long as the master, the bus, and the slave make identical assumptions on the data format, a single request and a single acknowledge signal will be adequate to control the transfer of data.

The numbering of wires within a signal vector depends on the bus. The documentation of the bus should be consulted to determine the name of the least significant bit of a word. As an example, Table 10.4 illustrates the signal naming for a bus slave input under various bus schemes.

While endianess can be configured on most buses, it is a static selection, and dynamic switching of endianess is not supported by AMBA, Coreconnect, Avalon-MM or Wishbone. The additional hardware and complexity introduced in the bus system does not justify the benefit. Indeed, as illustrated in Fig. 10.7, heterogeneous endianess can be resolved while interconnecting bus components to the bus system.

A bus system will also need to provide a mechanism for *transfer sizing*: selecting what part of a given word belongs to the actual data transfer. In a 32-bit data-bus for example, it is useful to be able to transfer a single byte or a halfword (16 bit). For example, this would allow a C program to write a single char (8 bit) to memory. Transfer sizing is expressed using byte-enable signals, or else by directly encoding the size of the transfer as part of the bus control signals. The former method, using byte-enable signals, is slightly more general than the latter, because it allows one to cope with un-aligned transfers.

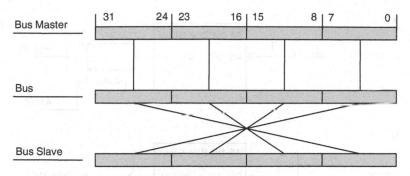

**Fig. 10.7** Connecting a big-endian slave to little-endian master

To see the difference between the two, consider the difference in performance for a processor running the following C program.

```
int main() {
  unsigned i;
  char a[32], *p = a;

  for (i=0; i<32; i++)
    *p++ = (char) (i + 4);

  return 0;
}
```

As the processor moves through all iterations of the i-loop, it will generate *byte-aligned* write operations to all addresses occupied by the a array. Assume that this happens in a system with a 32-bit data bus, and that the third byte of a 32-bit word needs to be written during a particular iteration. When the bus does not support unaligned data transfers, the processor will first need to *read* the word that contains the byte, update the word by modifying a single byte, and write it back to memory. On the other hand, when the bus does support unaligned data transfers, the processor can directly write to the third byte in a word. Therefore, the example program will complete quicker on systems that support unaligned transfers. Note that unaligned transfers can also lead to exceptions. For example, processors with a word-level memory organization do not support transfer of unaligned words. If a programmer attempts to perform such a transfer, an exception will result, which usually halts the execution of the program.

Endianess and byte-transfer sizing help bus components to deal with the ordering of individual bytes within a bus word. In some cases, we may run into the situation where masters and slaves of different physical width need to be interconnected. For example, a bus slave could have an 8-bit data bus but needs to be connected to a 32-bit bus. Or, a bus master could have a 64-bit data bus but needs to be connected to a 32-bit bus. These cases require additional hardware.

Figure 10.8 shows how a 64-bit bus slave and a 16-bit bus slave can be connected to a 32-bit bus. In the case of the 64-bit bus slave, a data-write will transfer only

**a**

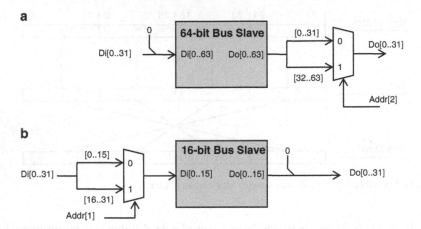

**b**

**Fig. 10.8** (a) Connecting a 64-bit slave to a 32-bit bus. (b) Connecting a 16-bit slave to a 32-bit bus

32 bits at a time; the upper 32 bits are wired to zero. Additionally, the interconnect may use the byte-enable signals to mask off non-written bytes to the slave. In the case of a data-read, one of the address lines, Addr[2], will be used to multiplex the 64 bits of data produced by the bus slave. The net effect of the multiplexing is that the bus slave appears as a continuous memory region when data is read.

The case of the 16-bit bus slave is opposite: the 32-bit bus system can deliver more data than the 16-bit bus slave can handle, and an additional address bit, Addr[1] is used to determine which part of the 32-bit bus will be transferred to the 16-bit bus slave. If the interconnect supports *dynamic* transfer sizing, it will expand each master-side transaction (read or write) into multiple slave-side transactions.

In summary, busses are able to deal with varying wordlength requirements by the introduction of additional control signals (byte-select signals), and by adding additional multiplexing hardware around the bus slaves or bus masters. A designer also needs to be aware of the endianess assumptions made by the on-chip bus, the bus master, and the bus slave.

### 10.2.3 *Improved Bus Transfers*

Each bus data transfer includes multiple phases. First, the bus master has to negotiate a bus access with the bus arbiter. Next, the bus master has to issue a bus address and a bus command. Third, the bus slave has to acknowledge the data transfer. Finally, the bus master has to terminate the bus transfer and release control over the bus. Each of these activities takes a finite amount of time to complete. Moreover, all of these activities are sequential, so that the overall system is limited by the speed of the slowest component. For high-speed buses, this is too slow.

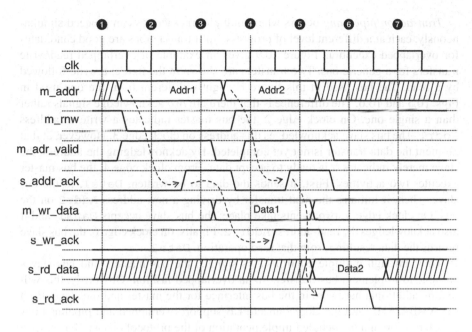

**Fig. 10.9** Example of pipelined read/write on a generic bus

On-chip buses use three mechanisms to speed up these transfers. The first mechanism, *transaction-splitting*, separates each bus transaction in multiple phases, and allows each phase to complete separately. This prevents locking up the bus over an extended period of time. The second mechanism, *pipelining*, introduces overlap in the execution of bus transfer phases. The third mechanism, *burstmode operation*, enables transfer of multiple data items, located at closely related addresses, during a single bus transaction.

A bus may use one or several of these mechanisms at the same time. Two of them, transaction splitting and pipelined transfers, often occur together. We will discuss them simultaneously.

### 10.2.3.1  Transaction Splitting and Pipelined Transfers

In *transaction splitting*, a single bus transfer is decomposed into separate phases. For example, the high-speed bus implementation of AMBA (AHB) and CoreConnect (PLB) treat the transfer of an address as a separate transaction from the transfer of data. Each of these transfers has a different acknowledge signal. The rationale is that a bus slave will need some time after the reception of an address in order to prepare for the data transfer. Hence, after the transfer of an address, the bus master should be released from waiting for the bus slave, and carry on.

*Transaction pipelining* occurs when multiple transactions can proceed simultaneously, each at a different level of progress. Split transactions are good candidates for overlapped execution. Figure 10.9 gives an example of overlapped read/write transfers for a generic bus. Two transfers are shown in the figure: a write followed by a read. The bus used in this figure is slightly different from the one used in Figs. 10.5 and 10.6. The difference is that there are *two* acknowledge signals rather than a single one. On clock edge 2, the bus master indicates a write to address Addr1. The bus slave acknowledges this address on clock edge 3. However, at that moment the data transfer is not yet completed. By acknowledging the address, the slave merely indicates it is ready to accept data. From clock edge 4, the bus master executes two activities. First, it sends the data to be written, Data1, to the bus slave. Then, it initiates the next transfer by driving a new address Addr2 on the bus. On clock edge 5, two events take place: the bus slave accepts Data1, and it also acknowledges the read-address Addr 2. Finally, on clock edge 6, the bus slave returns the data resulting from that read operation, Data2.

Thus, through multiple control/status signals, the bus masters and bus slaves are able to implement bus transfers in an overlapped fashion. Obviously, this will require additional hardware in the bus interface for the master and the slave. Both the AMBA/AHB and the Coreconnect/PLB support overlapped and pipelined bus transfers, although the detailed implementation of the protocol on each bus system is different. The Avalon bus and the Wishbone bus support pipelined transfers, but there are no separate acknowledge signals for address and data transfer.

### 10.2.3.2   Burstmode Transfers

The third technique to improve bus transaction performance is to use *burstmode transfers*. This will transfer multiple data items from closely related addresses in one bus transaction.

Burstmode transfers are useful when transfers of closely-related data items are required. The main memory accesses made by a processor with cache memory are an example. When there is cache miss on the processor, an entire cache *line* needs to be replaced. Assume that there would be 32 bytes in a cache line, then a cache miss implies reading 32 consecutive bytes from memory.

Burst-mode transfers can have a fixed or a variable length. In a fixed-length burst-mode scheme, the bus master will negotiate the burst properties at the start of the burst transfer, and next perform each transfer within the burst. In a variable-length scheme, the bus master (or the bus slave) has the option of terminating the burst after every transfer. The addresses within a burst are usually incremental, although there also applications where the address needs to remain constant, or where the address increments with a modulo operation. Thus, as part of the burst specification, a bus may allow the user to specify the nature of the burst address sequence.

Finally, the address step will depend on the size of the data within the burst: bytes, halfwords and words will increment addresses by 1, 2, and 4 respectively. Obviously, all of these options involve adding extra control-signals on the bus, at

**Table 10.5** Burst transfer schemes

| Burst property | CoreConnect/OPB | AMBA/APB |
|---|---|---|
| Burst length | Fixed (2 .. 16) or variable | Fixed (4, 8, 16) or variable |
| Address sequence | Incr | Incr/mod/const |
| Transfer size | Fixed by bus | Byte/halfword/word |
| Burst property | Avalon-MM | Wishbone |
| Burst length | Fixed (1 .. $2^{11}$) | Variable |
| Address sequence | Incr/mod/const | Incr/const |
| Transfer size | Byte/halfword/word | Byte/halfword/word |

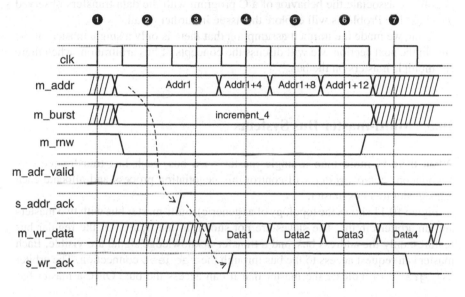

**Fig. 10.10** A four-beat incrementing write burst

the side of the master as well as the slave. Table 10.5 shows the main features for burst-mode support on Coreconnect, AMBA, Avalon, and Wishbone.

An example of a burst-mode transfer is shown in Fig. 10.10. This transfer illustrates a burst transfer of four words in adjacent address locations. Besides the commands discussed before (m_addr, m_rnw, m_adr_valid), a new command m_burst is used to indicate the type of burst transfer performed by the master. In this generic bus, we assume that one of the burst types is encoded as increment_4, meaning a burst of four consecutive transfers with incrementing address. On clock edge 3, the slave accepts this transfer, and after that the master will provide four data words in sequence. The address information provided by the master after the first address is, in principle, redundant. The addresses are implied from the burst type (increment_4) and the address of the first transfer. Figure 10.10 assumes that the wordlength of each transfer is 4 bytes (one word).

Therefore, the address sequence increments by 4. The scheme in this figure is similar to the scheme used by AMBA/AHB. The Coreconnect/PLB system is slightly more general (and as a consequence, more complicated), although the ideas of burst transfers are similar to those explained above.

This completes our discussion on improved bus data transfers. As demonstrated in this section, there are many variations and enhancements possible for data transfer over a bus. Optimal bus performance requires both the master as well as the slave to be aware of all features provided by a bus protocol. For the hardware-software codesigner, understanding the bus protocols is useful to observe the hardware-software communication at its lowest abstraction level. For example, it is very well possible to associate the behavior of a C program with the data transfers observed on a bus. The Problems will explore this issue in further detail.

So far, we made the implicit assumption that there is only a single master on the bus. In the next section, we will discuss the concepts of bus arbitration, when there are multiple masters on the bus.

## 10.3   Multi-master Bus Systems

When there is more than a single master on a bus, each bus transfer requires negotiation. A *bus arbiter* will control this negotiation process and allocate each transfer slot to a bus master.

Figure 10.11 shows the topology of a (generic) multi-master bus with two masters and an arbiter circuit. The slaves are not shown in the figure. Of the regular bus features, only the address bus, and a transfer-acknowledge signal are visible. Each master can request access to the bus through the request connection. The arbiter uses grant to indicate the master that it can access the bus. Once a master has

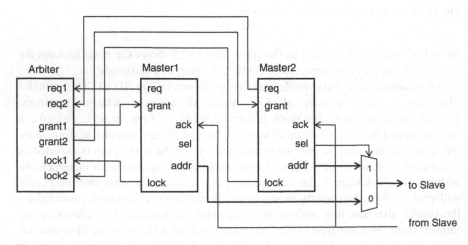

**Fig. 10.11** Multi-master arbitration

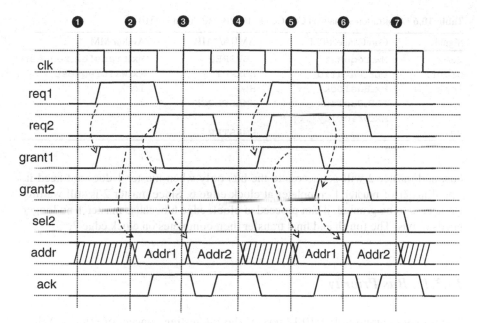

**Fig. 10.12**  Multi-master arbitration timing

control over the bus, it will proceed through one of the regular bus transfer schemes as discussed before. The `lock` signals are used by a master to grab exclusive control over the bus, and will be clarified later.

Figure 10.12 shows how two masters compete for the bus over several clock cycles. On clock edge 2, master 1 requests the bus through `req1`. Since master 2 does not need the bus at that moment, the arbiter will grant the bus to master 1. Note that the `grant` signal comes as an immediate response to the `request` signal. This means that the bus negotiation process can complete within a single clock cycle. In addition, it implies that the arbiter will need to use combinational logic to generate the `grant` signal based on the `request` signal.

After clock edge 2, master 1 drives an address onto the address bus and completes a regular bus transfer. We assume that the slave acknowledges the completion of this transfer on clock edge 3, by pulling `ack` high. The earliest time when the next arbitration for a bus transfer takes place is clock edge 3. This is called an *overlapping* arbitration cycle, because the arbitration of the next transfer happens at the same moment as the completion of the current transfer. The second transfer is granted to master 2, and completes on clock edge 4.

Between clock edges 4 and 5, the bus sits idle for one cycle, because no master has requested access to the bus. On clock edge 5, both masters 1 and 2 request access to the bus. Only one master is allowed to proceed, and this means that there is a *priority resolution* implemented among the masters. In this case, master 1 has fixed priority over master 2, which means that master 1 will always get access to the

**Table 10.6** Arbitration signals on CoreConnect/OPB and AMBA/AHB

| Signal | CoreConnect/PLB | AMBA/AHB | Avalon-MM |
|--------|-----------------|----------|-----------|
| reqx   | Mx_request      | HBUSREQ  | Mx_transfer_request |
| grantx | PLB_PAValid     | HGRANT   | Mx_waitrequest |
| lock   | Mx_Buslock      | HLOCK    | lock |
|        | PLB_Buslock     | HMASTLOCK |  |
|        | Mx_priority[..] |          |  |
| sel    |                 | HMASTER[..] |  |

bus, and master 2 will get access to the bus only when master 1 does not need it. The transfer of master 1 completes at clock edge 6. Since master 2 is still waiting for access to be granted, it can proceed at clock edge 6 because master 1 no longer needs to bus. The fourth and final transfer then completes on clock edge 7.

## 10.3.1  Bus Priority

The timing diagram in Fig. 10.12 reveals the interesting concept of priority. When multiple masters attempt to access the bus at the same time, only a single master is allowed to proceed based on priority resolution. The simplest priority scheme is to allocate a fixed priority, strictly increasing, to every master. While this is easy to implement, it is not necessarily the best solution. When a high-priority master continuously accesses the bus, other low-priority masters can be denied bus transfers for extended amounts of time.

Often, multiple masters on a bus should have equal priority. For example, when symmetrical multi-processors processors on a single bus access the same memory, no processor should have priority over the other. In this situation, priority resolution is implemented using *round-robin* scheme. Each master takes turns to access to the bus. When two masters request the bus continuously, then the bus transfers of master 1 and master 2 will be interleaved. Another possible solution is a *least-recently-used* scheme, in which the master who was waiting for the bus for the longest time will get access first. Equal-priority schemes such as round-robin or least-recently-used avoid *starvation* of the bus masters, but, they also make the performance of the bus unpredictable. When working with latency-critical applications, this can be a problem. To address this, designers can use a mixed scheme that combines multiple levels of priority with an equal-priority scheme to allow several masters to share the same priority level.

The priority algorithm used by the bus arbiter is not part of the definition of the bus transfer protocol. Therefore, Coreconnect/PLB, AMBA/AHB and Avalon-MM only describe the arbitration connections, but not the priority schemes. The Wishbone bus is special in that it does not define special bus request/grant signals. Instead, Wishbone leaves the design of the bus topology to the designer.

Table 10.6 makes a comparison between the generic bus arbtriation signals defined above, and those of CoreConnect, AMBA and Avalon. The table also lists a few arbitration signals that are unique to each individual bus protocol. The bus locking signals will be explained shortly. The other signals have the following meaning.

- Mx_priority[..] allows a PLB master to select its priority for each transfer. This scheme allows the master to change its priority level dynamically depending on the needs of the bus transfer.
- HMASTER[..] is an encoding of the identity of the master that was granted bus access by the arbiter. The signal is used to drive the bus address multiplexer.

### 10.3.2 Bus Locking

The final concept in multi-master bus schemes is *bus locking*: the exclusive allocation of a bus to a single master for the duration of multiple transfers. There are several reasons why bus locking may be needed. First, when large blocks of data need to be transferred with strict latency requirements, exclusive access to the bus may be required. While burst-mode transfers can help a master to complete these transfers quickly, these transfers can still be interrupted by another master with higher priority. By locking the bus, the master can be sure this will not happen.

The second need for bus locking is when a master needs to have guaranteed, exclusive access to consecutive transfers, typically a read transfer followed by a write transfer. This is needed, for example when implementing a *test-and-set* instruction. The test-and-set instruction is used to create a mutex, a well-known software primitive to implement mutual exclusion. A mutex is similar to a semaphore (discussed in Sect. 9.2.2), but it's not identical to it. A mutex implies ownership: once a mutex is locked, it can only be unlocked by the same entity that locked it. An example where a mutex can be used is to control the access of two bus masters to a single, shared region of memory. Bus masters will control access using a mutex, implemented through a test-and-set instruction as discussed next.

An example implementation of test-and-set is shown below. This C program runs on each of two processors (bus masters) attached to the same bus. They share a memory location at address 0x8000. By calling testandset, a processor will try to read this memory location, and write into it during a single locked-bus operation. This means that the function test_and_set() cannot be interrupted: only one processor will be able to read the value of the mutex when its value is low. The two processors use this function as follows. Before accessing the shared resource, the processors will call enter(), while they will call leave(). The shared resource can be anything that needs exclusive access by one of the processors.

```
int *mutex = (int *) 0x8000; // location of mutex
```

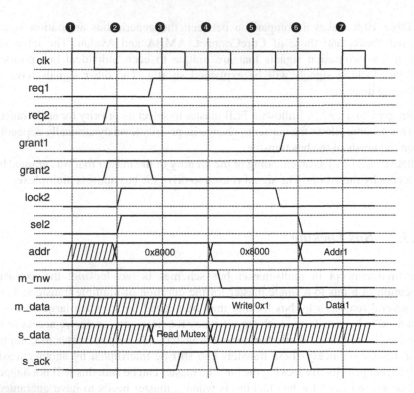

**Fig. 10.13** Test-and-set operation with bus locking

```
int test_and_set() {
  int a;
  lock_bus();
  a = *mutex;
  *mutex = 1;
  unlock_bus();
  return a;
}

void leave() {
  *mutex = 0;
}

void enter() {
  while (test_and_set()) ;
}
```

Figure 10.13 shows an example of test-and-set with bus-locking. On clock edge 2, master 2 requests access to the bus using req2. This access is granted by the arbiter through grant2. After clock edge 2, this master grabs the bus using sel2 and locks it using lock2. Master 2 will now perform a test-and-set operation, which involves a read of a memory address immediately followed by a write to the same memory address. The read operation starts on clock edge 3 and completes on clock edge 4. On clock edge 3, the master drives an address onto the bus and signals a read operation (m_rnw). On clock edge 4, the slave delivers the data stored at this address, and completes the transfer using s_ack.

Meanwhile, another master requested bus access starting on clock edge 3 (using req1). However, because master 2 has locked the bus, the arbiter will ignore this request. Master 2 will be granted further use of the bus until it release the lock. This access is guaranteed even if master 2 has a lower priority than other masters requesting the bus.

After performing the reading part of the test-and-set instruction, master 2 will now write a '1' into the same location. At clock edge 5, master 2 puts the address and the data on the bus, and at clock edge 6 the slave accepts the data. The lock can be released after clock edge 5. Note that, should the write operation to the slave fail, then the complete test-and-set instruction has failed. We assume however that the write operation completes correctly. As soon as master 2 releases lock2, control will go to master 1 because of the pending request on req1. As a result, starting with clock edge 6, a new bus transfer can start, which is allocated to master 1.

Wrapping up, when multiple masters are attached to a single bus, individual bus transfers need to be arbitrated. In addition, a priority scheme may be used among masters to ensure latency requirements for particular masters. Finally, bus-locking can be used to implement guaranteed access for an extended amount of time. Since all of these techniques are implemented in hardware, at the level of a bus transfer, they are very fast and efficient. As such, bus systems play an important role in building efficient hardware-software communication.

## 10.4 Bus Topologies

The bus topology is the logical and physical organization of bus components in a network. So far, we have used linear bus topologies: busses that consist of bus segments interconnected by bus bridges.

The bus topology has a major impact on system performance. Let's first consider a single bus segment. All modules connected to the same bus segment are sharing the same communication resource. This means that the communication among these modules needs to be sequentialized. Two bus masters on the same bus segment cannot initiate parallel bus transfers; they need bus arbitration to sequentialize their access.

Bus bridges can split busses in multiple segments, and are used to group bus components of similar performance. You can think of them as partitioning a road in multiple lanes, to accommodate fast traffic as well as slow traffic. However, bus bridges do not solve the issue of parallel bus transfers. Bus bridges introduce an

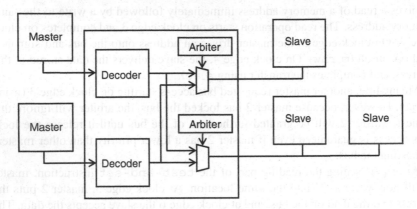

**Fig. 10.14** A multi-layer bus

implicit hierarchy among bus segments: a bus bridge is a master on one side and a slave on the other. In many cases, for example in multi-processor architectures, a hierarchy among the processors is not wanted or even counter-productive.

Besides the logical constraints of a bus segments, there are also significant technological issues. Implementing very long wires on a chip is hard, and distributing high-frequency signals and clocks using such wires is even harder. The power consumption of a wire is proportional to the length of the wire and the switching frequency of the signals on that wire. Hence, long wires will consume significantly more power than small, local wires. Chips which are organized using a linear bus topology will consume more energy for the same task than chips organized using a distributed topology.

Clearly, building on-chip bus communications with segmented busses has its limitations. Customizing the bus topology, and matching it to the application, is a logical improvement to this.

## 10.4.1  Bus Switches

A bus segment creates a static association between the bus masters and bus slaves attached to that segment. Let's see why a static assignment of bus masters to bus segments can be a problem. Assume that a master is connected to bus segment A. All slaves that need to communicate with this master will need to attach to A as well, or else to a segment which is directly bridged from A. Furthermore, all masters that need to talk to *any* of the slaves attached to A or any bridged segment on it, need to connect to bus segment A as well. Hence, all masters and all slaves in the system will tend to cluster to the same bus segment A. In simple, single-master systems, this is no problem. In heterogeneous architectures or architectures with multiple cores, however, a single segmented bus quickly becomes a performance bottleneck.

The solution to this problem is to use *bus switching* or *bus layering*, which makes the association between masters and slaves flexible. Figure 10.14 is an example of a two-layer bus. The two bus masters each connect to a bus switch with two output segments. There are three slaves connect to the output segments. The masters can connect to either of the slaves on the output segments. A master transfer is initially decoded to determine what output segment should receive it. All requests for the same output segment are then merged and arbitrated. Simultaneous transfers to the same output segment will be sequentialized: the master at the input segment will not be able to complete the transfer until the master at the other input segment is done.

Multiple masters can still share the same input segment on a bus switch, which makes bus switching compatible with multi-master bus arbitration. Bus systems that have implemented bus switching are for example AMBA/AHB and an improved version of Avalon, Merlin.

The most evolved form of bus switch is an implementation in which each master has its own input segment on the switch, and each slave has its own output segment. Such an implementation is called a *cross-bar*. The cross-bar is a highly parallel, but very expensive implementation of on-chip interconnect. And, while it addresses the logical limits of bus segments, it does not address the electrical issues of them. A cross-bar is a global interconnect system, and it is not scalable.

## 10.4.2 Network On Chip

Bus switches support a dynamic association of masters to slaves, but they have limited scalability. The fundamental issue with bus switches is that they maintain a tight association between a master and a slave. Indeed, each bus transaction is directly between a master and a slave, and the implementation of a bus transaction requires the interconnection network to create a pathway to complete it. Early telephone systems used the same concept: to connect a caller to a callee, a permanent circuit was established between the two parties for the duration of their telephone call.

The notion of a fixed and permanent pathway is however not scalable. Instead, the bus transfer itself must be implemented dynamically, in several steps. First, a master wraps a request into a packet. The master delivers the packet to the interconnect, and the interconnect finds a path from the master to the slave. Finally, the slave accepts the packet from the interconnect. When the slave is ready to issue a response, the opposite will happen: the slave delivers a packet to the interconnect, the interconnect returns the packet to the master, and the master accepts the response packet from the interconnect. Note the change of terminology from transaction to request/response and packet. The communication paradigm has changed from transactions to sending and receiving of packets. This new communication paradigm is called *Network-on-Chip*.

Figure 10.15 presents the concept of a Network-on-Chip. The computational elements of the chip and their interconnections are organized in a geometrical pattern, such as a matrix or a ring. The interconnections are made between the *tiles*

**Fig. 10.15** A generic
network-on-chip

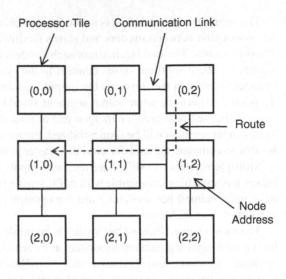

of the network on chip. In Fig. 10.15, every tile can directly communicate with its neighbouring tiles. In addition, every tile has an address, symbolically indicated by the matrix indices. This allows any tile to communicate with any other tile. A *route* for the communication is selected, and a data packet travels through a number of *hops*, from a source tile to a destination tile over a number of intermediate tiles. Figure 10.15 illustrates a route from tile (0,2) to tile (1,0).

The design of a network on-chip, and its operation, introduces a challenging set of problems. At the basic level, the communication and data representation is very different from the approach used in on-chip buses. In a network-on-chip, data items are encapsulated in a *packet* before being transmitted. Once a packet leaves a source tile and travels to a destination tile, it needs to find a route. As can be seen in Fig. 10.15, a route is not unique. Between a given source tile and destination tile, many different routes are possible. Hence, one needs a routing algorithm that is able to select interconnection segments so that the overall level of congestion will remain small. The design of Network-on-Chip has been an intensive area of research during the past decade.

The CELL processor is a well known multiprocessor device that relies on network-on-chip technology to implement on-chip communications. The CELL combines eight regular processing components called *SPE* (synergistic processing element). In addition, there is a control processor called *PPE* (power processor element) and an off-chip interface unit. All of these components are connected to the same network on chip, called the *EIB* (element interconnect bus). As illustrated in Fig. 10.16, the EIB consists of four ring structures, each 16 bytes wide. The communication model of the CELL processors assumes that processors will work using local memory, and that communication is implemented by moving blocks of data from one local memory location to the other. A *Direct Memory Access* (DMA) unit is used to handle communication between the bus interface and the local memory.

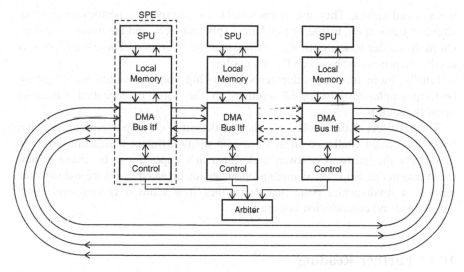

**Fig. 10.16** On-chip network in the CELL processor

The four rings run in opposite directions, so that each SPE can directly talk to its neighbors. Communication with other SPE is possible by taking several hops over the communication bus. Each time an SPE wants to transmit a block of data over EIB, it will send an appropriate request to the central on-chip arbiter. The arbiter will schedule all outstanding requests over the four rings. Up to three transfers can be concurrently scheduled over each ring, provided that these transfers use different segments on the ring.

The resulting data bandwidth on the CELL processor is impressive. In a 3.2 GHz CELL chip, the interface from the SPE to the rest of the chip supports 25.6 GB/s data bandwidth in each direction. The downside is that the CELL must be programmed in a very particular manner, as a set of concurrent programs that pass messages to one another. Optimizing the performance of a parallel CELL program is complex, and requires attention to a large collection of details, such as the granularity of tasks and message blocks, the synchronization mechanisms between processors, and the locality of data.

## 10.5 Summary

In this chapter, we discussed the concepts of on-chip interconnection buses, the lowest abstraction level where software and hardware meet. On-chip buses are shared communication resources to connect bus master components with bus slave components.

We discussed several examples, including the AMBA, CoreConnect, Avalon, and Wishbone bus. These busses support basic read/write transfers between bus masters and slaves. They use enhancements to improve the performance of on-chip communication, including pipelining, split-transfers, and burstmode transfers. On multi-master buses, each transfer needs to be arbitrated. These transfers may need to be prioritized based on the bus master.

Finally, more recent developments in on-chip communication with System-on-Chip emphasize distributed solutions in the form of switched busses or network-on-chip.

The net effect of this evolution is that the design of on-chip communication has become a design challenge on its own, with many different abstraction layers to tackle. For the hardware-software codesigner, it's important to be aware of new evolutions in this field. The communication latency between hardware and software remains a fundamental issue that determines if a solution is communication-constrained and computation-constrained.

## 10.6    Further Reading

The best reference to study on-chip bus systems is obviously the documentation from the vendors themselves. The AMBA bus specification can be obtained online from ARM (2009a). Likewise, the CoreConnect bus specification can be obtained online from IBM (2009), and the Avalon specification is available from Altera (2011). An in-depth discussion of contemporary on-chip bus systems, including AMBA and CoreConnect, is available from Pasricha and Dutt (2008). The same book also reviews ongoing research topics for on-chip bus systems.

Recently research efforts have focused on network-on-chip. An overview of the design principles may be found in De Micheli's book (Micheli and Benini 2006). A recent special issue of IEEE Design and Test Magazine has reviewed several proposals and open research issues (Ivanov and De Micheli 2005).

## 10.7    Problems

**Problem 10.1.** The timing diagram in Fig. 10.17 illustrates a write operation on the AMBA peripheral bus, AMBA APB. A *memory-mapped* register is a register which is able to intercept bus transfers from a specific address. In this case, we wish to create logic which will write PWDATA into a register when a write to address 0x2000 occurs. Develop a logic expression for the logic module shown in Fig. 10.17. Assume a 16-bit address.

**Listing 10.1** Program for Problem 10.2

```
#include <stdio.h>
void main() {
  int i, a[0x40];
```

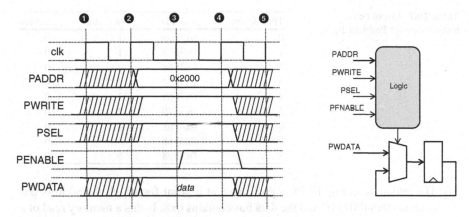

**Fig. 10.17** Timing diagram and schematic for Problem 10.1

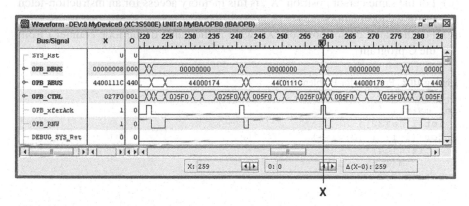

**Fig. 10.18** Timing diagram for Problem 10.2

```
for (i=0; i< 0x40; i++)
  if (i > 0x23)
      a[i] = a[i-1] + 1;
  else
    a[i] = 0x5;
}
```

**Problem 10.2.** While debugging a C program on a 32-bit microprocessor (shown in Listing 10.1), you capture the bus transfer shown in Fig. 10.18. The microprocessor is attached to an off-chip memory that holds the program and the data. The text and data segment both are stored in an off-chip memory starting at address 0x44000000. The array a[] starts at address 0x44001084. The instructions from the body of the loop start at address 0x44000170. Observe closely the timing diagram in Fig. 10.18 and answer the questions below.

**Table 10.7** List of bus
transactions for Problem 10.3

| Transaction | m_addr | m_data | m_rnw |
|---|---|---|---|
| 0 | 0x3F68 | 1 | 0 |
| 1 | 0x3F78 | 1 | 0 |
| 2 | 0x3F68 | 2 | 0 |
| 3 | 0x3F74 | 2 | 0 |
| 4 | 0x3F68 | 3 | 0 |
| 5 | 0x3F70 | 3 | 0 |
| 6 | 0x3F68 | 4 | 0 |
| 7 | 0x3F68 | 5 | 0 |
| 8 | 0x3F68 | 6 | 0 |

(a) The cursor X in Fig. 10.18 is positioned at a point for which the address bus contains 0x4400111C and the data bus contains 0x8. Is this a memory read of a memory write?
(b) For the same cursor position 'X', is this memory access for an instruction-fetch or for a data-memory read?
(c) For the same cursor position 'X', what is the value of the loop counter i from the C program?

**Listing 10.2** Program for Problem 10.3

```
int main() {
  int i, a[32];
  for (i=1; i<32; i++)
    if (i<4)
      a[4-i] = i;
  return 0;
}
```

**Problem 10.3.** You are debugging an embedded system by recording bus transactions between a microprocessor and an on-chip memory. The program you're analyzing is shown in Listing 10.2. Table 10.7 lists a sequence of bus transactions you observe when m_rnw is low. The signals in the Table follow the naming convention of the generic bus (defined in Table 10.2). Analyze the transactions and answer the questions below.

(a) What is the memory address of variable i?
(b) What is the memory address of variable a[0]?

**Problem 10.4.** The timing diagram in Fig. 10.19 shows the arbitration process of two masters, M1 and M2, requesting access to a shared bus. Answer the questions below using the information provided in the timing diagram.

(a) Based on the timing diagram, which master has the highest priority for bus transfers: M1, M2 or impossible to tell?
(b) Which master has control over the address bus between clock edge 3 and clock edge 4: M1, M2, or impossible to tell?

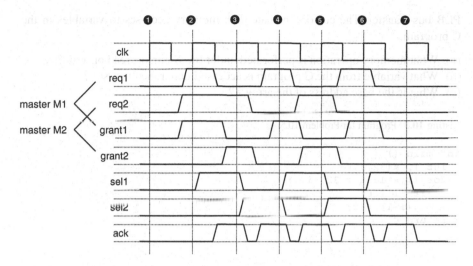

**Fig. 10.19** Timing diagram for Problem 10.4

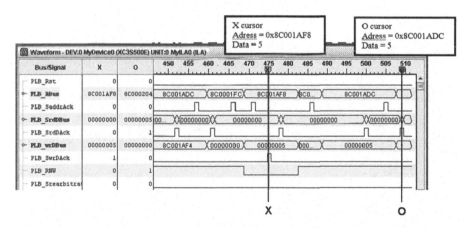

**Fig. 10.20** Timing diagram for Problem 10.5

(c) What type of component determines the value of the grant*x* signals: a bus master, a bus arbiter, or a bus slave?

(d) What type of component determines the value of the ack signal: a bus master, an bus arbiter, or a bus slave?

**Problem 10.5.** The C program in Listing 10.3 is running on a processor attached to a PLB bus (Fig. 10.20). The program and data are stored in an off-chip memory starting at 0x8C00000. During program execution, a PLB trace is captured in Chipscope, and the result is shown in Fig. 10.20. Observe closely the 'X' and 'O' cursor. Both cursors are positioned on top of the data-acknowledge phase of a

PLB bus transfer. The cursors indicate data memory accesses to variables in the C program.

(a)  What memory operation is performed under the X cursor: read or write?
(b)  What variable from the C program is accessed under the O cursor ?
(c)  What is the base address of the array a?

**Listing 10.3**  Program for Problem 10.5

```
int main () {
  int i, a[32];
  for (i = 1; i < 32; i++)
     if (i < 8)
        a[i+1] = i;
  return 0;
}
```

# Chapter 11
# Microprocessor Interfaces

## 11.1 Memory-Mapped Interfaces

A memory-mapped interface allocates part of the address space of a processor for communication between hardware and software. The memory-mapped interface is the most general, most wide-spread type of hardware/software interface. This is no surprise: memory is a central concept in software, and it's supported at the level of the programming language through the use of pointers. In this section, we look into the operation of memory-mapped registers, and into extended concepts such as mailboxes, queues, and shared memory. We also show how memory-mapped interfaces are modeled in GEZEL.

### 11.1.1 The Memory-Mapped Register

A memory-mapped interface can be as simple as a register which can be read and written through bus transfers on an on-chip bus. Figure 11.1 illustrates the generic setup of a memory-mapped register.

The register will be accessed when a bus transfer is made to a specific memory address, or within a specific memory address range. The memory address, and the related bus command, is analyzed by an address decoder. This decoder will generate a read pulse or a write pulse for the register. A full decoder will generate these pulses for a single address value. However, the complexity of the decoder is proportional to the number of bits that must be decoded. Therefore, it may be cheaper to build a decoder for a range of memory addresses (See Problem Section). The result of such a multi-address decoder is that a register is *aliased* at multiple memory locations. That is, the register will be read or written upon a transfer from or to any address within the range of addresses captured by the address decoder.

A memory-mapped register works as a shared resource between software and hardware. A write-conflict may occur if the hardware and the software attempt to

P.R. Schaumont, *A Practical Introduction to Hardware/Software Codesign*,
DOI 10.1007/978-1-4614-3737-6_11, © Springer Science+Business Media New York 2013

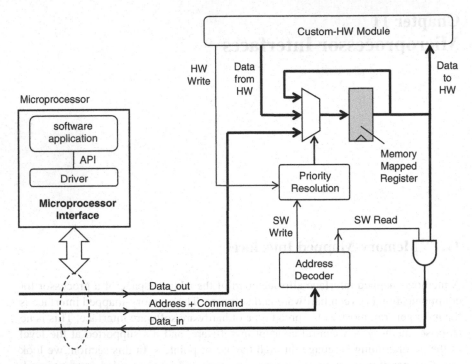

**Fig. 11.1** A memory-mapped register

write into the register during the same clock cycle. To resolve this case, a priority decoder will allow hardware or software to proceed. Note that it seldom makes sense to sequentialize the write operations into the register, since one value would overwrite the other.

In software, a memory-mapped register can be accessed using an initialized pointer, as shown below. The hardware abstraction layer of a particular microprocessor may provide wrapper functions to write into and read from memory-mapped registers.

```
volatile unsigned int *MMRegister = (unsigned int *) 0x8000;

// write the value '0xFF' into the register
*MMRegister = 0xFF;

// read the register
int value = *MMRegister;
```

Figure 11.2 explains why the pointer must be a `volatile` pointer. A memory-mapped register is integrated into the memory hierarchy of a processor, at the level of main-memory. When a processor instruction will read from or write into that register, it will do so through a memory-load or memory-store operation. Through the use of the `volatile` qualifier, the C compiler will treat the memory hierarchy slightly different.

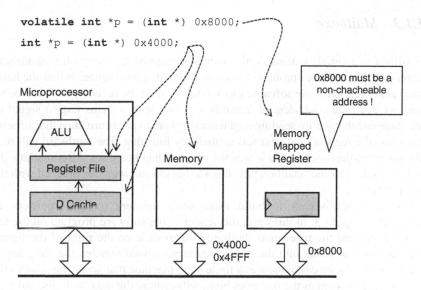

**Fig. 11.2** Integrating a memory-mapped register in a memory hierarchy

- When using normal pointer operations, the processor and the compiler will attempt to minimize the number of operations to the main memory. This means that the value stored at an `int *` can appear in three different locations in the memory hierarchy: in main memory, in the cache memory, and in a processor register.
- By defining a register as a `volatile int *`, the compiler will avoid maintaining a copy of the memory-mapped register in the processor registers. This is essential for a memory-mapped register, because it can be updated by a custom-hardware module, outside of the control of a microprocessor.

However, defining a memory-mapped register with a `volatile` pointer will not prevent that memory address from being cached. There are two approaches to deal with this situation. First, the memory addresses that include a memory-mapped register could be allocated into a non-cacheable memory area of a processor. This requires a processor with a configurable cache. For example, the Microblaze processor uses this technique. A second approach is to use specific cache-bypass instructions on the microprocessor. Such instructions are memory-access operations on the same address as normal load/store instructions, but they do not relay on the processor data cache. The Nios-II processor, for example, has `ldio` and `stio` instructions that serve as cache-bypass counterparts to the normal load (`ld`) and store (`st`) instructions.

Building on the principle of a memory-mapped register, we can create communication structures to tie hardware and software together.

## 11.1.2   Mailboxes

A mailbox is a simple extension of a memory-mapped register with a handshake mechanism. The obvious problem with a memory-mapped register is that the hardware cannot tell when the software has written or read the register, and vice versa. Thus, we need the equivalent of a mailbox: a box with a status flag to signal its state. Suppose that we are sending data from software to hardware, then the software writes into the register, and next sets a 'mailbox full' flag. The hardware will read the value from the register after it sees the mailbox flag being set. After reading, the hardware will clear the 'mailbox full' flag so that the software can proceed with the next transfer.

This construct is easy to build using three memory mapped registers, as illustrated in Fig. 11.3. In this case, the sender is the software program on the left of the figure, and the receiver is the hardware module on the right of the figure. After writing fresh data into the data memory-mapped register, the req flag is raised. The hardware component is a finite state machine that scans the state of the req flag and, as soon as the flag goes high, will capture the data, and raise the ack flag in response. Meanwhile, the software program is waiting for the ack flag to go high. Once both ack and req are high, a similar sequence is followed to reset them again.

The entire protocol thus goes through four phases: req up, ack up, req down, ack down. Because both parties work through the protocol in an interlocked fashion, the communication automatically adapts to the speed of the slowest component. The protocol has two synchronization points: once just after both ack and req have transitioned high, and a second time just after both ack and req are low. This means that it is quite easy to double the throughput of the protocol in Fig. 11.3.

A mailbox based on memory-mapped registers has a high overhead, in terms of design cost as well as in terms of performance. The frequent synchronization of hardware and software through handshakes has two disadvantages. First it requires a fine-grained interaction in the execution flow of hardware and software. Keep in mind that each four-phase handshake implies two synchronization points. The second disadvantage is that frequent synchronization implies additional bus transfers. For example, consider the while statements in the C program in Fig. 11.3. Each iteration in the while loop generates one read from a volatile pointer, resulting in one bus transfer.

Both of these problems – tight coupling and extra bus transfers – can be solved by improving the buffer mechanism between hardware and software. We will discuss two examples. The first is to use *FIFO queues* instead of mailboxes to uncouple hardware and software and to remove the need for interlocked read/write of the memory-mapped interface. The second is to use *shared memory*. This can reduce the need for synchronization by increasing the granularity of the data transfer, from a single word to an entire memory block.

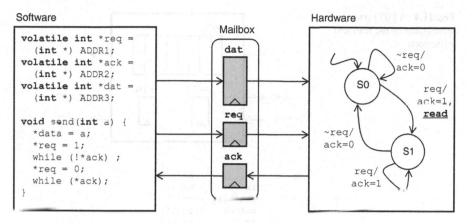

**Fig. 11.3** A mailbox register between hardware and software

### 11.1.3  First-In First-Out Queues

When a handshake protocol is used to implement a mailbox, the write and read operations are interleaved. This is inconvenient when the nature of write-operations and read-operations is very different. For example, writing into the FIFO could be bursty, so that several tokens are written in rapid succession, while reading from the FIFO could be very regular, so that tokens are read at a steady pace. The role of the FIFO is to store the extra tokens during write operations, and to gradually release them during read operations. Of course, in the long term, the average rate of writing into the FIFO must be equal to the average reading rate.

FIFO-based communications can be implemented using handshake signals too. Instead of having a single pair of request/acknowledge signals, we will now have *two* pairs. One pair controls the write operations into the FIFO, while the second pair controls the read operations into the FIFO. Figure 11.4 illustrates a FIFO queue with individual handshakes for read and write operations into the queue. In this case, we have assumed a FIFO with eight positions. The GEZEL code on the left of Fig. 11.4 shows the register-transfer level operations. In this case, the handshake operations are tied to incrementing a read-pointer and a write-pointer on a dual-port memory. The increment operations are conditional on the state of the FIFO, and the level of the request inputs. The state of the FIFO can be empty, full or non-empty, and this condition is evaluated based on comparing the read and write pointer values. Encoding the status of the FIFO using only the value of the read and write pointer values has a negative side-effect: the code shown in Fig. 11.4 requires the FIFO buffer to maintain at least one empty space at all times. By introducing a separate full flag, all spaces in the buffer can be utilized (See Problem Section).

**Fig. 11.4** A FIFO with
handshakes on the read and
write ports

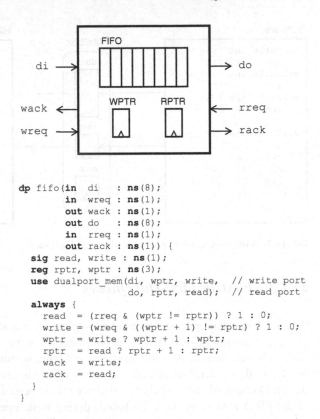

```
dp fifo(in  di   : ns(8);
         in  wreq : ns(1);
         out wack : ns(1);
         out do   : ns(8);
         in  rreq : ns(1);
         out rack : ns(1)) {
    sig read, write : ns(1);
    reg rptr, wptr : ns(3);
    use dualport_mem(di, wptr, write,   // write port
                     do, rptr, read);   // read port
    always {
      read  = (rreq & (wptr != rptr)) ? 1 : 0;
      write = (wreq & ((wptr + 1) != rptr) ? 1 : 0;
      wptr  = write ? wptr + 1 : wptr;
      rptr  = read ? rptr + 1 : rptr;
      wack  = write;
      rack  = read;
    }
}
```

## 11.1.4  Slave and Master Handshakes

The FIFO shown in Fig. 11.4 has two *slave* interfaces: one for writing and one
for reading. A slave interface waits for a control signal and reacts to it. Thus,
the acknowledge signals will be set in response to the request signals. There is a
matching master protocol required for a slave protocol. In the hardware/software
interface of Fig. 11.3, the software interface uses a master-protocol and the hardware
interface uses a slave-protocol.

By building a FIFO section with a slave input and a master output, multiple
sections of FIFO can be connected together to build a larger FIFO. An example
of this scheme is shown in Fig. 11.5. In this implementation, we use a FIFO with
a single storage location, implemented as a register. The updates of this register
are under control of a finite state machine, which uses the request/acknowledge
handshake signals as inputs. Note the direction of the request/acknowledge arrows
on the input port and the output port of the FIFO. At the input port, the request
signal is an input, while at the output port, the request signal is an output.

A master handshake interface can only be connected to a slave handshake
interface, otherwise the handshake protocol does not work. The FIFO in Fig. 11.5

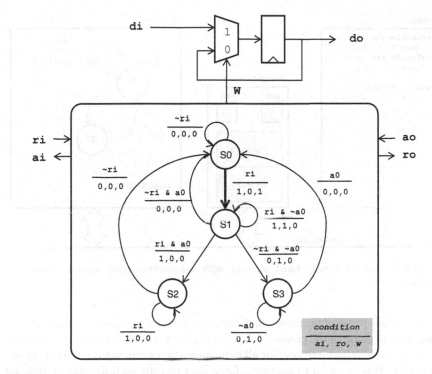

**Fig. 11.5** A one-place FIFO with a slave input handshake and a master output handshake

operates as follows. Initially, the FSM is in state S0, waiting for the input request signal to be set. Once it is set, it will write the value on the input port into the register and transition to state S1. In state S1, the FSM sets the request signal for the output port, indicating that the FIFO stage is non-empty. From state S1, three things can happen, depending upon which handshake (input or output) completes first. If the input handshake completes (ri falls low), the FSM goes to state S3. If the output handshake responds (ao raises high), the FSM goes to state S2. If both handshakes complete at the same time, the FSM directly goes back to S0.

## 11.1.5 Shared Memory

Instead of controlling the access on a single register, a single handshake can also be used to control access to a region of memory. In that case, a shared-memory scheme is obtained (Fig. 11.6).

In this example, a memory module is combined with two memory-mapped registers. The registers are used to implement a two-way handshake. The memory is split up in two different regions. The handshake protocol is used to control access to these

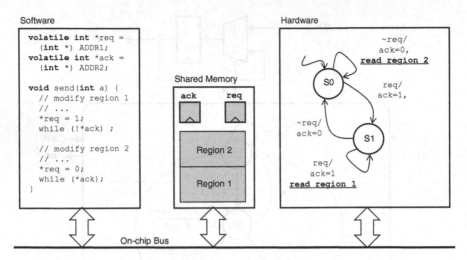

**Fig. 11.6** A double-buffered shared memory with a memory-mapped request/acknowledge handshake

regions. In one phase of the protocol, changes are allowed to region 1 of the memory. In the second phase of the protocol, changes are allowed in region 2 of the memory. This way, the protocol ensures that all data values in a given region of memory are consistent. This is useful to exchange large data records such as images, internet packets, file headers, and so on. Distributed memory organizations in System-on-Chip (Chap. 8) often make use of the shared-memory concept to implement communication channels. In addition, multiprocessor systems often make use of shared-memory structures to support inter-processor communications. A second observation is that, in some applications, clever organization of the read/write access patterns into a shared memory may lead to substantial performance improvements. This is especially true for applications that require data-reorganization in between processing stages.

## *11.1.6   GEZEL Modeling of Memory-Mapped Interfaces*

To conclude our discussion on memory-mapped interfaces. We describe the modeling of memory-mapped interfaces in GEZEL. A memory-mapped interface is represented using a dedicated simulation primitive called an `ipblock`. There is a separate primitive for a read-interface and a write-interface, and each is mapped to a user-specified memory address.

The following example shows the modeling of a memory-mapped interface for a coprocessor that evaluates the Greatest Common Divisor Algorithm. The coprocessor uses a single input and a single output, and is controlled using memory-mapped registers. Listing 11.1 illustrates the design of the coprocessor. Lines 1–28 contain five ipblock, modeling the software processor and the memory-mapped hardware/software interfaces. These interfaces are not modeled as actual registers, but as modules capable of decoding read operations and write operations to a given memory bus. The decoded addresses are given as a parameter to the ipblock. For example, the request-signal of the handshake is mapped to address 0x80000000, as shown on lines 6–10. The coprocessor kernel is shown on lines 30–43, and is a standard implementation of the greatest-common-divisor algorithm similar to the one used in earlier examples in this book. The hardware-software interface logic is embedded in the interface module, included on lines 45–101, which links the ipblock with this datapath. This module is easiest to understand by inspecting the FSM description. For each GCD computation, the hardware will go through two complete two-way handshakes. The first handshake (lines 84–89) provides the two operands to the GCD hardware. These operands are provided sequentially, over a single input port. After the first handshake, the computation starts (line 92). The second handshake (lines 93–100) is used to retrieve the result. This approach of tightly coupling the execution of the algorithm with the hardware/software interface logic has advantages and disadvantages: it results in a compact design, but it also reduces the flexibility of the interface.

**Listing 11.1**  A memory-mapped coprocessor for GCD

```
 1  ipblock my_arm {
 2    iptype "armsystem";
 3    ipparm "exec=gcddrive";
 4  }
 5
 6  ipblock m_req(out data : ns(32)) {
 7    iptype "armsystemsource";
 8    ipparm "core=my_arm";
 9    ipparm "address=0x80000000";
10  }
11
12  ipblock m_ack(in data : ns(32)) {
13    iptype "armsystemsink";
14    ipparm "core=my_arm";
15    ipparm "address=0x80000004";
16  }
17
18  ipblock m_data_out(out data : ns(32)) {
19    iptype "armsystemsource";
20    ipparm "core=my_arm";
21    ipparm "address=0x80000008";
22  }
23
24  ipblock m_data_in(in data : ns(32)) {
25    iptype "armsystemsink";
26    ipparm "core=my_arm";
27    ipparm "address=0x8000000C";
28  }
```

```
29
30   dp euclid(in  m_in, n_in : ns(32);
31            in  go         : ns( 1);
32            out ready       : ns( 1);
33            out gcd         : ns(32)) {
34     reg m, n : ns(32);
35     sig done : ns(1);
36
37     always { done  = ((m==0) | (n==0));
38              ready = done;
39              gcd   = (m > n) ? m : n;
40              m     = go ? m_in : ((m > n) ? m - n : m);
41              n     = go ? n_in : ((n >= m) ? n - m : n);
42            }
43   }
44
45   dp tb_euclid {
46     sig m, n  : ns(32);
47     sig ready : ns(1);
48     sig go    : ns(1);
49     sig gcd   : ns(32);
50     use euclid(m, n, go, ready, gcd);
51     use my_arm;
52     sig req, ack, data_out, data_in : ns(32);
53
54     use m_req(req);
55
56     use m_ack(ack);
57     use m_data_out(data_out);
58     use m_data_in(data_in);
59
60     reg r_req   : ns(1);
61     reg r_done  : ns(1);
62     reg r_m, r_n : ns(32);
63
64     always  { r_req  = req;
65              r_done = ready;
66              data_in = gcd;
67              m      = r_m;
68              n      = r_n;
69            }
70     sfg ack1 { ack = 1;          }
71     sfg ack0 { ack = 0;          }
72     sfg getm { r_m = data_out; }
73     sfg getn { r_n = data_out; }
74     sfg start{ go  = 1;          }
75     sfg wait { go  = 0;          }
76   }
77
78   fsm ctl_tb_euclid(tb_euclid) {
79     initial s0;
80     state s1, s2, s3, s4, s5, s6;
81
```

```
82    @s0 (ack0, wait) -> s1;
83
84    // read m
85    @s1 if (r_req)   then (getm, ack1, wait) -> s2;
86                     else (ack0, wait)       -> s1;
87    // read n
88    @s2 if (~r_req)  then (getn, ack0, wait) -> s3;
89                     else (ack1, wait)       -> s2;
90
91    // compute
92    @s3 (start, ack0) -> s4;
93    @s4 if (r_done)  then (ack0, wait)        -> s5;
94                     else (ack0, wait)        -> s4;
95
96    // output result
97    @s5 if (r_req)   then (ack1, wait)        -> s6;
98                     else (ack0, wait)        -> s5;
99    @s6 if (~r_req)  then (ack0, wait)        -> s1;
100                    else (ack1, wait)        -> s6;
101  }
102
103  system S {
104    tb_euclid;
105  }
```

Listing 11.2 shows a C driver program that matches the coprocessor design of Listing 11.1. The program evaluates the GCD operation of the numbers 80 and 12, followed by the GCD of the numbers 80 and 13. Note the difference between a master handshake protocol, as shown in the functions sync1() and sync0(), and a slave handshake protocol, as illustrated in the FSM transitions in Listing 11.1. In a master handshake, the request signals are written first, and followed by reading and testing of the acknowledge signals. In a slave handshake, the request signals are tested first, followed by a response on the acknowledge signals.

Executing this cosimulation is easy. We first cross-compile the C program to an executable. Next, we run the cosimulator on the executable and the GEZEL program.

```
> arm-linux-gcc -static  gcddrive.c -o gcddrive
> gplatform  gcd.fdl
core my_arm
armsystem: loading executable [gcddrive]
armsystemsink: set address 2147483652
armsystemsink: set address 2147483660
gcd(80,12) = 4
gcd(80,13) = 1
Total Cycles: 11814
```

In conclusion, memory-mapped interfaces are a general-purpose mechanism to create hardware/software interfaces. The principle of a two-way handshake is applicable to many different situations. In this case, hardware and software are synchronized using simple shared storage locations. Because memory-mapped interfaces rely on a general-purpose on-chip bus, they easily become a bottleneck

**Listing 11.2**  A C driver for the GCD memory-mapped coprocessor

```
1   #include <stdio.h>
2   volatile unsigned int *req = (unsigned int *) 0x80000000;
3   volatile unsigned int *ack = (unsigned int *) 0x80000004;
4
5   void sync1() {
6       *req = 1;  while (*ack == 0) ;
7   }
8
9   void sync0() {
10      *req = 0;  while (*ack == 1) ;
11  }
12
13  int main() {
14      volatile unsigned int *di  = (unsigned int *) 0x80000008;
15      volatile unsigned int *ot  = (unsigned int *) 0x8000000C;
16
17      *di = 80;
18      sync1();
19      *di = 12;
20      sync0();
21      sync1();
22      printf("gcd(80,12) = %d\n", *ot);
23      sync0();
24
25      *di = 80;
26      sync1();
27      *di = 13;
28      sync0();
29      sync1();
30      printf("gcd(80,13) = %d\n", *ot);
31      sync0();
32
33      return 0;
34  }
```

when throughput requirements increase. In addition, because the on-chip bus is shared with other components, a memory-mapped interface will also show a variable latency. For cases that require a dedicated, tightly-controlled link, we will need a more efficient hardware/software interface, such as the one discussed next.

## 11.2   Coprocessor Interfaces

In cases where high data-throughput between the software and the custom hardware is needed, the memory-mapped interface may become inadequate. Instead, we can use a coprocessor interface, a dedicated processor interface specifically created to attach custom hardware modules. As demonstrated in Fig. 11.7, a coprocessor

**Fig. 11.7** Coprocessor
interface

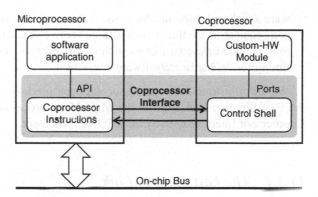

interface does not make use of the on-chip bus. Instead, it uses a dedicated port on the processor. This port is driven using a set of specialized instructions, the coprocessor instructions. The coprocessor instruction set, as well as the specific coprocessor interface, depends on the type of processor. A classic coprocessor example is a floating-point calculation unit, which is attached to an integer core. Not all processors have a coprocessor interface.

The choice of specializing a custom hardware module towards a specific coprocessor interface is an important design decision. Indeed, it locks the custom hardware module into a particular processor, and therefore it limits the reusability of that custom hardware module to systems that also use the same processor. In comparison, a custom hardware module with a memory-mapped interface is specific to a bus system, and not to a particular processor type. For example, a hardware module for a CoreConnect bus can be used in any system that uses a CoreConnect bus. Hence, the use case of the coprocessor model is more specific than the use case of the memory-mapped module model.

The main advantages of a coprocessor interface over an on-chip bus are higher throughput and fixed latency. We consider each of these aspects.

- Coprocessor interfaces have a higher throughput than memory-mapped interfaces because they are not constrained by the wordlength of the bus. For example, coprocessor ports on 32-bit CPUs may support 64- or 128-bit interfaces, allowing them to transport four words per coprocessor instruction. Hardware/software interfaces based on coprocessor instructions may also be implemented more efficiently than load/store instructions. A coprocessor instruction typically uses two source operands and one destination operand. In contrast, a load/store instruction will specify only a single operand. A complete hardware/software interaction over a coprocessor interface may thus be specified with fewer instructions as the same interaction over an on-chip bus.
- A coprocessor interface can also maintain fixed latency, so that the execution timing of software and hardware is precisely known. Indeed, a coprocessor bus is a dedicated connection between hardware and software, and it has a stable, predictable timing. This, in turn, simplifies the implementation of hardware/soft-

ware synchronization mechanisms. In contrast, an on-chip bus interface uses a communication medium which is shared between several components, and which may include unknown factors such as bus bridges. This leads to unpredictable timing for the hardware/software interaction.

We will illustrate the coprocessor interface model by means of two examples: the Fast Simplex Link used in the Microblaze processor, and the Floating Point Coprocessor Interface used by the LEON-3 processor.

### 11.2.1 The Fast Simplex Link

The Microblaze processor, a soft-core processor that can be configured in a Xilinx FPGA, is configurable with up to eight Fast Simplex Link (FSL) interfaces. An FSL link consists of an output port with a master-type handshake, and an input port with a slave-type handshake. The *simplex* part of FSL refers to the direction of data, which is either output or input. The Microblaze processor has separate instructions to write to the output port and read from the input port.

Figure 11.8 shows an FSL interface. In between the hardware coprocessor and the Microblaze, FIFO memories can be added to loosen the coupling between hardware and software. Data going from the Microblaze to the FIFO goes over a master interface consisting of the signals data, write, and full. Data going from the FIFO to the Microblaze goes over a slave interface which includes the signals data, exists, and read. Each of the master interface and slave interface implements a two-way handshake protocol. However, the labeling of handshake signals is slightly different compared to what we discussed before: write and exists correspond to req, while full and read correspond to ack.

Figure 11.9 shows the operation of the FSL interface for a FIFO with two positions. The timing diagram demonstrates the activities of writing three tokens into the FIFO, and reading them out again. The operation will be familiar because of the two-way handshake implemented in the FSL protocol. On clock edge 2, the first data item is written into the FIFO. The exists flag goes up because the FIFO is non-empty. On clock edge 3, a second data item is written into the FIFO. We assume that the FIFO holds no more than two places, so that this second write will fill the FIFO completely, and the full flag is raised as a result. In order to write more tokens into the FIFO, at least one read operation needs to complete first. This happens on clock edge 5. The third data item is written into the FIFO on clock edge 6, which completely fills the FIFO again. The FIFO is emptied by read operations on clock edges 7 and 8. From clock edge 9 on, the FIFO is empty.

The read and write operations on an FSL interface are controlled using dedicated instructions on the Microblaze processor. The basic read and write operations are of the form:

```
put   rD, FLSx   // copy register rD to FSL interface FSLx
get   rD, FSLx   // copy FSL interface FSLx to register rD
```

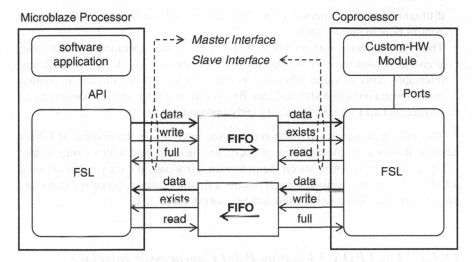

**Fig. 11.8** The fast simplex link interface

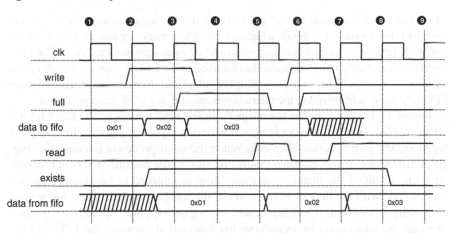

**Fig. 11.9** The FSL handshake protocol

There are many variants of these instructions, and we only discuss the main features.

- The instruction can be configured as a blocking as well as a non-blocking operation. When a non-blocking instruction is unable to complete a read operation or a write operation, it will reset the carry flag in the processor. This way, a conditional jump instruction can be used to distinguish a successful transfer from a failed one.
- The FSL I/O instructions can also read a control status flag directly from the hardware interface: The *data* bus shown in Fig. 11.8 includes a 32-bit data word and a single-bit control flag. An exception can be raised if the control bit if

different from the expected value. This allows the hardware to influence the control flow in the software.

- The FSL I/O instructions can be configured as *atomic* operations. In that case, a group of consecutive FSL instructions will run as a single set, without any interrupts. This is useful when the interface to a hardware module is created using several parallel FSL interfaces. By disallowing interrupts, the designer can be sure that all FSL interfaces are jointly updated.

The FSL interface is a popular coprocessor interface in the context of FPGA designs. It uses a simple hardware protocol, and is supported with a configurable, yet specialized instruction set on the processor. However, it's only a data-moving interface. In the next section, we will discuss a floating-point coprocessor interface. Such an interface has a richer execution semantics.

## 11.2.2   The LEON-3 Floating Point Coprocessor Interface

The interface in this section is a tightly-coupled interface to attach a floating-point unit (FPU) to a processor. While a high-end desktop processor has the FPU built-in, embedded processors often configure this module as an optional extension. The FPU interface discussed in this section is the one used in the LEON-3 processor, designed by Gaisler Research. It is a tightly-coupled interface: instructions executed on the FPU remain synchronized to the instructions executing on the main processor.

Figure 11.10 illustrates the main signals in the FPU interface. The LEON-3 32-bit microprocessor includes an integer-instruction pipeline, a set of floating-point registers, and an instruction-fetch unit. When the microprocessor fetches a floating-point instruction, it will dispatch that instruction to the floating-point coprocessor. After the result of the floating point operation is returned to the microprocessor, it is merged with the flow of instructions in the integer pipeline. There are several interesting issues with this scheme, and the signals on the coprocessor interface can best be understood by examining the interaction between the FPU and the microprocessor in detail.

The FPU contains two different datapaths. One is a *linear* pipeline with three pipeline stages. The second is a *non-linear* pipeline, and it consists of a pipeline with feedback, so that one pipeline stage remains in use for several cycles. FPU operations such as add, subtract, and multiply are handled by the linear pipeline, while operations such as divide and square-root are handled by the non-linear pipeline. FPU instructions through the linear pipeline have a latency of three clock cycles and a throughput of one instruction per clock cycle. However, FPU instructions through the non-linear pipeline can have a latency of up to 24 clock cycles, and their throughput can be as low as one instruction every 24 clock cycles.

The challenge of the coprocessor interface is to maintain the instruction sequence in the FPU synchronized with the microprocessor. This is non-trivial because the latency and throughput of instructions in the FPU is irregular. Indeed, results must

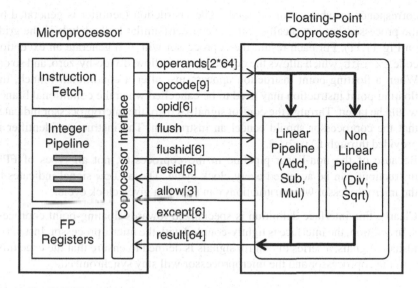

**Fig. 11.10** GRFPU floating point coprocessor interface

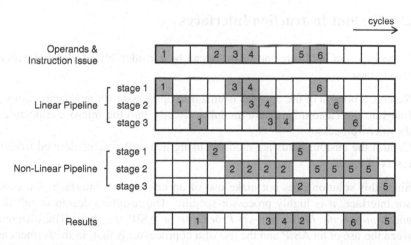

**Fig. 11.11** GRFPU instruction issue and result return

be merged in the microprocessor in the same order as their operands are dispatched to the FPU. Figure 11.11 demonstrates that, due to the complex pipeline architecture of the FPU however, results may arrive out-of-order. The GRFPU interface handles this problem.

- Each operation for the FPU is provided as a set of operands, with a given opcode, and an instruction identifier opid. When the FPU operation finishes, it returns the result, together with a corresponding instruction identifier resid. By examining the value of resid, the processor can determine what result

corresponds to what set of operands. The instruction identifier is generated by
the processor, but is typically a simple counter (similar to the labels in the grids
on Fig. 11.11). For each result, the coprocessor will also generate an exception
code except, which allows the detection of overflow, divide-by-zero, and so on.

• When a floating-point instruction appears just after a conditional branch, the
floating-point instruction may need to be cancelled when the conditional branch
would be taken. Through the control signal flush, the microprocessor indicates
that the coprocessor should cancel an instruction. The instruction identifier is
provided through flushid.

• Because of the non-linear pipeline in the coprocessor, not all types of FPU
instructions can be accepted every clock cycle. The allow signal indicates to
the microprocessor what instructions can start at a given clock cycle.

Clearly, this interface definition is specialized towards floating-point coprocessors. In addition, the interface is tightly-coupled with the micro-processor. Instead of
handshake signals, a series of control signals is defined to ensure that the execution
flow of the coprocessor and the microprocessor will stay synchronized.

## 11.3   Custom-Instruction Interfaces

The integration of hardware and software can be considerably accelerated with the
following idea.

1. Reserve a portion of the opcodes from a microprocessor for new instructions
2. Integrate the custom-hardware modules directly into the micro-architecture of
the micro-processor
3. Control the custom-hardware modules using new instructions derived from the
reserved opcodes.

Since this solution does not make use of an on-chip bus interface or a coprocessor interface, it is highly processor-specific. The resulting design is called an
*Application-Specific Instruction-set Processor* or ASIP for short. The difference
between the use of an ASIP and the use of a coprocessor is that, in the former case,
the design of the instruction set is defined by the application, while in the latter case,
the coprocessor instruction set is defined as part of the micro-processor.

Design with ASIP is experienced by designers as an easier form of hardware-
software codesign, because it automates some of the more difficult aspects of
hardware/software codesign. First, the instruction-fetch and dispatch mechanism in
the micro-processor ensures that custom-hardware and software remain synchronized. Second, design of an ASIP proceeds in an incremental fashion. A solution
can be made with one custom instruction, two custom instructions, or more. The
incremental nature of the design avoids drastic changes to the system architecture.
In contrast, the traditional hardware design process is bottom-up, exact and rigorous.

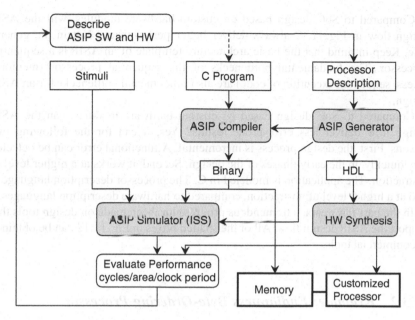

**Fig. 11.12** ASIP design flow

## 11.3.1 ASIP Design Flow

When creating an ASIP, a designer captures the application initially as a C program. After the performance of this program on the processor is evaluated, adjustments to the program and the processor architecture can be made. Such adjustments include, for example, new hardware in the processor datapath, and new processor instructions. After the processor hardware is adjusted, the C program can be tuned as well to make use of these instructions. This leads to a design flow as illustrated in Fig. 11.12.

The design starts with a C program and a description of the processor. The *processor description* is not a hardware description in terms of FSMD, but a specialized description of processor resources. It enumerates the instructions supported by the processor, the configuration and size of register files, and the memory architecture. Using the processor description, an *ASIP generator* will create design components for the ASIP. This includes a software development toolkit (compiler, assembler, linker, debugger) as well as a synthesizable hardware description of the processor. Using the software development toolkit, the application program in C can be compiled, simulated and evaluated. The hardware description can be technology-mapped onto gates or FPGA, which yields the processor implementation as well as technology metrics such as area and achievable processor clock. Based on the performance evaluation, the processor description can be reworked to obtain a better performance on a given application. This may also require rework of the application in C.

Compared to SoC design based on custom-hardware modules, will the ASIP design flow in Fig. 11.12 always deliver better performance? Not in the general case. Keep in mind that the basic architecture template of an ASIP is a sequential processor. The fundamental bottlenecks of the sequential processor (memory-access, sequential execution of code) are also fundamental bottlenecks for an ASIP design.

Compared to SoC design based on custom-hardware modules, can the ASIP design flow deliver less error-prone results? Yes, it can for the following two reasons. First, the design process is incremental. A functional error can be detected very quickly, in the early phases of the design. Second, it works at a higher level of abstraction. The application is modeled in C. The processor description language is also at a higher level of abstraction, compared to hardware description languages.

In the past few years, a tremendous progress has been made on design tools that support the ASIP design flow. All of the shaded boxes in Fig. 11.12 can be obtained as commercial tools.

### 11.3.2   Example: Endianness Byte-Ordering Processor

We describe an example of ASIP design, as well as GEZEL modeling of the custom-instruction. The application is an endianness byte-ordering processor.

Figure 11.13 illustrates the design problem. Processors have a chosen byte-ordering or endianness. When a processor transmits data words over a network, the byte order of the transmitted packets is converted from the processor byte-order into network byte-order, which is big-endian in practice. For example, when a processor is little endian, it will store the word 0x12345678 with the lowest significant byte in the lowest memory address. However, when that word is transmitted over a network, the packet byte-order must follow a big-endian convention that will send 0x78 first and 0x12 last. The communication protocol stack on the processor will therefore convert each word from little-endian format to big-endian format before handing it off to the network buffer on the Ethernet card.

For a 32-bit processor, endianness conversion involves byte-level manipulation using shifting and masking. The following is an example of a function that converts little-endian to big-endian (or vice versa) in C.

```
for (i=0; i<4096; i++)
  ot[i] = ( ((in[i] & 0x000000ff) << 24) |
            ((in[i] & 0x0000ff00) <<  8) |
            ((in[i] & 0x00ff0000) >>  8) |
            ((in[i] & 0xff000000) >> 24));
```

On a StrongARM processor (which is little-endian), this loop requires 13 cycles per iteration (assuming no cache misses). Examining the assembly, we would find that there are 11 instructions inside of the loop, leaving two cycles of pipeline stalls per iteration – one for the branch and one data-dependency (ldr instruction).

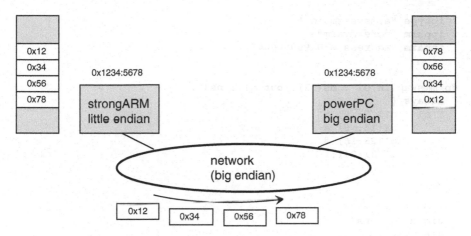

**Fig. 11.13** Endianness byte-ordering problem

```
.L6:
    ldr r3, [r4, ip, asl #2]     ; read in[i]
    and r2, r3, #65280           ; evaluate ot[i]
    mov r2, r2, asl #8           ;   ...
    orr r2, r2, r3, asl #24      ;   ...
    and r1, r3, #16711680        ;   ...
    orr r2, r2, r1, lsr #8       ;   ...
    orr r2, r2, r3, lsr #24      ;   ...
    str r2, [r0, ip, asl #2]     ;   and write back
    add ip, ip, #1               ; next iteration
    cmp ip, lr                   ;
    ble .L6
```

Let's consider possible design improvements. In hardware, the endianness-conversion is obviously very simple: it is a simple wiring pattern. The hardware-software codesign problem, therefore, essentially revolves around the problem of moving data around. Before designing an ASIP, let's try to solve this problem with a memory-mapped coprocessor.

The memory-mapped coprocessor looks as follows. It uses two memory-mapped registers, one for input and one for output.

```
ipblock myarm {
  iptype "armsystem";
  ipparm "exec=byteswap";
}

ipblock port1(out data     : ns(32)) {
  iptype "armsystemsource";
  ipparm "core=myarm";
  ipparm "address = 0x80000000";
}

ipblock port2(in data      : ns(32)) {
```

```
  iptype "armsystemsink";
  ipparm "core=myarm";
  ipparm "address = 0x80000004";
}

dp mmswap(in d1 : ns(32); out q1 : ns(32)) {
  always {
    q1 = d1[ 7: 0] #
         d1[15: 8] #
         d1[23:16] #
         d1[31:24];
  }
}

dp top {
  sig d, q : ns(32);
  use myarm;
  use port1(d);
  use port2(q);
  use mmswap(d, q);
}

system S {
  top;
}
```

The C driver program for this memory-mapped coprocessor looks as follows.

```
volatile unsigned int * mmin = (unsigned int *) 0x80000000;
volatile unsigned int * mmot = (unsigned int *) 0x80000004;
for (i=0; i<4096; i++) {
  *mmin = in[i];
  ot[i] = *mmot;
}
```

The equivalent StrongARM assembly program for this loop looks as follows.

```
.L21:
    ldr r3, [r0, ip, asl #2]    ; load in[i]
    str r3, [r4, #0]            ; send to coprocessor
    ldr r2, [r5, #0]            ; read from coprocessor
    str r2, [r1, ip, asl #2]    ; write ot[i]
    add ip, ip, #1              ; next iteration
    cmp ip, lr
    ble .L21
```

The execution time of the loop body now takes ten clock cycles per iteration (seven instructions, one stall for branching, two stall for data-dependencies). This is a gain of three cycles. We assume that we have single-cycle access to the memory-mapped coprocessor, which is hard to achieve in practice. Hence, the gain of three cycles will probably be lost in a real design. The problem with this design is apparent from the assembly program: each data element travels *four* times over the memory bus in order to be converted.

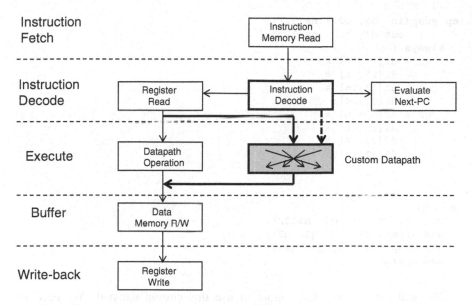

**Fig. 11.14** Single-argument custom-datapath in an ASIP

Using an ASIP, this wasteful copying over the memory bus can be avoided: we can retrieve in[i] once, convert it inside of the processor, and write back the converted result to ot[i]. In order to do this as part of an instruction, we need to modify (extend) the datapath of the processor with a new operation. Figure 11.14 illustrates how this works: the execution stage of the pipeline is extended with a new datapath (endianness conversion), and a new instruction is integrated into the instruction decoder.

GEZEL supports experiments with custom datapaths in a StrongARM processor, by using several unused opcodes of that processor. In particular, GEZEL supports 2-by-2 and 3-by-1 custom datapaths in a StrongARM. A 2-by-2 custom datapath reads two register operands and produces two register operands. A 3-by-1 custom datapath reads three register operands and produces a single register operand. The following GEZEL program shows a 2-by-2 custom datapath for endianness conversion. An ipblock of type armsfu2x2 is used to represent a custom instruction.

```
ipblock myarm {
  iptype "armsystem";
  ipparm "exec=nettohost_sfu";
}

ipblock armsfu(out d1, d2 : ns(32);
               in  q1, q2 : ns(32)) {
  iptype "armsfu2x2";
  ipparm "core = myarm";
  ipparm "device=0";
}
```

```
dp swap(in   d1, d2 : ns(32);
        out  q1, q2 : ns(32)) {
  always {
    q1 = d1[ 7: 0] #
         d1[15: 8] #
         d1[23:16] #
         d1[31:24];
    q2 = d2[ 7: 0] #
         d2[15: 8] #
         d2[23:16] #
         d2[31:24];
  }
}

dp top {
  sig d1, d2, q1, q2: ns(32);
  use armsfu(d1, d2, q1, q2);
  use swap (d1, d2, q1, q2);
  use myarm;
}
```

We will now write a C program to use this custom datapath. We need to make use of a custom opcode to trigger execution of the custom datapath. The GEZEL armsfu interface relies on the smullnv opcode, which is unused by the C compiler (arm-linux-gcc 3.2). Since smullnv cannot be written directly in C, its opcode is inserted in a regular program by making use of inline assembly. The following C snippet shows how to define an inline assembly macro, and how to call this instruction for a single-argument and a double-argument instruction.

```
#define OP2x2_1(D1, D2, S1, S2) \
  asm volatile ("smullnv %0, %1, %2, %3" : \
                "=&r"(D1), "=&r"(D2) : \
                "r"(S1), "r"(S2));

// use as a single-argument instruction
for (i=0; i<4096; i++) {
  OP2x2_1(ot[i], dummy1, in[i], dummy2);
}

// use as a dual-argument instruction
for (i=0; i<2048; i++) {
  OP2x2_1(ot[2*i], ot[2*i+1], in[2*i], in[2*i+1]);
}
```

The resulting assembly for the single-argument case now looks as follows. The loop requires eight cycles, used by six instructions and two stalls. This is a gain of two clock cycles compared to the previous case. Equally important is that the program now only needs half the bus transfers, since the coprocessor is integrated *inside* of the StrongARM.

**Table 11.1** Performance summary for the endianness byte-ordering processor

| Type | Software | Memory-mapped | ASIP (1x1) | ASIP (2x2) |
|------|----------|---------------|------------|------------|
| Instructions | 11 | 7 | 6 | 7 |
| Pipeline stalls | 2 | 3 | 2 | 2 |
| Cycles/iteration | 13 | 10 | 8 | 9 |
| Cycles/word | 13 | 10 | 8 | 4.5 |

```
.L38:
    ldr     r3, [r4, lr, asl #2]    ; load in[i]
    smullnv r2, r7, r3, ip          ; perform conversion
    str     r2, [r1, lr, asl #2]    ; write ot[i]
    add     lr, lr, #1              ; next iteration
    cmp     lr, r0
    ble     .L38
```

The dual-argument design is even more efficient, because the loop management code is now shared over two endianness conversions. We have nine cycles per loop iteration: seven instructions and two stalls. However, each iteration of the loop performs two conversions, so that the effective cycle cost is 4.5 cycles per endianness conversion (compared to eight cycles in the previous case).

```
.L53:
    ldr     r1, [r5], #4            ; read in[2*i], adjust pointer
    ldr     r2, [r5], #4            ; read in[2*i+1], adjust pointer
    smullnv r0, ip, r1, r2          ; perform conversion
    str     r0, [r4], #4            ; store ot[2*i], adjust pointer
    subs    lr, lr, #1             ; next iteration
    str     ip, [r4], #4            ; store ot[2*i+1], adjust pointer
    bne     .L53
```

Summing up, by converting an all-software design to an ASIP-type design, the cycle cost for endianness conversion on a StrongARM reduces from 13 cycles per word to 4.5 cycles per word, an improvement of 2.9 times. Table 11.1 summarizes the performance of different configurations.

What is the limiting factor of the final design (ASIP 2x2)? Based on the code listing, 4 of the 7 instructions are load and store instructions. In other words, the majority of the execution time is spent in moving data between memory and the processor. This illustrates a point we made earlier: the strength of an ASIP design is also its weakness. An instruction-set architecture is convenient to build extensions, but bottlenecks in the instruction-set architecture will also be bottlenecks in the resulting hardware/software codesign.

## 11.3.3  Example: The Nios-II Custom-Instruction Interface

The Nios-II processor has an custom-instruction interface to attach hardware modules. Following our earlier definition of ASIP vs. coprocessor interface, we

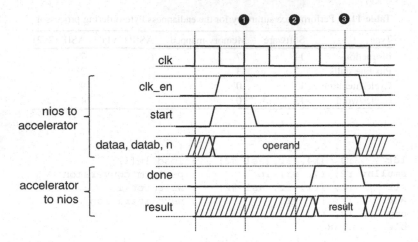

**Fig. 11.15**  Nios-II custom-instruction interface timing

should call this interface a coprocessor interface, since the Nios-II has predefined instructions to move operands through it. However, the look and feel of the interface closely resembles that of a custom-instruction in an ASIP. For this reason, the interface is introduced as a custom-instruction interface.

Figure 11.15 illustrates the timing of the interface. A custom instruction has three input operands: two 32-bit inputs `dataa` and `datab`, and an 8-bit input `n`. The result of the custom instruction is captured in a 32-bit output `result`. The interface supports variable-length execution of custom instructions through a two-way handshake mechanism. As shown in Fig. 11.15, the execution of the custom instruction starts with the Nios processor sending arguments to the custom hardware module at clock edge 1. The custom hardware uses a `done` output pin to indicate completion of the custom instruction at clock edge 3. The `clk_en` input is used to mask off the clock to the custom hardware when the instruction is not active.

In software, the custom-instruction is executed through a dedicated instruction, `custom`. For example, `custom 0x5, r2, r3, r5` would assign the value `0x5` to the `n` input, and associate the ports `dataa`, `datab`, and `result` with registers `r2`, `r3` and `r5` respectively. Figure 11.16a shows the result.

The `n` input is part of the instruction opcode, and is not a value in a register file. The purpose of the `n` input is to multi-plex different custom instruction implementations within the custom hardware module.

The Nios custom instruction interface also supports the use of local register files in the custom hardware module. This can be used to store local parameters for the hardware, and it can reduce the need for processor registers. Figure 11.16b illustrates this case, and shows how the custom-instruction interface needs to be extended. Only the case for the first input operand (a) is shown. An additional control signal, `reada`, selects if the custom hardware needs to accept data from the Nios register file or else from the local register file. In the former case, the input `dataa` contains the value of the processor register to be used as operand. In the latter case, the input

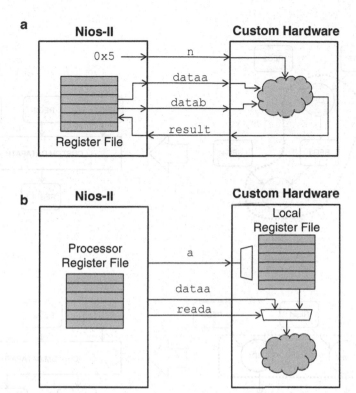

**Fig. 11.16** Nios-II custom-instruction integration. (**a**) With processor register file. (**b**) With local register file

a indicates the address of the register field in the local register file to use as operand. To the software programmer, selection between these two cases is seamless, and the custom-instruction mnemonic integrates both. Registers prefixed with r are located in the processor, while registers prefixed with c are located in the custom hardware. Thus, an opcode such as custom 0x5, c2, c3, r5 would take operand a and b from the local register file, and store the result in the processor register file.

## 11.3.4  Finding Good ASIP Instructions

How do we identify good ASIP instructions? Most likely, the application domain itself will suggest what type of primitives are most frequently needed. For example, image processing often works with 8-bit pixels. Hence, instructions that support efficient 8-bit operations will be useful for an image-processing ASIP. Another example, coding and cryptography make use of modular arithmetic. Hence, in an ASIP for coding algorithms, it makes sense to have support for modular addition and multiplication.

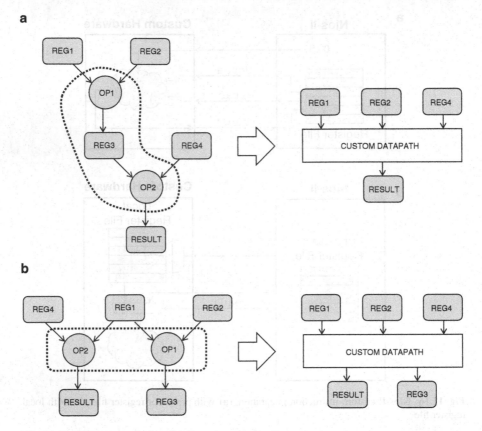

**Fig. 11.17** (**a**) Operator fusion. (**b**) Operator compounding

The study of automatic instruction definition is a research field on its own, and is of particular interest to compiler developers. We will describe two basic techniques that work directly at the level of assembly language, and that are not specific to a given application domain. The first technique is called *operator fusion*, and the second technique is called *operator compounding*.

In operator fusion, we define custom instructions as a combination of dependent operations. The dependencies are found by means of data flow analysis of the code. After data flow analysis, we can cluster assembly operations together. When we group two operations together, the intermediate register storage required for the individual operations disappears. Figure 11.17a illustrates operator fusion. There are obviously a few limitations to the clustering process.

- All operations that are fused are jointly executed. Hence, they must be at the same loop hierarchy level, and they must be within the same branch of an if-then-else statement. Note that it is still possible to integrate an entire if-then-else statement as a custom instruction; see Problem 11.5.

- The number of input and output operands must be limited to ensure that the register-file bandwidth stays within bounds. Indeed, as the new custom instruction executes, it will require all the input arguments to be available at the same clock cycle, and it will produce all output arguments at the same clock cycle.
- The length of the chained operations must not be too long, since this adversely affects the critical path of the processor.

Figure 11.17b shows an example of operator compounding. In this case, we are combining multiple possibly unrelated operations together, especially when they share common inputs. The operator compounds are identified using data-flow analysis of the assembly code, and they have similar limitations as fused operators.

We'll now consider the endianness-conversion example once more and consider how much more it can be optimized beyond a single OP2x2 ASIP instruction for endianness conversion. In this case, we consider a complete C function as follows. Notice how this code has been optimized with incremental pointer arithmetic.

```
void byteswap(unsigned *in, unsigned *ot) {
  int i;
  int d1, d2, d3, d4;
  unsigned *q1 = in;
  unsigned *q2 = ot;
  for (i=0; i<2048; i++) {
    d1 = *(q1++);
    d2 = *(q1++);
    OP2x2_1(d3, d4, d1, d2);
    *(q2++) = d3;
    *(q2++) = d4;
  }
}
```

The assembly code for this function looks as follows. The loop body contains nine instructions. Only a single instruction performs the actual byte swap operation! Four other instructions are related to moving data into and out of the processor (ldr, str), two instructions do loop counter management (cmp, add), one instruction implements a conditional return (ldmgtfd) and one instruction is a branch. There will be two pipeline stalls: one for the branch, and one for the second memory-load (ldr).

```
byteswap:
      stmfd   sp!, {r4, r5, lr}     ; preserve registers
      mov     ip, #0                ; init loop counter
      mov     lr, #2032             ; loop counter limit
      add     lr, lr, #15
.L76:
      ldr     r2, [r0], #4          ; load in[i]
      ldr     r3, [r0], #4          ; load in[i+1]
      smullnv r4, r5, r2, r3        ; endianness conversion
      str     r4, [r1], #4          ; store ot[i]
      str     r5, [r1], #4          ; store ot[i+1]
      add     ip, ip, #1            ; increment loop counter
```

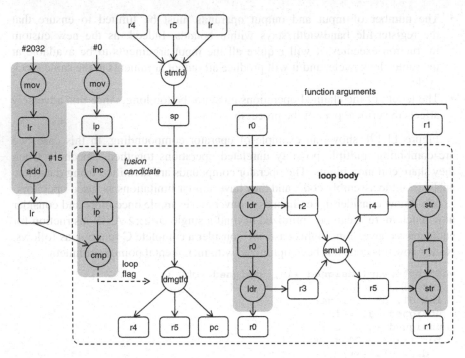

**Fig. 11.18**  Analysis of the `byteswap` function for ASIP instructions

```
cmp        ip, lr                    ; test
ldmgtfd    sp!, {r4, r5, pc}         ; and conditionally return
b          .L76
```

To optimize this assembly code with additional ASIP instructions, we can construct a data-flow diagram for the assembly code, and investigate the opportunities for operation fusion and compounding. Figure 11.18 shows the dataflow analysis diagram. The boxes represent registers, while the circles represent operations. A distinction is made between operations inside of the loop and those outside of it; recall that fusion and compounding only work within a single loop level. The shaded clusters are examples of possible operation fusion.

- A set of fused operations could merge the loop counter logic into two new operations. The first operation initializes two registers. The second operation increments one register, compares it to the next, and sets a loop flag as a result of it. We will call these new operations `initloop` and `incloop` respectively.
- The second set of fused operations is trickier: they involve memory access (`str` and `ldr`), so they cannot be implemented with modifications to the execution stage of the RISC pipeline alone. In addition, since the StrongARM has only a single memory port, fusing these operations into a single operation will not provide performance improvement if there is not a similar modification to the memory architecture as well. Thus, we cannot define new instructions for these

fusion candidates. However, we may still reduce the footprint of the loop body by collapsing the store and load instructions in store-multiple and load-multiple instructions.

The resulting assembly code after these transformations would therefore look as follows. The loop body contains six instructions, but will have three stalls (one for ldm, one for the data dependency from ldm to smullnv, and one for the branch). A loop round-trip now costs 9 cycles, or 4.5 cycles per word. This is as fast as we found before, but in this version of the code, function call overhead is included.

```
byteswap:
        stmfd      sp!, {r4, r5, lr}    ; preserve registers
        initloop   ip, lr, #0, #2047    ; loop-counter instruction 1
.L76:
        ldm        r0!, {r2, r3}        ; load in[i], in[i+1]
        smullnv    r4, r5, r2, r3       ; endianness conversion
        stm        r1!, {r4, r5}        ; store ot[i], ot[i+1]
        incloop    ip, lr               ; loop counter instruction 2
        ldmgtfd    sp!, {r4, r5, pc}    ; and conditionally return
        b          .L76
```

This concludes our discussion of the third type of hardware/software interface, the custom instruction. The design of customized processors is an important research topic, and optimizations have been investigated far beyond the examples we discussed above. For the hardware/software codesigner, it's important to understand that ASIP design takes a top-down view on the codesign problem: one starts with a C program, and next investigates how to improve its performance. In a classic coprocessor design, the view is bottom-up: one starts with a given kernel in hardware, and next investigates how to integrate it efficiently into a system. Both approaches are viable, and in both cases, a codesigner has to think about interfacing hardware and software efficiently.

## 11.4 Summary

Microprocessor interfaces are at the core of hardware/software codesign. We made a distinction between three classes of interfaces, each with a different integration within the System on Chip architecture.

A memory-mapped interface reuses the addressing mechanism of an on-chip bus to reserve a few slots in the address space for hardware/software communication. Single memory locations can be implemented using memory-mapped registers. A range of memory locations can be implemented using a shared-memory. Specialized communication mechanisms, such as FIFO queues, further improve the characteristics of the hardware/software communication channel.

The coprocessor interface is a second type of hardware/software interface. It requires a dedicated port on a microprocessor, as well as a few predefined instructions to move data through that port. We discussed two examples of this interface,

including the Fast Simplex Link interface, and a floating-point coprocessor interface
for the LEON-3 processor.

The final hardware/software interface is the custom-instruction, created by
modifying the instruction-set architecture of a micro-processor. This interface
requires a rather detailed understanding of micro-processor architecture, and is often
supported with a dedicated toolchain.

As a hardware/software codesigner, it is useful to spend time thinking about
the breadth of this design space, which is enormous. Probably the most important
point is to realize that there is no single silver bullet to capture all the variations
of interconnections for hardware and software. There are many trade-offs to
make for each variation, and different solutions can tackle different bottlenecks:
computational power, data bandwidth, power consumption, design cost, and so on.

Also, keep in mind that no system is free of bottlenecks: the objective of
hardware/software codesign is not to remove bottlenecks, but rather to locate and
understand them. In the next chapter, we will focus on the hardware-side of custom-
hardware modules, and describe how hardware can be controlled from within
software through a hardware/software interface.


## 11.5  Further Reading

Early research in hardware/software codesign suggested that much of the hardware/-
software interface problem can be automated. Chapter 4 in Micheli et al. (2001)
describes some of the work in this area. To date however, no standards for creating
hardware/software interfaces exist, and the design of such interfaces largely remains
an ad-hoc process.

Yet, design support is critical to ensure error-free design. Designers often build
virtual platforms of a chip during the implementation. A virtual platform is a
complete simulation of the entire chip, emphasizing a detailed representation of
the hardware/software interactions.

Memory-mapped interfaces are ubiquitous in the design of System-on-Chip
architectures. One can consult the datasheet of a typical micro-controller and
find that all peripherals are programmed or configured using memory-mapped
input/output. For example, check the datasheet of Atmel AT91SAM7L128, an
ARM-based microcontroller with numerous on-chip peripherals (Atmel 2008).

In contrast to general hardware/software interfaces, the literature on custom
processor design is rich and detailed. In fact, the ASIP approach is one of the most
successful models for hardware/software codesign when practical implementations
are considered. A comprehensive treatment of the ASIP design process is provided
by Rowen in (2004). Leupers and Ienne discuss customized processor architectures
in Leupers and Ienne (2006). There are numerous publications on design applications based on ASIP, and as many conferences that cover them (a major event is
Embedded Systems Week, grouping three conferences together on compiler design,
on system design, and on embedded software implementation).

## 11.6 Problems

**Problem 11.1.** Build a memory-mapped register for the address bus described in Table 10.2. Evaluate the complexity of two different designs. What is the general recommendation for the design of memory-mapped address decoders you can make?

(a) A decoder that maps the register to any address of the range 0x3F000000 - 0x3F00FFFF.
(b) A decoder that maps the register to the single address 0x3F000000.

**Problem 11.2.** Modify the design of Fig. 11.3 so two integers can be transferred for each full request/acknowledge cycle. Thus, show that you can transfer twice as much data without changing the number of phases in the two-way handshake protocol.

**Problem 11.3.** Modify the design of Fig. 11.4 so that all positions of the FIFO are used before the full flag is set high.

**Problem 11.4.** Design a GEZEL implementation for a loopback interface on an Fast Simplex Link. Figure 11.19 illustrates the problem. The module has a single register, loopreg, that can hold the output of a single putfsl instruction. The module has a master interface for writing into the register, and a slave interface for reading from the register. Design the GEZEL program and a suitable testbench (in GEZEL).

**Problem 11.5.** Consider the C program in Listing 11.3, and the corresponding ARM assembly code in Listing 11.4.

(a) Study the following C program and the corresponding assembly code out of it.
(b) Define a custom instruction max rx, ry, which compares ry to the current value of rx and replaces that value if ry is bigger than rx. Assume that rx and ry are unsigned, 32-bit values.

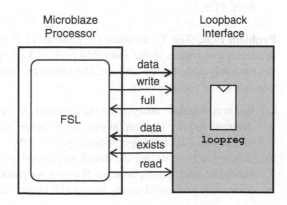

Fig. 11.19 The fast simplex link interface loopback interface for Problem 11.4

**Listing 11.3** C program for Problem 11.5

```
int findmax(unsigned int data[1024]) {
  unsigned int max = 0;
  int i;
  for (i=0; i<1024; i++)
    if (max < data[i])
      max = data[i];
  return max;
}
```

**Listing 11.4** Assembly program for Problem 11.5

```
findmax:
        mov       r2, #0
        mov       r1, #1020
        mov       ip, r2
        add       r1, r1, #3
.L7:
        ldr       r3, [r0, r2, asl #2]
        cmp       ip, r3
        strcc     ip, [r0, r2, asl #2]
        add       r2, r2, #1
        cmp       r2, r1
        movgt     r0, #0
        movgt     pc, lr
        b         .L7
```

**Listing 11.5** C program for Problem 11.6

```
int absmax(int v, int w) {
  return (v * 6 + w * 4);
}
```

(c) Design a GEZEL datapath for this custom instruction, following the example in Sect. 11.3.2.

**Problem 11.6.** The C function of Listing 11.5 was compiled to a Microblaze processor and results in the assembly code of Listing 11.6. The input arguments of the assembly code are r5 and r6; the return argument is r3; the return instruction was left out of the assembly.

(a) Perform dataflow analysis on the assembly code and draw the data dependency diagram below. Use rectangles to indicate registers and circles to indicate operations. Label the operations 'op1' to 'op6'.
(b) When you have drawn the dataflow dependency diagram, define several ASIP candidate instructions (at least 2) using operation fusion. Indicate clearly which operations you could merge into ASIP instructions.

**Listing 11.6** Assembly program for Problem 11.6

```
// input arg: r5, r6
// return arg: r3
addk      r3,r5,r5      // op1
addk      r3,r3,r5      // op2
addk      r3,r3,r3      // op3
addk      r6,r6,r6      // op4
addk      r6,r6,r6      // op5
addk      r3,r3,r6      // op6
```

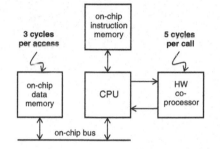

```
int a[1000], r[1000];
void transform() {
    int i;
    int *ra = a;
    int *rr = r;
    for (i=0; i<1000; i++)
        *rr++ = call_coproc(*ra++);
}
```

**Fig. 11.20** Architecture and C program for Problem 11.7

**Problem 11.7.** The system-on-chip in Fig. 11.20 combines a coprocessor, a processor, and on-chip data- and instruction-memory. The processor will copy each element of an array a[] to the coprocessor, each time storing the result as an element of an array r[]. The C program that achieves this is shown on the right of Fig. 11.20. All the operations in this architecture take *zero* time to execute, apart from the following two: accessing the on-chip data memory takes three cycles, and processing a data item on the coprocessor takes five cycles.

(a) Find the resulting execution time of the C program.
(b) Show how you can rewrite the C program so that the resulting execution time becomes smaller than 8,000 cycles. As a hint, assume that you can rewrite the y = call_coproc(x); function call as two separate calls. The first call is send_coproc(x);, and it issues an argument from software to the coprocessor. The second call is y = get_coproc();, and it takes a result from the coprocessor.

**Problem 11.8.** An ASIP processor performs operations on a stream of samples. The samples appear at fixed rate. The samples are processed using an algorithm A. Which of the following optimizations will reduce the energy consumption E needed to process a single sample? The answer for each question is one of: yes, no, impossible to decide.

(a) Rewrite the algorithm A so that it requires fewer MOPS from the processor (MOPS = Million Operations per Second). Does this reduce the energy consumption E?

(b) Add a custom instruction B that will make algorithm A complete in only half the clock cycles. You can assume the power consumed by added hardware is negligible. Does this reduce the energy consumption E?

(c) Increase the clock frequency of the ASIP. Does this reduce the energy consumption E?

(d) Lower the voltage of the ASIP. Does this reduce the energy consumption E?

# Chapter 12
# Hardware Interfaces

## 12.1 The Coprocessor Hardware Interface

A hardware interface connects a custom hardware module to a coprocessor bus or an on-chip bus. The hardware interface steers the input- and output ports of the custom hardware module. This may affect many different activities of the custom hardware module, including data transfer as well as control. Figure 12.1 shows the location of the hardware interface in the overall integration of a microprocessor with a custom-hardware module.

### 12.1.1 Functions of the Coprocessor Hardware Interface

The design of the hardware interface is a classic hardware/software codesign problem, that matches the flexibility of custom-hardware design to the realities of the hardware/software interface. For example, the hardware interface typically takes are of any of the following.

- **Data Transfer**: The hardware interface handles data transfer between software and hardware. This includes read/write transactions on the on-chip bus, using a master-protocol or a slave-protocol. In other cases, the hardware interface implements handshake sequences for a coprocessor bus. A hardware interface can be optimized for high-throughput and bursty data transfers, for example with a dedicated Direct Memory Address Controller.
- **Wordlength Conversion**: The hardware interface converts data operands of the custom-hardware module, which can be arbitrary in size and number, into data structures suitable for on-chip bus communication. For example, in a 32-bit system, software can deliver no more than 32-bits of data at a time, while the custom hardware module may be using a much wider 1,024-bit input bus. In that case, the hardware interface needs to support the conversion of a single

P.R. Schaumont, *A Practical Introduction to Hardware/Software Codesign,*
DOI 10.1007/978-1-4614-3737-6_12, © Springer Science+Business Media New York 2013

**Fig. 12.1** The hardware
interface maps a
custom-hardware module to a
hardware-software interface

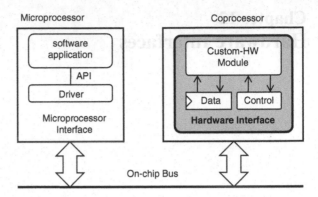

1,024-bit operand into an array of 32-bit words. Furthermore, the hardware interface takes care of the bit- and byte-level organization of the data in case conversions between software and hardware are needed.

- **Operand Storage**: The hardware interface provides local and/or temporary storage for arguments and parameters for the custom-hardware component. Besides arguments and parameters, the hardware interface can also include local buffering to support the on-chip interface. In fact, the distinction between arguments and parameters is very important. Arguments are updated *every time* the custom-hardware module executes. Parameters, on the other hand, may be updated only infrequently. Hence, to minimize the hardware-software communication bandwidth, parameters are transmitted only once, and then held in a local memory in the hardware interface.

- **Instruction Set**: The hardware interface defines the software-view of a custom-hardware component in terms of instructions and data structures. The design of instruction-sets for custom hardware modules is a particularly interesting and important problem. A carefully-designed custom instruction-set can make the difference between an efficient coprocessor, and a confusing blob of logic.

- **Local Control**: The hardware interface implements local control interactions with the custom hardware component, such as sequencing a series of micro-operations in response to a single software command.

## 12.1.2  *Layout of the Coprocessor Hardware Interface*

Figure 12.2 illustrates the layout of a generic coprocessor hardware interface, which connects a custom hardware module to an on-chip bus interface or coprocessor interface. Since the hardware interface itself is a user-defined hardware component, it can have an arbitrary architecture. The following components are commonly found in a hardware interface.

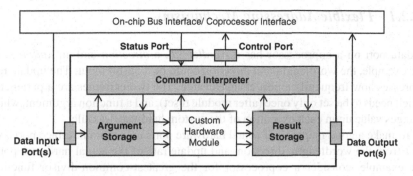

**Fig. 12.2** Layout of a coprocessor hardware interface

- A data input buffer for *Argument Storage*.
- A data output buffer for *Result Storage*.
- A *Command Interpreter* to control the data buffers and the custom hardware module based on commands from software.

The hardware interface has several *ports*, addressable inputs/outputs of the co-processor. For example, an on-chip bus interface uses an address decoder, while a coprocessor interface may have dedicated ports. From the perspective of the custom hardware module, it is common to partition the collection of ports into data input/output ports and control/status ports.

The separation of control and data is an important aspect in the design of coprocessors. Indeed, in a software program on a micro-processor, control and data are tightly connected through the instruction-set of the micro-processor. In custom-hardware however, the granularity of interaction between data and control is chosen by the designer. Observe that in Fig. 12.2, control signals and data signals are orthogonal: control flows vertically, and data flows horizontally.

Despite the relatively simple organization of the hardware interface, the design space of the data buffers and the command interpreter is rich and deep. In the following, we will discuss mechanisms to help us design the hardware interface efficiently. We will differentiate between *data-design* and *control-design*. Data-design implements data dependencies between software and hardware. Control-design implements control dependencies between software and hardware. We will discuss both of them.

## 12.2  Data Design

*Data Design* is the implementation of a mapping between the hardware interface ports and the custom hardware ports. Typically, this includes the introduction of buffers and registers, as well as the creation of an address map.

## 12.2.1   Flexible Addressing Mechanisms

A data port on a coprocessor has a *wordlength*, a *direction* and an *update rate*. For example, the wordlength and direction could be a 32-bit input. The update rate expresses how frequently a port changes value. The two extremes are a parameter, which needs to be set only once (after module reset), and a function argument, which changes value upon each execution of the custom hardware module.

To make a good mapping of actual hardware ports to custom interface ports, we start from the wordlength, direction, and update rate of the actual hardware ports. For example, consider a coprocessor for the greatest-common-divisor function, gcd. The high-level specification of this function would be:

```
int gcd(int m, int n);
```

The hardware module equivalent of gcd has two input ports m and n, which are 32-bit wide. The module also has a single output port of 32-bit. These three ports are the actual ports of the hardware module. When we implement this module as a memory-mapped coprocessor, the ports of the hardware interface will be implemented as memory-mapped registers.

A straightforward approach is to map each actual hardware port to a different memory-mapped register. This makes each port of the hardware module independently addressable from software. However, it may not always be possible to allocate an arbitrary number of memory-mapped ports in the hardware interface. In that case, we need to *multiplex* the custom-hardware module ports over the hardware interface ports.

## 12.2.2   Multiplexing and Masking

There are several occasions when the ports of the hardware module need to be multiplexed over the ports of the hardware interface.

- The hardware interface may have insufficient ports available to implement a one-to-one mapping between hardware-module ports and control-shell ports.
- Some hardware-module ports need to be programmed only once (or very infrequently), so that it is inefficient to allocate a separate hardware interface port for each of them.

Multiplexing will increase the control complexity of the hardware interface slightly. In addition, careless multiplexing will reduce the available input/output bandwidth of the hardware module. Thus, there is a risk that the module becomes communication-constrained because of the multiplexing process.

Multiplexing can be implemented in different ways. The first is *time-multiplexing* of the hardware module ports. The second to use an *index register* in the hardware interface. Figure 12.3 shows an example of a time-multiplexed port for the GCD

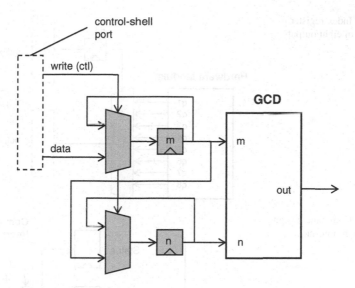

**Fig. 12.3** Time-multiplexing of two hardware-module ports over a single control-shell port

coprocessor. In this case, the arguments need to be provided by writing the value of m and n sequentially to the hardware interface port.

The index-register technique works as follows. Several ports (say $N$) on the hardware module are mapped into two ports on the hardware interface. One port on the hardware interface is a data port of sufficient width to hold any single hardware module port. The second port is an index port and has width $log_2N$ bits. The index port controls the mapping of the data port of the hardware interface to one of the ports on the hardware module. Figure 12.4 shows an example of the index-register technique to merge eight outputs to a single data output. The index register technique is more flexible than time-multiplexing, because the interface can freely choose the readout order of the hardware-module output ports. At the same time, it also requires double the amount of interface operations. Hence, multiplexing with index-registers is most useful for ports that update very infrequently, such as parameters.

Multiplexing is also useful to handle operands with very long wordlengths. For example, if the hardware module uses 128-bit operands, while the hardware-interface ports are only 32-bit, then the operands can be provided one word at-a-time by means of time-multiplexing.

Finally, here's a technique to work with very short operands, such as single-bit arguments. In this case, it is expensive to allocate a single hardware interface port for each single-bit hardware-module port. Instead, several hardware-module ports can be grouped together in a hardware interface port. To enable sharing of the hardware interface port among logically independent hardware-module port, an additional *mask register* is used. The mask register indicates which bits of a hardware interface port should be taken into account when updating the

**Fig. 12.4** Index-register to select one of eight output ports

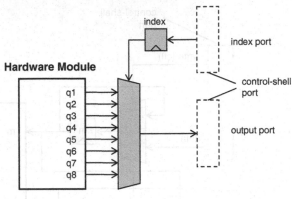

**Fig. 12.5** Command design of a hardware interface

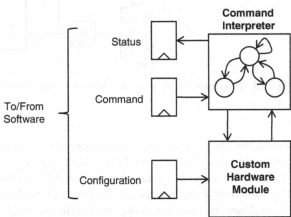

hardware-module ports. The updated value is obtained by simple bit-masking of the previous value on the hardware ports with the new value of the hardware interface port.

```
new_hardware_port = (old_hardware_port & ~mask_value) |
                    (control_shell_value & mask_value);
```

## 12.3 Control Design

Control design in a coprocessor is the collection of activities to generate control signals and to capture status signals. The result of control design is a set of *commands* or instructions that can be executed by the coprocessor. These commands are custom-tailored for the design.

Figure 12.5 shows a generic architecture to control a custom hardware module through software. It includes a *command interpreter* which accepts commands from software and which returns status information. The command interpreter is the

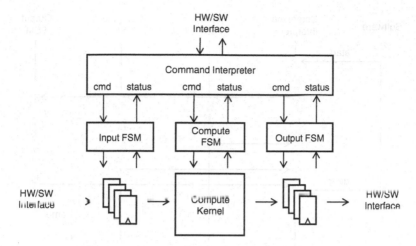

**Fig. 12.6** Hierarchical control in a coprocessor

top-level controller in the coprocessor, and it communicates directly with software. We also make a distinction between a *command* and a *configuration*. A command is a one-time control operation. A configuration is a value which will affect the execution of the coprocessor over an extended period of time, possibly over multiple commands.

The following sections discuss several architectural techniques that can be used to optimize the performance of the coprocessor.

## 12.3.1  Hierarchical Control

Figure 12.6 shows the architecture of a coprocessor that can achieve communication/computation overlap. The coprocessor has a hierarchy of controllers, which allow independent control of the input buffering, the computations, and output buffering. The command interpreter analyzes each command from software and splits it up into commands for the lower-level FSMs. In the simplest form, these commands are simple start/done handshakes. Thus, for each command of software, the command interpreter runs a combination of lower-level FSM sequences. Often, a single level of command decoding is insufficient. For example, we may want to use a coprocessor which has an addressable register set in the input or output buffer. In that case, we can embed the register address into the command coming from software. To implement these more complicated forms of subcontrol, the start/done handshakes need to be replaced with more complex command/status pairs.

A control hierarchy simplifies the design of control, as is shown in Fig. 12.7. The command interpreter can easily adapt to the individual schedules from the input FSM, compute FSM and output FSM. On the other hand, the method of

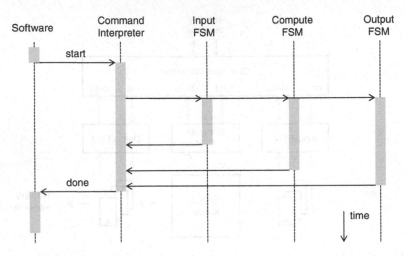

**Fig. 12.7** Execution overlap using hierarchical control

**Table 12.1** Command Set to control block-level pipelining

| Command | Input FSM | Compute FSM | Output FSM |
|---|---|---|---|
| pipe_load1 | Start | | |
| pipe_load2 | Start | Start | |
| pipe_continue | Start | Start | Start |
| pipe_flush1 | | Start | start |
| pipe_flush2 | | | Start |

using start/done pulses is inconvenient when working with pipelined submodules, since a done pulse only appears after the pipeline latency of the submodule. This will require a modification to the start/done scheme, which will be discussed later.

First, we examine how a hierarchical controller as illustrated in Fig. 12.6 can help in achieving computation/communication overlap. The basic principle, of course, is well known: we need to pipeline the input/compute/output operations within the coprocessor. Table 12.1 illustrates a set of five software commands to achieve pipelined execution within the coprocessor. This scheme is called *block-level pipelining*. The commands obtain precise pipeline startup and shutdown. The first three commands (pipe_load1, pipe_load2, pipe_continue) require the software to send an argument to the coprocessor. The last three commands (pipe_continue, pipe_flush1, pipe_flush2) require the software to retrieve a result from the coprocessor. Once the pipeline is filled up through the sequence of pipe_load1 and pipe_load2, the software can repeat the command pipe_continue as often as needed.

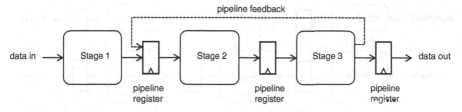

**Fig. 12.8** Pipeline terminology

## 12.3.2 Control of Internal Pipelining

When a custom hardware module has internal pipeline stages, hierarchical control becomes more intricate. There are two issues of concern in the design of a hardware interface. First, we need to find a way to generate control signals for the pipeline stages. Next, we need to define a proper mechanism to interface these control signals to the higher layers of control. Indeed, a simple start/done handshake is insufficient for a pipeline, because it does not reflect the pipeline effect. In this section we will address both of these aspects.

Figure 12.8 introduces some terminology on pipeline architectures. A pipeline consists of a number of pipeline stages separated by pipeline registers. The latency of a pipeline is the number of clock cycles it takes for an operand to move from the input of the pipeline to the output. The throughput of a pipeline measures the number of results produced per clock cycle. If a pipeline accepts a new operand each clock cycle, its throughput is one (per cycle). If, on the other hand, a pipeline accepts a new operand every $N$ cycles, its throughput is $1/N$. In a *linear pipeline* architecture, there are no feedback connections. For such a pipeline, the latency equals the number of pipeline stages, and the throughput equals 1. In a *non-linear pipeline* architecture, there are feedback connections. This happens when certain stages of a pipeline are reused more than a single time for each data token that enters the pipeline. In a non-linear pipeline, the latency can be higher than the number of pipeline stages, and the throughput can be lower than 1.

In any pipelined coprocessor, the pipeline control signals are eventually under the control of a higher-level controller. Pipeline control signals will be required in two cases.

- When a pipeline needs to perform more than a single operation, there needs to be a way to send control information into the pipeline. This control information will determine the operation of individual pipeline stages.
- In a non-linear pipeline architecture, multiplexers may be needed between the pipeline stages in order to feed operands from multiple locations within the pipeline. These multiplexers need additional control signals.

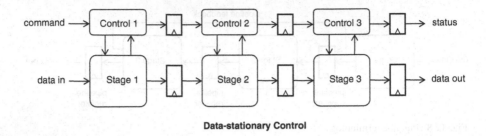

**Data-stationary Control**

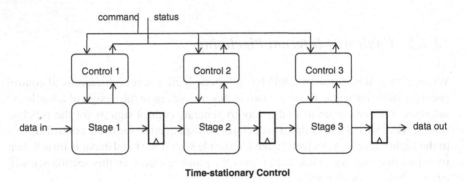

**Time-stationary Control**

**Fig. 12.9** Data-stationary and time-stationary pipeline control

### 12.3.2.1   Control of Linear Pipelines

We first discuss two techniques to control multi-function pipelines. The first is called *data-stationary* control, while the second is called *time-stationary* control. Figure 12.9 illustrates the differences between these two schemes.

- In a data-stationary scheme, control signals will travel along with the data through the pipeline. At each pipeline stage, the control word is decoded and transformed into appropriate control signals for that stage.
- In a time-stationary scheme, a single control word will control the entire pipeline for a single clock cycle. Because the pipeline contains fragments of different data items, each in a different stage of processing, a time-stationary control scheme will specify the operations to be performed on several data elements at the same time.

From a programmer's perspective, a data-stationary approach is more convenient because it hides the underlying pipeline structure in the program. A RISC instruction-set, for example, uses data-stationary encoding. On the other hand, time-stationary control makes the underlying machine structure explicitly visible in the control signals. Time-stationary control is therefore suitable for tasks that require access to the entire pipeline at once, such as exception handling. In addition,

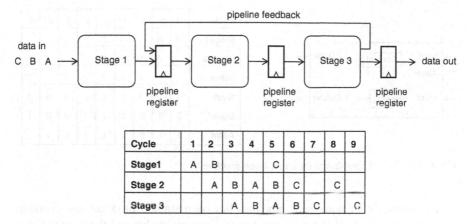

Fig. 12.10  A reservation table for a non-linear pipeline

non-linear pipeline architectures may be easier to control with a time-stationary approach then with a data-stationary approach. Indeed, generating the control signals for multiplexers between pipeline stages requires an overview of the entire pipeline.

#### 12.3.2.2  Control of Non-linear Pipelines

When a pipeline has feedback connections, the pipeline stages are reused multiple times per data item that enters the pipeline. As a result, the throughput of a pipeline can no longer be 1.

Figure 12.10 illustrates the operation of a non-linear pipeline structure with three stages. Each data item that enters the pipeline will use stages 2 and 3 two times. When a new data item enters the pipeline, it will occupy stage 1 in the first cycle, stage 2 in the second, and stage 3 in the third cycle. The data item is then routed back to stage 2 for the fourth cycle of processing, and into stage 3 for the fifth and final cycle of processing. We can thus conclude that this non-linear, three-stage pipeline has a latency of 5. The diagram below Fig. 12.10 is a *reservation table*, a systematic representation of data items as they flow through the pipeline, with stages corresponding to rows and clock cycles corresponding to columns. The table demonstrates the pipeline processing of three data items A, B and C. From the diagram, we can see that data items A and B are processed in subsequent clock cycles. However, item C cannot immediately follow item B: the pipeline is busy and will occupy stages 2 and 3. Hence, item C will need to wait. This situation is called a pipeline *conflict*. In cycle 5, the pipeline is available again and item C can start. We conclude that the pipeline is able to process two elements (A and B) in four clock cycles. Therefore, the throughput of this pipeline is 0.5.

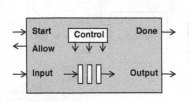

| Cycle | 1 | 2 | 3 | 4 | 5 | 6 | 7 | 8 | 9 |
|---|---|---|---|---|---|---|---|---|---|
| Stage1 | A | B | | | C | | | | |
| Stage 2 | | A | B | A | B | C | | C | |
| Stage 3 | | | A | B | A | B | C | | C |
| Start | 1 | 1 | 0 | 0 | 1 | 0 | 0 | 0 | 0 |
| Done | 0 | 0 | 0 | 0 | 1 | 1 | 0 | 0 | 1 |
| Allow | 1 | 1 | 0 | 0 | 1 | 1 | 0 | 1 | 1 |

**Fig. 12.11** Control handshakes for a non-linear pipeline

The operation of non-linear pipelines has been studied in detail, for example in the seminal work of Peter Kogge, but this material is beyond the scope of this chapter. Instead, we will consider the impact of pipelining on the generation of control handshake signals.

### 12.3.2.3   Control Handshakes for Pipelines

Earlier in this section we used a simple start/done handshake to implement hierarchical control for an iterated component. How do these handshake signals have to be modified for pipeline architectures?

In an iterated structure, a single done signal is sufficient to mark two distinguished but coinciding events. The first event is when a result is available at the output. The second event is when a new argument can be accepted at the input. In a pipeline structure however, input and output activities do not coincide. Indeed, in a pipeline, the latency does not have to be the reciprocal of the throughput. The case of a linear pipeline is still easy: the latency is equal to the number of pipeline stages, and the throughput is equal to 1. The case of a non-linear pipeline is more complex, and the latency as well as the throughput can both be different from 1.

To distinguish input events from output events, we will use *two* handshake-acknowledge signals. The first one, done, indicates when the pipeline produces valid data. The second one, allow, indicates when the input is able to accept new arguments.

Figure 12.11 illustrates the relation of the handshake signals to the operation of the pipeline. The left side of the figure illustrates the interface to the pipeline, while the right side shows the values for start, done and allow over several clock cycles. The beginning and end of a pipeline instruction are marked through start and done. The allow signal indicates if a new instruction can be started at the *next* clock cycle. If allow is zero, this means that starting an instruction will cause a pipeline conflict. You can observe that done is a delayed version of start, with the delay equal to the pipeline latency. The format of the allow signal is more

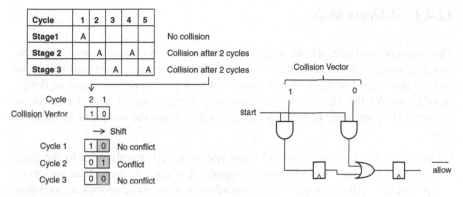

Fig. 12.12 Generation of the done signal

complex, because it depends on the exact pattern of pipeline interconnections. For example, allow must be zero in cycles 3 and 4 because the second pipeline stage is occupied by instruction A and B.

Nevertheless the allow signal is easy to generate, as demonstrated in Fig. 12.12. The allow signal indicates when the pipeline is occupied and cannot accept a new instruction. For the reservation table shown in Fig. 12.12, this happens two clock cycles after a pipeline instruction starts. This duration of two clock cycles is called a *forbidden latency*. A *collision vector* is a bit-vector where forbidden latencies are marked by means of a bit. The index of this bit corresponds to the forbidden latency. For the reservation table shown, the collision vector equals 10, since the only forbidden latency is two clock cycles. The allow signal can now be generated using the collision vector and a shift register, as shown on the right of Fig. 12.12. Each time a new instance of a pipeline instruction starts, a copy of the collision vector is added to the shift register. The last bit of this shift register indicates if the current clock cycle coincides with a forbidden latency. Hence, this bit is the inverse of the allow bit.

## 12.4 Programmer's Model = Control Design + Data Design

The previous two sections highlighted two aspects that affect control shell design. *Data design* is concerned with moving data from software to the encapsulated hardware module and back. *Control design* is concerned with generating control signals for the encapsulated hardware module.

In this section, we consider the impact of these design decisions on the software driver. The software view of a hardware module is defined as the *programmer's model*. This includes a collection of the memory areas used by the custom hardware module, and a definition of the commands (or instructions) understood by the module.

## 12.4.1   Address Map

The *address map* reflects the organization of software-readable and software-writable storage elements of the hardware module, as seen from software. The address map is part of the design of a memory-mapped coprocessor, and its design should consider the viewpoint of the software designer rather than the hardware designer. Here are some of the considerations that affect the design of the address map.

- To a software designer, *read* and *write* operations commonly refer to the *same* memory location. For a hardware designer, it is easy to route read and write operations to different registers, since read strobes and write strobes are available on a bus as separate registers. This practice should be avoided because it goes against the expectations of the software designer. A given memory-mapped address should always affect the same hardware registers.
- In the same spirit, a hardware designer can create memory-mapped registers that are read-only, write-only or read-write registers. By default all memory-mapped registers should be read/write. This matches the expectations of the software designer. *Read/write* memory-mapped registers also allow a software designer to conveniently implement bit-masking operations (such as flipping a single bit in a memory-mapped register). In some cases, *read-only* registers are justified, such as for example to implement registers that reflect hardware status information or sampled-data signals. However, there are very few cases that justify a *write-only* register.
- In software, read/write operations always handle aligned data. For example, extracting bits number 5–12 out of a 32-bit word is more complicated than extracting the second byte of the same word. While a hardware designer may have a tendency to make everything as compact as possible, this practice may result in an address map that is very hard to handle for a software designer. Hence, the address map should respect the alignment of the processor.

## 12.4.2   Instruction Set

The *instruction set* of a custom-hardware module defines how software controls the module. The design of a good instruction-set is a hard problem; it requires the codesigner to make the proper trade-off between flexibility and efficiency. Instructions that trigger complex activities in the hardware module may be very efficient, but they are difficult to use and understand for a software designer. The design of an instruction-set strongly depends on the function of the custom-hardware module, and therefore the number of generic guidelines is limited.

- One can distinguish *three classes* of instructions: one-time commands, on-off commands, and configurations. One-time commands trigger a single activity in the hardware module (which may take multiple clock cycles to complete). Pipeline control commands, such as discussed in Sect. 12.3.1, fall in the same category. On-Off commands come in pairs, and they control a continuous activity on the hardware module. Finally, configurations provide a parameter to an algorithm. They affect the general behavior of the hardware module. Making the proper choice between these is important in order to minimize the amount of control interaction between the software driver and the hardware module.

- *Synchronization* between software and hardware is typically implemented at multiple levels of abstraction. At the lowest level, the hardware/software interfaces will ensure that data items are transferred correctly from hardware to software and vice versa. However, additional synchronization may be needed at the algorithm level. For example, a hardware module with a data-dependent execution time could indicate completion to the driver software through a status flag. In this case, a status flag can support this additional layer of synchronization.
- Another synchronization problem is present when multiple software users share a single hardware module. In this case, the different users all see the same hardware registers, which may be undesirable. This synchronization issue can be handled in several ways. Coprocessor usage could be serialised (allowing only a single user at a time), or else a context switch can be implemented in the hardware module.
- Finally, *reset design* must be carefully considered. An example of flawed reset design is when a hardware module can only be initialized by means of full system reset. It makes sense to define one or several instructions for the hardware module to handle module initialization and reset.

## 12.5 Summary

In this chapter we discussed design techniques to encapsulate hardware modules onto a predefined hardware/software interface. These techniques are collected under the term *hardware interface design*. A hardware interface implements the connection mechanisms between low-level software and a hardware module, including the transfer of operands to and from the hardware module, and the generation of control signals for the hardware module. This requires, in general, the addition of data input/output buffers, and the addition of a controller.

Optimizing the communication between hardware and software is an important objective. One optimization technique is to improve the overlap between hardware/software communication and computation. This can be achieved by means of block-level pipelining and/or internal pipelining. Both forms of pipelining provide improved system-level performance, at the expense of additional hardware and increased complexity in system control.

Design of a good hardware interface is a challenging task; the application section of this book includes several in-depth examples.

## 12.6   Further Reading

Similar to hardware/software interface design, the implementation of coprocessor hardware interfaces is an ad-hoc design process.

The classic work on optimization of pipelined architectures is by Kogge (1981), and its ideas on scheduling of pipelined architectures are still relevant.

## 12.7   Problems

**Problem 12.1.** A C function requires 8 programmable constants that change very infrequently, and 1 data input that changes upon each call. Design a memory-mapped interface that minimizes the amount of memory locations required, while at the same time introducing minimal impact on the runtime performance of the design. Be precise: show a CPU bus on one side, and $8 + 1$ registers on the other side, with your memory-mapped interface design in between. Assume the interface is mapped starting at address 0x100.

**Problem 12.2.** The Motorola DSP56000 processor is a pipelined processor. One of its assembly instructions look as follows.

```
MPY   x0, Y0, A      X:  (R0)+, X0    Y:  (R4)+, Y0
```

There are no comments in the above line – everything on that line is part of the instruction! This instruction multiplies register X0 with Y0 and places the product in the A accumulator. At the same time, the value of register X0 is updated with the memory location pointed to by registers R0, and Y0 is updated with the memory location pointed to by register R4. Does the Motorola DSP56000 use time-stationary control or does it use data-stationary control?

**Problem 12.3.** Listing 12.1 is a design of a control shell for a median module, which evaluates the median of three values. Study the operation of the coprocessor by studying the GEZEL code listing, and answer the questions below.
(a) How many data-input and data-output ports does the coprocessor have?
(b) Is the median module communication-constrained or computation-constrained with respect to this hardware/software interface?
(c) Describe how software should operate the coprocessor (write values to it, and retrieve the result from it).
(d) Write a C program that uses this coprocessor to evaluate the median value of the numbers 36, 99, and 58.

**Problem 12.4.** Listing 12.2 is a hardware implementation of a vector generator module. Given a set of input coordinates (ix, iy), the module will generate all integer coordinates lying on the straight line between (0,0) and (ix, iy). The algorithm implemented by the module, the Bresenham algorithm, is used to draw lines on raster-scan displays. The Listing 12.2 assumes that (ix, iy) lies

**Listing 12.1** A hardware interface for a median module (Problem 12.3)

```
dp median(in  a, b, c : ns(32);
          out q        : ns(32)) {
  sig q1, q2 : ns(32);
  always {
   q1 = (a > b) ? a : b;
   q2 = (a > b) ? b : a;
   q  = (c > q1) ? q1 : (c < q2) ? q2 : c;
  }
}

ipblock myarm {
  iptype "armsystem";
  ipparm "exec=median_driver";
}

ipblock b_datain(out data : ns(32)) {
  iptype "armsystemsource";
  ipparm "core=myarm";
  ipparm "address=0x80000000";
}

ipblock b_dataout(in data : ns(32)) {
  iptype "armsystemsink";
  ipparm "core=myarm";
  ipparm "address=0x80000004";
}

dp medianshell {
  reg a1, a2, a3 : ns(32);
  sig q : ns(32);
  use median(a1, a2, a3, q);

  sig v_in, v_out : ns(32);
  use b_datain(v_in);
  use b_dataout(v_out);

  use myarm;

  reg old_v : ns(32);
  always {
    old_v = v_in;
    a1   = ((old_v == 0) & (v_in > 0)) ? v_in : a1;
    a2   = ((old_v == 0) & (v_in > 0)) ? a1   : a2;
    a3   = ((old_v == 0) & (v_in > 0)) ? a2   : a3;
    v_out = q;
  }
}

system S {
  medianshell;
}
```

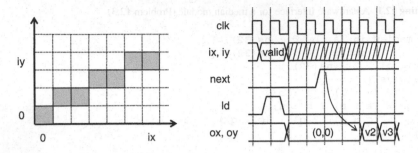

**Fig. 12.13** Bresenham vector generation module (Problem 12.4)

in the first quadrant, so that `ix` and `iy` are always positive. The timing diagram on the right of Fig. 12.13 shows how to operate the module. New target coordinates can be entered using the `ld` control input. Loading new coordinates also resets the output coordinates to `(0,0)`. After that, the `next` control input can be used to retrieve new output points from the vector. The ouput will be refreshed on the second clock edge after `next` is high.

(a) Design a hardware interface for this module implemented as a coprocessor with a single 32-bit input port, and a single 32-bit output port. Optimize your design to take advantage of the fact that the coordinates of the module are 12-bit. Assume a memory-mapped interface. The design of the hardware interface includes: definition of the coprocessor instruction set, design of the hardware interface hardware, and design of sample driver software.

(b) How would you modify the design of (a) when the coprocessor needs to be connected through an 8-bit bus? Describe the required modifications to the instruction-set, the control-shell hardware, and the sample driver software.

**Listing 12.2** A vector generator

```
dp bresen(in   ix, iy   : ns(12);
          in   ld, next : ns( 1);
          out  ox, oy   : ns(12)) {
  reg x, y              : ns(12);  // current position
  sig nx, ny            : ns(12);
  reg e                 : tc(12);
  reg eol               : ns(1);
  reg rix, riy          : ns(12);  // current target
  reg einc1, einc2      : tc(12);
  reg xinc, yinc        : ns(1);
  reg ldr, nextr        : ns(1);

  always {
    ldr   = ld;
    nextr = next;
    rix   = ld ? ix : rix;
    riy   = ld ? iy : riy;
```

```
        ox    = x;
        oy    = y;
    }

    sfg init {
      einc1 = (rix > riy) ? (riy - rix) : (rix - riy);
      einc2 = (rix > riy) ? riy          : rix;
      xinc  = (rix > riy) ? 1 : 0;
      yinc  = (rix > riy) ? 0 : 1;
      e     = (rix > riy) ? 2 * riy - rix : 2 * rix - riy;
      x     = 0;
      y     = 0;
    }

    sfg loop {
      nx    = (e >= 0) ? x + 1     : x + xinc;
      ny    = (e >= 0) ? y + 1     : y + yinc;
      e     = (e >= 0) ? e + einc1 : e + einc2;
      x     = nx;
      y     = ny;
      eol   = ((nx == rix) & (ny == riy));
    }

    sfg idle { }
}
fsm f_bresen(bresen) {
  initial s0;
  state s1, s2;
  @s0 (init)                          -> s1;
  @s1 if (ldr) then (init)            -> s1;
      else if (eol)    then (idle) -> s2;
      else if (nextr) then (loop) -> s1;
      else (idle)                   -> s1;

  @s2 if (ldr) then (init)            -> s1;
      else (idle)                   -> s2;
}

// testbench
dp test_bresen {
  reg ix, iy   : ns(12);
  reg ld, next : ns(1);
  sig ox, oy   : ns(12);
  use bresen(ix, iy, ld, next, ox, oy);

  always   { $display("<",$cycle, ">", $dec, " ox ", ox, " oy ",
                oy); }
  sfg init { ld = 1; next = 0; ix = 11; iy = 7; }
  sfg step { ld = 0; next = 1; ix =  0; iy = 0; }
}
fsm ctl_test_bresen(test_bresen) {
  initial s0;
```

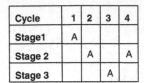

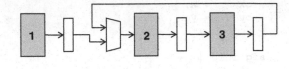

Fig. 12.14   A non-linear pipeline (Problem 12.5)

```
    state s1;
    @s0 (init) -> s1;
    @s1 (step) -> s1;
}

system S {
    test_bresen;
}
```

**Problem 12.5.** Figure 12.14 shows a non-linear pipeline architecture with three stages. The shaded blocks labeled 1, 2, and 3 represent combinational logic. Pipeline stage 2 iterates two times over each data item entered. As a result, this architecture can only process one data item every clock cycle.

(a) Find the forbidden latencies for this pipeline.
(b) Can this pipeline accept new data inputs at regularly spaced interfaces? If not, how could you modify this architecture so that this becomes possible?

# Part IV
# Applications

The final part of the book describes three complete applications of hardware/software codesign and provides solutions to selected exercises.

Each application example also includes an introduction on the application background, so that the system-level specification and the design refinement process can be fully understood. The examples are implemented as prototypes in Field Programmable Gate Arrays (FPGA), and the reader has access to the full source code of these designs. The application examples include a co-processor for the Trivium stream-cipher, a coprocessor for the AES block cipher, and a coprocessor for the evaluation of digital rotation functions (CORDIC).

# Chapter 13
# Trivium Crypto-Coprocessor

## 13.1 The Trivium Stream Cipher Algorithm

The Trivium stream cipher algorithm was proposed by Christophe De Canniere and Bart Preneel in 2006 in the context of the eSTREAM project, a European effort that ran from 2004 to 2008 to develop a new stream ciphers. In September 2008, Trivium was selected as a part of the official eSTREAM portfolio, together with six other stream cipher algorithms. The algorithm is remarkably simple, yet to this date it remains unbroken in practice. We will clarify further what it means to *break* a stream cipher. In this section, we discuss the concept of a stream cipher, and the details of the Trivium algorithm.

### 13.1.1 Stream Ciphers

There are two types of symmetric-key encryption algorithms: stream ciphers, and block ciphers. Trivium is a stream cipher. AES, discussed in the next chapter, is a block cipher. The left of Fig. 13.1 illustrates the difference between a stream cipher and a block cipher. A stream cipher is a state machine with an internal state register of $n$ bits. The stream cipher kernel will initialize the state register based on a key, and it will update the state register while producing the keystream.

In contrast to a stream cipher, a block cipher is a state-less function that combines an $m$ bit key with a block of $n$ bits of plaintext. Because there is no state, the encryption of one block of plaintext bits is independent of the encryption of the previous block of bits. Of course, many hardware implementations of block ciphers contain registers. These registers are an effect of sequentializing the block cipher algorithm over multiple clock cycles. It is perfectly feasible to implement block ciphers without any registers.

The cryptographic properties of the stream cipher are based on the highly non-linear functions used for state register initialization and state register update.

P.R. Schaumont, *A Practical Introduction to Hardware/Software Codesign*,
DOI 10.1007/978-1-4614-3737-6_13, © Springer Science+Business Media New York 2013

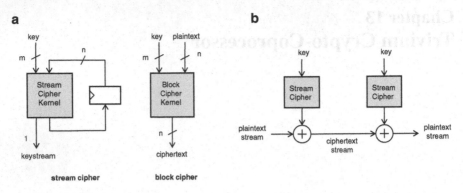

**Fig. 13.1** (**a**) Stream cipher and block cipher. (**b**) Stream cipher encryption/decryption

These non-linear functions ensure that the keystream cannot be predicted even after a very large number of keystream bits has been observed. *Breaking* a stream cipher means that one has found a way to predict the output of the stream cipher, or even better, one has found a way to reveal the contents of the state register. For a state register of $n$ bits, the stream cipher can be in $2^n$ possible states, so the total length of the key stream is on the order of $2^n$ bits. Practical stream ciphers have state register lengths between 80 bits and several 100 bits.

A stream cipher by itself does not produce ciphertext, but only a stream of keybits. The right of Fig. 13.1 illustrates how one can perform encryption and decryption with a stream cipher. The keystream is combined (xor-ed) with a stream of plaintext bits to obtain a stream of ciphertext bits. Using an identical stream cipher that produces the same keystream, the stream of ciphertext bits can be converted back to plaintext using a second xor operation.

A stream cipher algorithm produces a stream of bits. In a typical implementation of a stream cipher, we may be interested in using multiple key bits at a time. This is so because of two reasons. First, the encryption may require multiple bits at a time. Each letter from a message, for example, may be encoded with multiple bits (e.g. ASCII), so that encrypting a message letter needs multiple key bits as well. Second, depending on the target architecture, producing a single key bit at a time may be inefficient. On a RISC processor, for example, it makes sense to represent a stream as a sequence of 32-bit words. Therefore, depending on the computer architecture, we would have a key-stream formatted as single bits, as bytes, as 32-bit words, and so on.

One way to obtain a wider keystream is to run the stream cipher kernel at high speed and perform a serial-to-parallel conversion of the output. An alternative is illustrated in Fig. 13.2: the stream cipher can be easily parallelized to produce multiple keystream bits per clock cycle. This is especially useful when the stream cipher kernel is a very simple function, as is the case with Trivium.

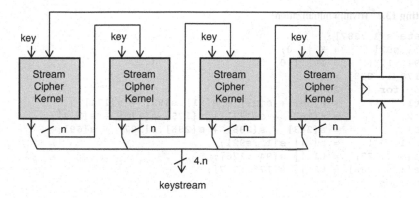

**Fig. 13.2** Parallel stream cipher implementation

**Listing 13.1** Trivium round

```
state s[1..288];
loop
  t1         = s[66] + s[93];
  t2         = s[162] + s[177];
  t3         = s[243] + s[288];
  z          = t1 + t2 + t3;
  t1         = t1 + s[91].s[92] + s[171];
  t2         = t2 + s[175].s[176] + s[264];
  t3         = t3 + s[286].s[287] + s[69];
  s[1..93]   = t3 || s[1..s92];
  s[94..177] = t1 || s[94..176];
  s[178..288] = t2 || s[178..287];
end loop
```

## *13.1.2 Trivium*

Trivium is a stream cipher with a state register of 288 bits. The state register is initialized based on an 80-bit key and an 80-bit initial value (IV). After initialization, Trivium produces a stream of keybits. The specification of Trivium keystream generation is shown in Listing 13.1. Each iteration of the loop, a single output bit z is generated, and the state register s is updated. The addition and multiplication (+ and .) are taken over $GF(2)$. They can be implemented with exclusive-or and bitwise-and respectively. The double-bar operation ( | | ) denotes concatenation.

The initialization of the state register proceeds as follows. The 80-bit key K and the 80-bit initial value IV are loaded into the state register, and the state register is updated 1,152 times (4 times 288) without producing keybits. After that, the state register is ready to produce keystream bits. This is illustrated in the pseudocode of Listing 13.2.

**Listing 13.2** Trivium initialization

```
state s[1..287];
s[1..s92]     = K || 0;
s[94..177]    = IV || 0;
s[178 .. 288] = 3;
loop for (4 * 288)
  t1          = s[66] + s[93] + s[91].s[92] + s[171];
  t2          = s[162] + s[177] + s[175].s[176] + s[264];
  t3          = s[243] + s[288] + s[286].s[287] + s[69];
  s[1..93]    = t3 || s[1..s92];
  s[94..177]  = t1 || s[94..176];
  s[178..288] = t2 || s[178..287];
end loop
```

These listings confirm that, from a computational perspective, Trivium is a very simple algorithm. A single state register update requires nine single-bit xor operations and three single-bit and operations. We need two additional single-bit xor operations to produce the output bit z.

## 13.1.3   Hardware Mapping of Trivium

A straightforward hardware mapping of the Trivium algorithm requires 288 registers, 11 xor gates, and 3 and gates. Clearly, the largest cost of this algorithm is in the storage. Figure 13.3 shows how Trivium is partitioned into hardware modules.

- The trivium module calculates the next state. We will use the term Trivium *kernel* to indicate the loop body of Listing 13.1, without the state register update.
- The keyschedule module manages state register initialization and update. The keyschedule module has a single control input ld to control state register initialization. In addition, keyschedule has a single status bit e that indicates when the initialization has completed, and thus when the output keystream z is valid. This partitioning between keyschedule and trivium kernel was done with loop unrolling in mind (Fig. 13.2).

Based on this partitioning and the Trivium specification given earlier, it is straightforward to create a GEZEL description of Trivium. Listing 13.3 shows the implementation of a 1 bit per cycle Trivium. The control signals in the keyschedule module are created based on a counter which is initialized after a pulse on the ld control input.

To create a bit-parallel keystream, we need to modify the code as follows. First, we need to instantiate the trivium module multiple times, and chain the state input and output ports together as shown in Fig. 13.2. Second, we need to adjust the key schedule, since the initialization phase will take less than 4 times 288 clock cycles. As an example, Listing 13.4 shows how to unroll Trivium eight times, thus

**Listing 13.3** 1-bit-per-cycle Trivium

```
1  dp trivium(in  si : ns(288);   // state input
2                out so : ns(288);  // state output
3                out z  : ns(1)) {  // crypto bit out
4     sig  t1,  t2,  t3 : ns(  1);
5     sig t11, t22, t33 : ns(  1);
6     sig saa           : ns( 93);
7     sig sbb           : ns( 84);
8     sig scc           : ns(111);
9     always {
10       t1  = si[ 65] ^ si[ 92];
11       t2  = si[161] ^ si[176];
12       t3  = si[242] ^ si[287];
13       z   = t1 ^ t2 ^ t3;
14       t11 = t1 ^ (si[ 90] & si[ 91]) ^ si[170];
15       t22 = t2 ^ (si[174] & si[175]) ^ si[263];
16       t33 = t3 ^ (si[285] & si[286]) ^ si[ 68];
17       saa = si[  0: 92] # t33;
18       sbb = si[ 93:176] # t11;
19       scc = si[177:287] # t22;
20       so  = scc # sbb # saa;
21    }
22 }
23
24 dp keyschedule(in  ld : ns(1);     // reload key & iv
25                  in  iv : ns(80);    // initialization vector
26                  in key : ns(80);    // key
27                  out  e : ns(1);     // output valid
28                  in  si : ns(288);   // state input
29                  out so : ns(288)) { // state output
30    reg s        : ns(288);           // state register
31    reg cnt      : ns(11);            // initialization counter
32    sig saa      : ns( 93);
33    sig sbb      : ns( 84);
34    sig scc      : ns(111);
35    sig cte      : ns(111);
36    always {
37       saa  = ld ? key        : si[  0: 92];
38       sbb  = ld ? iv         : si[ 93:176];
39       cte  = 7;
40       scc  = ld ? (cte << 108) : si[177:287];
41       s    = scc # sbb # saa;
42       so   = s;
43       cnt  = ld ? 1152 : (cnt ? cnt - 1 : cnt);     1152 = 4 * 288
44       e    = (cnt ? 0 : 1);
45    }
46 }
```

**Fig. 13.3** Hardware mapping of Trivium

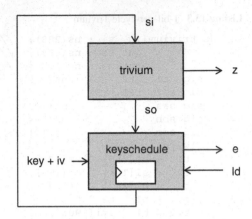

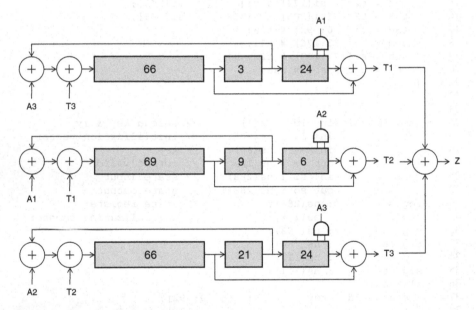

**Fig. 13.4** Trivium round structure

obtain a stream cipher that generates 1 byte of keystream per clock cycle. In this case, the initialization completes 8 times faster, after 143 clock cycles (line 33).

**Listing 13.4** 1-byte-per-cycle Trivium

```
1  dp trivium(in  si : ns(288);   // state input
2             out so : ns(288);   // state output
3             out z  : ns(1)) {   // crypto bit out
4     // ...
5  }
6  dp trivium2 : trivium
7  dp trivium3 : trivium
```

```
 8  dp trivium4 : trivium
 9  dp trivium5 : trivium
10  dp trivium6 : trivium
11  dp trivium7 : trivium
12  dp trivium8 : trivium
13
14  dp keyschedule(in  ld : ns(1);     // reload key & iv
15                  in  iv : ns(80);   // initialization vector
16                  in key : ns(80);   // key
17                  out  e : ns(1);    // output valid
18                  in  si : ns(288);  // state input
19                  out so : ns(288)) { // state output
20     reg s      : ns(288);           // state register
21     reg cnt    : ns(11);            // initialization counter
22     sig saa    : ns( 93);
23     sig sbb    : ns( 84);
24     sig scc    : ns(111);
25     sig cte    : ns(111);
26     always {
27       saa = ld ? key           : si[  0: 92];
28       sbb = ld ? iv            : si[ 93:176];
29       cte = 7;
30       scc = ld ? (cte << 108) : si[177:287];
31       s   = scc # sbb # saa;
32       so  = s;
33       cnt = ld ? 143 : (cnt ? cnt - 1 : cnt);    │143 = 4 * 288 / 8 - 1│
34       e   = (cnt ? 0 : 1);
35     }
36  }
37
38  dp triviumtop(in  ld : ns(1);       // reload key & iv
39                in  iv : ns(80);      // initialization vector
40                in key : ns(80);      // key
41                out  z : ns(8);       // encrypted output
42                out  e : ns(1)) {     // output valid
43     sig si, so0, so1, so2, so3, so4, so5, so6, so7 : ns(288);
44     sig z0, z1, z2, z3, z4, z5, z6, z7 : ns(1);
45     use keyschedule(ld, iv, key, e, si, so0);
46     use trivium     (so0, so1, z0);
47     use trivium2    (so1, so2, z1);
48     use trivium3    (so2, so3, z2);
49     use trivium4    (so3, so4, z3);
50     use trivium5    (so4, so5, z4);
51     use trivium6    (so5, so6, z5);
52     use trivium7    (so6, so7, z6);
53     use trivium8    (so7, si,  z7);
54     always {
55       z = z0 # z1 # z2 # z3 # z4 # z5 # z6 # z7;
56     }
57  }
```

What is the limiting factor when unrolling Trivium? First, notice that unrolling the algorithm will not increase the critical path of the Trivium kernel operations as long as they affect different state register bits. Thus, as long as the state register bits read are different from the state register bits written, then all the kernel operations are independent. Next, observe that a single Trivium round consists of three circular shift registers, as shown in Fig. 13.4. The length of each shift register is indicated inside of the shaded boxes. To find how far we can unroll this structure, we look for the *smallest* feedback loop. This loop is located in the upper circular shift register, and spans 69 bits. Therefore, we can unroll Trivium at least 69 times before the critical path will increase beyond a single and-gate and two xor gates. In practice, this means that Trivium can be easily adjusted to generate a key-stream of double-words (64 bits). After that, the critical path will increase each 69 bits. Thus, a 192 bit-parallel Trivium will be twice as slow as a 64 bit-parallel Trivium, and a 256 bit-parallel Trivium will be roughly three times as slow.

### 13.1.4  A Hardware Testbench for Trivium

Listing 13.5 shows a hardware testbench for the Trivium kernel. In this testbench, the key value is programmed to 0x80 and the IV to 0x0. After loading the key (lines 12–15), the testbench waits until the e-flag indicates the keystream is ready (lines 29–30). Next, each output byte is printed on the output (lines 19–22). The first 160 cycles of the simulation generate the following output.

```
> fdlsim trivium8.fdl 160
147 11001100 cc
148 11001110 ce
149 01110101 75
150 01111011 7b
151 10011001 99
152 10111101 bd
153 01111001 79
154 00100000 20
155 10011010 9a
156 00100011 23
157 01011010 5a
158 10001000 88
159 00010010 12
```

The key stream bytes produced by Trivium consists of the bytes 0xcc, 0xce, 0x75, 0x7b, 0x99, and so on. The bits in each byte are read left to right (from most significant to least significant). In the next sections, we will integrate this module as a coprocessor next to the processor.

**Listing 13.5** Testbench for a 1-byte-per-cycle Trivium

```
1   // testbench
2   dp triviumtest {
3     sig ld        : ns(1);
4     sig iv, key : ns(80);
5     sig e         : ns(1);
6     reg re        : ns(1);
7     sig z         : ns(8);
8     reg rz        : ns(8);
9     sig bs        : ns(8);
10    use triviumtop(ld, iv, key, z, e);
11    always    { rz  = z;
12                re  = e;     }
13    sfg init0 { iv  = 0;
14                key = 0x80;
15                ld  = 1;
16              }
17    sfg idle  { ld = 0; }
18    sfg bstuf { ld = 0; }
19    sfg show  { ld = 0;
20                bs = rz;
21                $display$(cycle, " ", $bin, bs, $hex, " ", bs);
22              }
23  }
24  fsm ft(triviumtest) {
25    initial s0;
26    state s10, s1, s2;
27    @s0  (init0)                  -> s10;
28    @s10 (init0)                  -> s1;
29    @s1  if (re) then (bstuf)     -> s2;
30               else (idle)        -> s1;
31    @s2  (show)                   -> s2;
32  }
```

## 13.2 Trivium for 8-bit Platforms

Our first coprocessor design will attach the Trivium stream cipher hardware to an 8-bit microcontroller. We will make use of an 8051 micro-controller. Like many other micro-controllers, it has several general-purpose digital input-output ports, which can be used to create hardware-software interfaces. Thus, we will be building a *port-mapped* control shell for the Trivium coprocessor. The 8051 micro-controller also has an external memory bus (XBUS), which supports a memory-space of 64 K. Such external memory busses are rather uncommon for micro-controllers. However, we will demonstrate the use of such a memory-bus in our design as well.

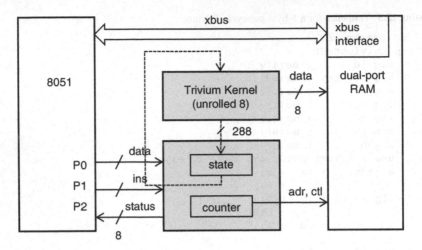

**Fig. 13.5** Trivium coprocessor integration on a 8051

## 13.2.1   Overall Design of the 8051 Coprocessor

Figure 13.5 illustrates the overall design. The coprocessor is controlled through three 8-bit ports (P0, P1, P2). They are used to transfer operands, instructions, and to retrieve the coprocessor status respectively. The Trivium hardware will dump the resulting keystream into a dual-port RAM module, and the contents of the keystream can be retrieved by the 8051 through the XBUS.

The system works as follows. First, the 8051 programs a key and an initialization vector into the Trivium coprocessor. Next, the 8051 commands the Trivium coprocessor to generate N keybytes, which will be stored in the shared RAM on the XBUS. Finally, the 8051 can retrieve the keybytes from the RAM. Note that the retrieval of the keybytes from RAM is only shown as an example; depending on the actual application, the keystream may be used for a different purpose. The essential part of this example is the control of the coprocessor from within the micro-controller.

To design the control shell, we will need to develop a command set for the Trivium coprocessor. Since the 8-bit ports of the 8051 don't include strobes, we need to introduce our own handshake procedure: a simple *idle* instruction is used to create a synchronization point between hardware and software. The command set for the coprocessor is shown in Table 13.1. All of the commands, except ins_enc, complete within a single clock cycle. The last command, ins_enc, takes up to 256 clock cycles to complete. The status port of the 8051 is used to indicate encryption completion. Figure 13.6 illustrates the command sequence for the generation of 10 bytes of keystream. Note that the status port becomes zero when the keystream generation is complete.

**Table 13.1**  Command set for Trivium coprocessor

| Value at P0 (Instruction) | Value at P1 (Data) | Value at P2 (Status) | Meaning |
|---|---|---|---|
| ins_idle | don't care | don't care | Idle Instruction |
| ins_key0 | Key Byte 0 | don't care | Program Key Byte |
| ins_key1 | Key Byte 1 | don't care | Program Key Byte |
| ins_key2 | Key Byte 2 | don't care | Program Key Byte |
| ins_key3 | Key Byte 3 | don't care | Program Key Byte |
| ins_key4 | Key Byte 4 | don't care | Program Key Byte |
| ins_key5 | Key Byte 5 | don't care | Program Key Byte |
| ins_key6 | Key Byte 6 | don't care | Program Key Byte |
| ins_key7 | Key Byte 7 | don't care | Program Key Byte |
| ins_key8 | Key Byte 8 | don't care | Program Key Byte |
| ins_key9 | Key Byte 9 | don't care | Program Key Byte |
| ins_iv0 | IV Byte 0 | don't care | Program IV Byte |
| ins_iv1 | IV Byte 1 | don't care | Program IV Byte |
| ins_iv2 | IV Byte 2 | don't care | Program IV Byte |
| ins_iv3 | IV Byte 3 | don't care | Program IV Byte |
| ins_iv4 | IV Byte 4 | don't care | Program IV Byte |
| ins_iv5 | IV Byte 5 | don't care | Program IV Byte |
| ins_iv6 | IV Byte 6 | don't care | Program IV Byte |
| ins_iv7 | IV Byte 7 | don't care | Program IV Byte |
| ins_iv8 | IV Byte 8 | don't care | Program IV Byte |
| ins_iv9 | IV Byte 9 | don't care | Program IV Byte |
| ins_init | don't care | don't care | Initializes state register |
| ins_enc | rounds | isready | Encrypts rounds |

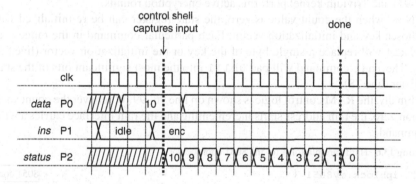

**Fig. 13.6**  Command sequence for encryption

## 13.2.2   Hardware Platform of the 8051 Coprocessor

We will now capture the hardware platform of Fig. 13.5 as a GEZEL program. Listing 13.6 shows the complete platform apart from the Trivium kernel (which was discussed in Sect. 13.1.3). The first part of the Listing captures all the 8051-specific interfaces. The Trivium coprocessor will be attached to these interfaces.

- Lines 1–6: The 8051 core my8051 will read in an executable called trivium. ihx. The executable is in Intel Hex Format, a common format for micro-controller binaries. The period parameter of the core is 1, meaning that the clock frequency of the 8051 core is the same as the hardware clock frequency. A traditional 8051 architecture uses 12 clock cycles per instruction. Thus, a period of 1 means that there will be a single instruction executing each 12 clock cycles.
- Lines 7–21: Three I/O ports of the 8051 are defined as P0, P1, and P2. A port is either configured as input or else as output by choosing its type to be i8051systemsource (e.g. lines 8, 13) or else i8051systemsink (e.g. line 18).
- Lines 22–30: A dual-port, shared-memory RAM module attached to the XBUS is modeled using an ipblock. The module allows to specify the starting address (xbus, line 28) as well as the amount of memory locations (xrange, line 29).

The triviumitf module integrates the Trivium hardware kernel (line 44) and the hardware/software interfaces. Several registers are used to manage this module, including a Trivium state register tstate, a round counter cnt, and a RAM address counter ramcnt (lines 50–53).

The key and initialization vector are programmed into the state register through a sequence of chained multiplexers (lines 56–86). This works as follows. First consider the update of tstate on line 92. If the counter value cnt is nonzero, tstate will copy the value so, which is the output of the Trivium kernel. If the counter value cnt is zero, tstate will instead copy the value of init, which is defined through lines 56–88. Thus, by loading a nonzero value into cnt (lines 90–91), the Trivium kernel performs active encryption rounds.

Now, when the count value is zero, the state register can be re-initialized with a chosen key and initialization vector. Each particular command in the range 0x1 to 0x14 will replace a single byte of the key or the initialization vector (line 56–86). The init command will pad 0b111 into the most significant bits of the state register (line 78).

Finally, the RAM control logic is shown on line 96–99. Whenever the count value is non-zero, the ram address starts incrementing and the ram interface carries a write command.

**Listing 13.6**  Hardware platform for the 8051 coprocessor

```
 1  ipblock my8051 {                                           8051 core
 2     iptype "i8051system";
 3     ipparm "exec=trivium.ihx";
 4     ipparm "verbose=1";
 5     ipparm "period=1";
 6  }
 7  ipblock my8051_data(out data : ns(8)) {              8051 interfaces
 8     iptype "i8051systemsource";
 9     ipparm "core=my8051";
10     ipparm "port=P0";
11  }
```

```
12   ipblock my8051_ins(out data : ns(8)) {
13     iptype "i8051systemsource";
14     ipparm "core=my8051";
15     ipparm "port=P1";
16   }
17   ipblock my8051_status(in data : ns(8)) {
18     iptype "i8051systemsink";
19     ipparm "core=my8051";
20     ipparm "port=P2";
21   }
22   ipblock my8051_xram(in   idata   : ns(8);
23                       out  odata   : ns(8);
24                       in   address : ns(8);
25                       in   wr      : ns(1)) {
26     iptype "i8051buffer";
27     ipparm "core=my8051";
28     ipparm "xbus=0x4000";
29     ipparm "xrange=0x100"; // 256 locations at address 0x4000
30   }
31
32   dp triviumitf {                              Trivium control shell
33     sig updata, upins, upstatus : ns(8);
34     use my8051_data   (updata  );
35     use my8051_ins    (upins   );
36     use my8051_status (upstatus);
37
38     sig ramadr, ramidata, ramodata : ns(8);
39     sig wr : ns(1);
40     use my8051_xram(ramidata, ramodata, ramadr, wr);
41
42     sig si, so : ns(288);
43     sig z : ns(8);
44     use trivium80(si, so, z);                    Trivium kernel
45
46     sig k0, k1, k2, k3, k4, k5, k6, k7, k8, k9 : ns(288);
47     sig v0, v1, v2, v3, v4, v5, v6, v7, v8, v9 : ns(288);
48     sig init : ns(288);
49
50     reg tstate : ns(288);
51     sig newcnt : ns(8);
52     reg cnt    : ns(8);
53     reg ramcnt : ns(8);
54
55     always {
56       k0 = (upins == 0x1) ? tstate[287: 8] # updata : tstate;
57       k1 = (upins == 0x2) ? k0[287: 16] # updata # k0[ 7:0]
                   :k0;
58       k2 = (upins == 0x3) ? k1[287: 24] # updata # k1[15:0]
                   :k1;
59       k3 = (upins == 0x4) ? k2[287: 32] # updata # k2[23:0]
                   :k2;
```

```
60      k4 = (upins == 0x5) ? k3[287: 40] # updata # k3[31:0]
              :k3;
61      k5 = (upins == 0x6) ? k4[287: 48] # updata # k4[39:0]
              :k4;
62      k6 = (upins == 0x7) ? k5[287: 56] # updata # k5[47:0]
              :k5;
63      k7 = (upins == 0x8) ? k6[287: 64] # updata # k6[55:0]
              :k6;
64      k8 = (upins == 0x9) ? k7[287: 72] # updata # k7[63:0]
              :k7;
65      k9 = (upins == 0xA) ? k8[287: 80] # updata # k8[71:0]
              :k8;
66
67      v0 = (upins == 0xB)  ? k9[287:101] # updata # k9[ 92: 0]
               : k9;
68      v1 = (upins == 0xC)  ? v0[287:109] # updata # v0[100: 0]
               : v0;
69      v2 = (upins == 0xD)  ? v1[287:117] # updata # v1[108: 0]
               : v1;
70      v3 = (upins == 0xE)  ? v2[287:125] # updata # v2[116: 0]
               : v2;
71      v4 = (upins == 0xF)  ? v3[287:133] # updata # v3[124: 0]
               : v3;
72      v5 = (upins == 0x10) ? v4[287:141] # updata # v4[132: 0]
               : v4;
73      v6 = (upins == 0x11) ? v5[287:149] # updata # v5[140: 0]
               : v5;
74      v7 = (upins == 0x12) ? v6[287:157] # updata # v6[148: 0]
               : v6;
75      v8 = (upins == 0x13) ? v7[287:165] # updata # v7[156: 0]
               : v7;
76      v9 = (upins == 0x14) ? v8[287:173] # updata # v8[164: 0]
               : v8;
77
78      init = (upins == 015) ? 0b111 # v9[284:0]  : v9;
79
80      newcnt    = (upins == 0x16) ? updata : 0;
81      cnt       = (cnt) ? cnt - 1 : newcnt;
82      tstate    = (cnt) ? so : init;
83      si        = tstate;
84      upstatus  = cnt;
85
86      ramcnt    = (cnt) ? ramcnt + 1 : 0;
87      ramadr    = ramcnt;
88      wr        = (cnt) ? 1 : 0;
89      ramidata  = z;
90    }
91  }
92
93  system S {
94    my8051;
95    triviumitf;
96  }
```

### 13.2.3   Software Driver for 8051

The software driver for the above coprocessor is shown in Listing 13.7. This C code is written for the 8051 and can be compiled with SDCC, the Small Devices C Compiler (http://sdcc.sourceforge.net). This compiler allows one to use symbolic names for specific memory locations, such as the names of the I/O ports P0, P1, and P2.

The program demonstrates the loading of a key and initialization vector (lines 21–43), the execution of the key schedule (lines 46–50), and the generation of a keystream of 250 bytes (lines 53–56). Note that the software driver does not strictly follow the interleaving of active commands with ins_idle. However, this code will work fine for the hardware model from Listing 13.6.

As discussed before, the key scheduling of Trivium is similar to the normal operation of Trivium. Key scheduling involves running Trivium for a fixed number of rounds while discarding the keystream. Hence, the key scheduling part of the driver software is, apart from the number of rounds, identical to the encryption part.

Finally, line 64 illustrates how to terminate the simulation. By writing the value 0x55 into port P3, the simulation will halt. This is an artificial construct. Indeed, the software on a real micro-controller will run indefinitely.

**Listing 13.7**   8051 software driver for the Trivium coprocessor

```
1   #include <8051.h>
2
3   enum {ins_idle, ins_key0, ins_key1,
4         ins_key2, ins_key3, ins_key4, ins_key5,
5         ins_key6, ins_key7, ins_key8, ins_key9,
6         ins_iv0,  ins_iv1,  ins_iv2,  ins_iv3,
7         ins_iv4,  ins_iv5,  ins_iv6,  ins_iv7,
8         ins_iv8,  ins_iv9,  ins_init, ins_enc};
9
10  void terminate() {
11    // special command to stop simulator
12    P3 = 0x55;
13  }
14
15  void main() {
16    volatile xdata unsigned char *shared =
17      (volatile xdata unsigned char *) 0x4000;
18    unsigned i;
19
20    // program key, iv
21    P1 = ins_key0; P0 = 0x80;
22    P1 = ins_key1; P0 = 0x00;
23    P1 = ins_key2; P0 = 0x00;
24    P1 = ins_key3; P0 = 0x00;
25    P1 = ins_key4; P0 = 0x00;
26    P1 = ins_key5; P0 = 0x00;
27    P1 = ins_key6; P0 = 0x00;
28    P1 = ins_key7; P0 = 0x00;
```

```
29      P1 = ins_key8;  P0 = 0x00;
30      P1 = ins_key9;  P0 = 0x00;
31      P1 = ins_iv0;   P0 = 0x00;
32      P1 = ins_iv1;   P0 = 0x00;
33      P1 = ins_iv2;   P0 = 0x00;
34      P1 = ins_iv3;   P0 = 0x00;
35      P1 = ins_iv4;   P0 = 0x00;
36      P1 = ins_iv5;   P0 = 0x00;
37      P1 = ins_iv6;   P0 = 0x00;
38      P1 = ins_iv7;   P0 = 0x00;
39      P1 = ins_iv8;   P0 = 0x00;
40      P1 = ins_iv9;   P0 = 0x00;
41
42      // prepare for key schedule
43      P1 = ins_init;
44
45      // execute key schedule
46      P0 = 143; P1 = ins_enc;
47      P1 = ins_idle;
48
49      // wait until done
50      while (P2) ;
51
52      // produce 250 stream bytes
53      P0 = 250; P1 = ins_enc;
54      P1 = ins_idle;
55
56      while (P2) ;  // wait until done
57
58      // read out shared ram and send to port P0, P1
59      for (i=0; i< 8; i++) {
60        P0 = i;
61        P1 = shared[i];
62      }
63
64      terminate();
65    }
```

We can now compile the software driver and execute the simulation. The following commands illustrate the output generated by the program. Note that the 8051 micro-controller does not support standard I/O in the traditional sense: it is not possible to use printf statements without additional I/O hardware and appropriate software libraries. The instruction-set simulator deals with this limitation by printing the value of all ports each time a new value is written into them. Hence, the four columns below correspond to the value of P0, P1, P2 and P3 respectively. The tool output was annotated to clarify the meaning of the sequence of values.

```
> sdcc trivium.c
> gplatform tstream.fdl
i8051system: loading executable [trivium.ihx]
0xFF     0x01     0x00     0xFF
0x80     0x01     0x00     0xFF                        Program Key
```

```
0x80      0x02      0x00      0xFF
0x00      0x02      0x00      0xFF
0x00      0x03      0x00      0xFF
0x00      0x04      0x00      0xFF
0x00      0x05      0x00      0xFF
0x00      0x06      0x00      0xFF
0x00      0x07      0x00      0xFF
0x00      0x08      0x00      0xFF
0x00      0x09      0x00      0xFF
0x00      0x0A      0x00      0xFF

0x00      0x0B      0x00      0xFF             Program IV

0x00      0x0C      0x00      0xFF
0x00      0x0D      0x00      0xFF
0x00      0x0E      0x00      0xFF
0x00      0x0F      0x00      0xFF
0x00      0x10      0x00      0xFF
0x00      0x11      0x00      0xFF
0x00      0x12      0x00      0xFF
0x00      0x13      0x00      0xFF
0x00      0x14      0x00      0xFF
0x00      0x15      0x00      0xFF
0x8F      0x15      0x00      0xFF
0x8F      0x16      0x00      0xFF

0x8F      0x00      0x7A      0xFF             Run key schedule

0xFA      0x00      0x00      0xFF
0xFA      0x16      0x00      0xFF

0xFA      0x00      0xE5      0xFF             Produce 250 bytes

0x00      0x00      0x00      0xFF

0x00      0xCB      0x00      0xFF             First output byte

0x01      0xCB      0x00      0xFF

0x01      0xCC      0x00      0xFF             Second output byte

0x02      0xCC      0x00      0xFF

0x02      0xCE      0x00      0xFF             Third output byte

0x03      0xCE      0x00      0xFF
0x03      0x75      0x00      0xFF
0x04      0x75      0x00      0xFF
0x04      0x7B      0x00      0xFF
0x05      0x7B      0x00      0xFF
0x05      0x99      0x00      0xFF
0x06      0x99      0x00      0xFF
0x06      0xBD      0x00      0xFF
0x07      0xBD      0x00      0xFF
0x07      0x79      0x00      0xFF

0x07      0x79      0x00      0x55             Terminate
Total Cycles: 13332
```

   The last line of output shows 13,332 cycles, which is a long time when we realize
that a single key stream byte can be produced by the hardware within a single clock
cycle. How hard is it to determine intermediate time-stamps on the execution of

this program? While some instruction-set simulators provide direct support for this, we will need to develop a small amount of support code to answer this question. We will introduce an additional coprocessor command which, when observed by the triviumitf module, will display the current cycle count. This is a debug-only command, similar to the terminate call in the 8051 software.

The modifications for such a command to the code are minimal. In the C code, we add a function to call when we would like to see the current cycle count.

```
void showcycle() {
   P1 = 0x20; P1 = 0x0;
}
```

In the GEZEL code, we extend the triviumitf with a small FSM to execute the new command.

```
dp triviumitf {
   reg rupins : ns(8);
   ...
   always {
      ...
      rupins    = upins;
   }
   sfg show {
      $display("Cycle: ", $cycle);
   }
   sfg idle { }
}
fsm f_triviumitf(triviumitf) {
   initial s0;
   state s1;

   @s0 if (rupins == 0x20) then (show)  -> s1;
                           else (idle)  -> s0;
   @s1 if (rupins == 0x00) then (idle)  -> s0;
                           else (idle)  -> s1;
}
```

Each time showcycle() executes, the current cycle count will be printed by GEZEL. This particular way of measuring performance has a small overhead (88 cycles per call to showcycle()). We add the command in the C code at the following places.

- In the main function, just before programming the first key byte.
- In the main function, just before starting the key schedule.
- In the main function, just before starting the key stream.

Figure 13.7 illustrates the resulting cycle counts obtained from the simulation. The output shows that most time is spent in startup (initialization of the microcontroller), and that the software-hardware interaction, as expected, is expensive in cycle-cost. For example, programming a new key and re-running the key schedule costs 1,416 cycles, almost ten times as long as what is really needed by the hardware

**Fig. 13.7** Performance measurement of Trivium coprocessor

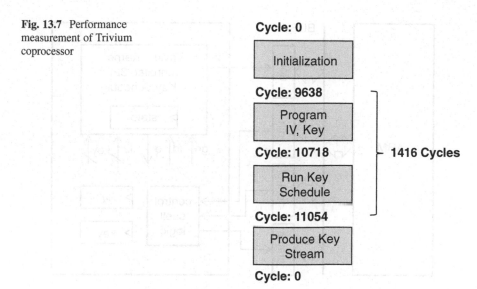

(143 cycles). This stresses once more the importance of carefully considering hardware-software interactions during the design.

## 13.3 Trivium for 32-bit Platforms

Our second Trivium coprocessor integrates the algorithm on a 32-bit StrongARM processor. We will compare two integration strategies: a memory-mapped interface and a custom-instruction interface. Both scenario's are supported through library modules in the GEZEL kernel. The hardware kernel follows the same ideas as before. By unrolling a trivium kernel 32 times, we obtain a module that produces 32 bits of keystream material per clock cycle. After loading the key and initialization vector, the key schedule of such a module has to execute for $4 * 288/32 = 36$ clock cycles before the first word of the keystream is available.

### 13.3.1 Hardware Platform Using Memory-Mapped Interfaces

Figure 13.8 shows the control shell design for a Trivium kernel integrated to a 32-bit memory-mapped interface. There are four memory-mapped registers involved: din, dout, control, and status. In this case, the key stream is directly read by the processor. The Trivium kernel follows the design we discussed earlier in Sect. 13.1.3. There is one additional control input, go, which is used to control the update of the state register. Instead of having a free-running Trivium kernel, the

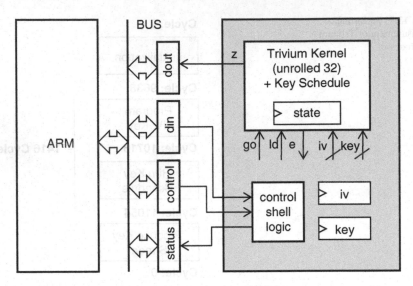

**Fig. 13.8**  Memory-mapped integration of Trivium on a 32-bit processor

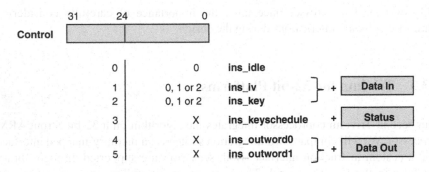

**Fig. 13.9**  Command set for a memory-mapped Trivium coprocessor

update of the state register will be strictly controlled by software, so that the entire keystream is captured using read operations from a memory-mapped interface.

As with other memory-mapped interfaces, our first task is to design a hardware interface to drive the Trivium kernel. We start with the command set. The command set must be able to load a key, an initialization vector, run the key schedule, and retrieve a single word from the key stream. Figure 13.9 illustrates the command set for this coprocessor.

The `control` memory-mapped register has a dual purpose. It transfers an instruction opcode as well as a parameter. The parameter indicates the part of the key or initial value which is being transferred. The parameter is 0, 1, or 2, since 3 words are sufficient to cover the 80 bits from the stream cipher key or the stream cipher initial value. The `ins_idle` instruction has the same purpose as before: it is used to synchronize the transfer of data operands with instructions. There are two

commands to retrieve keystream bits from the coprocessor: ins_outword0 and
ins_outword1. Both commands transfer a single word from the stream cipher
dout, and they are used alternately in order to avoid sending dummy ins_idle
to the coprocessor.

**Listing 13.8**  Hardware platform for the StrongARM coprocessor

```
1   ipblock myarm {                                              ARM Core
2     iptype "armsystem";
3     ipparm "exec=trivium";
4     ipparm "period=1";
5   }
6   ipblock armdout(in data : ns(32)) {                     ARM interfaces
7     iptype "armsystemsink";
8     ipparm "core=myarm";
9     ipparm "address=0x80000000";
10  }
11  ipblock armdin(out data : ns(32)) {
12    iptype "armsystemsource";
13    ipparm "core=myarm";
14    ipparm "address=0x80000004";
15  }
16  ipblock armstatus(in data : ns(32)) {
17    iptype "armsystemsink";
18    ipparm "core=myarm";
19    ipparm "address=0x80000008";
20  }
21  ipblock armcontrol(out data : ns(32)) {
22    iptype "armsystemsource";
23    ipparm "core=myarm";
24    ipparm "address=0x8000000C";
25  }
26
27  dp triviumitf(in   din    : ns(32);        Trivium hardware interface
28                out dout   : ns(32);
29                in   ctl    : ns(32);
30                out status : ns(32)) {
31    sig ld       : ns(1);
32    sig go       : ns(1);
33    sig iv, key  : ns(80);
34    sig e        : ns(1);
35    sig z        : ns(32);
36    use triviumtop(ld, go, iv, key, z, e);          Trivium kernel
37    reg ivr, keyr        : ns(80);
38    sig ivr0, ivr1, ivr2 : ns(32);
39    sig key0, key1, key2 : ns(32);
40    reg oldread : ns(3);
41
42    always {
43      iv  = ivr;
44      key = keyr;
45
```

```
46      // program new IV
47      ivr0= ((ctl[24:26] == 0x1) & (ctl[0:1] == 0x0)) ? din
                : ivr[31: 0];
48      ivr1= ((ctl[24:26] == 0x1) & (ctl[0:1] == 0x1)) ? din
                : ivr[63:32];
49      ivr2= ((ctl[24:26] == 0x1) & (ctl[0:1] == 0x2)) ? din
                : ivr[79:64];
50      ivr = ivr2 # ivr1 # ivr0;
51
52      // program new KEY
53      key0= ((ctl[24:26] == 0x2) & (ctl[0:1] == 0x0)) ? din
                : keyr[31: 0];
54      key1= ((ctl[24:26] == 0x2) & (ctl[0:1] == 0x1)) ? din
                : keyr[63:32];
55      key2= ((ctl[24:26] == 0x2) & (ctl[0:1] == 0x2)) ? din
                : keyr[79:64];
56      keyr= key2 # key1 # key0;
57
58      // control start
59      ld  = ((ctl[24:26] == 0x3) ? 1 : 0);
60
61      // read status
62      status = e;
63
64      // read output data
65      dout= z;
66
67      // trivium control
68      oldread = ((ctl[24:26]));
69      go  = ((ctl[24:26] == 0x4) & (oldread ==0x5)) |
70            ((ctl[24:26] == 0x5) & (oldread ==0x4)) |
71            ((ctl[24:26] == 0x3) & (oldread ==0x0));
72    }
73  }
74
75  dp triviumsystem {
76    sig din, dout, ctl, status : ns(32);
77    use myarm;
78    use triviumitf(din, dout, ctl, status);
79    use armdin(din);
80    use armdout(dout);
81    use armstatus(status);
82    use armcontrol(ctl);
83  }
84  system S {
85    triviumsystem;
86  }
```

Listing 13.8 shows the design of the hardware interface. The implementation of
the Trivium kernel is not shown, although a very similar design can be found in
Listing 13.4. The first part of Listing 13.8, lines 1–25, shows the memory-mapped
interface to the ARM core. This includes instantiation of the core (lines 1–5), and
four memory-mapped registers (line 6–25). The bulk of the code, lines 25–79,

contains the control shell for the Trivium kernel. The kernel is instantiated on line 36. The registers for key and initial value, defined on line 33, are programmed from software through a series of simple decode steps (line 46–62). The encoding used by the control memory mapped register corresponds to Fig. 13.9.

The control pins of the Trivium kernel (ld, go) are programmed by means of simple decoding steps as well (lines 65, 75–77). Note that the go pin is driven by a pulse of a single clock cycle, rather than a level programmed from software. This is done by detecting the exact cycle when the value of the control memory mapped interface changes. Note that the overall design of this control shell is quite simple, and does not require complex control or a finite state machine. Finally, the system integration consists of interconnecting the control shell and the memory-mapped interfaces (lines 81–89).

## 13.3.2  Software Driver Using Memory-Mapped Interfaces

A software driver for the memory-mapped Trivium coprocessor is shown in Listing 13.9. The driver programs the initial value and key, runs the key schedule, and next receives 512 words of keystream. The state update of the Trivium coprocessor is controlled by alternately writing 4 and 5 to the command field of the control memory-mapped interface. This is done during the key schedule (lines 28–32) as well as during the keystream generation (lines 34–39).

The code also contains an external system call getcyclecount(). This is a simulator-specific call, in this case specific to SimIt-ARM, to return the current cycle count of the simulation. By inserting such calls in the driver code (in this case, on lines 27, 33, 40), we can obtain the execution time of selected phases of the keystream generation.

To execute the system simulation, we compile the software driver, and run the GEZEL hardware module and the software executable in gplatform. The simulation output shows the expected keystream bytes: 0xcc, 0xce,0x75, ... The output also shows that the key schedule completes in 435 cycles, and that 512 words of keystream are generated in 10,524 cycles.

```
>arm-linux-gcc -static trivium.c cycle.s -o trivium
>gplatform trivium32.fdl
core myarm
armsystem: loading executable [trivium]
armsystemsink: set address 2147483648
armsystemsink: set address 2147483656
ccce757b ccce757b 99bd7920 9a235a88 1251fc9f aff0a655 7ec8ee4e
   bfd42128
86dae608 806ea7eb 58aec102 16fa88f4 c5c3aa3e b1bcc9f2 bb440b3f
   c4349c9f
key schedule cycles: 435   stream cycles: 10524
Total Cycles: 269540
```

**Listing 13.9** StrongARM software driver for the memory-mapped Trivium coprocessor

```
1  extern unsigned long long getcyclecount();
2
3  int main() {
4    volatile unsigned int *data   = (unsigned int *) 0x80000004;
5    volatile unsigned int *ctl    = (unsigned int *) 0x8000000C;
6    volatile unsigned int *output = (unsigned int *) 0x80000000;
7    volatile unsigned int *status = (unsigned int *) 0x80000008;
8
9    int i;
10   unsigned int stream[512];
11   unsigned long long c0, c1, c2;
12
13   // program iv
14   *ctl  = (1 << 24);           *data = 0;    // word 0
15   *ctl  = (1 << 24) | 0x1;                   // word 1
16   *ctl  = (1 << 24) | 0x2;                   // word 2
17
18   // program key
19   *ctl  = (2 << 24);           *data = 0x80; // word 0
20   *ctl  = (2 << 24) | 0x1; *data = 0;        // word 1
21   *ctl  = (2 << 24) | 0x2;                   // word 2
22
23   // run the key schedule
24   *ctl  = 0;
25   *ctl  = (3 << 24);  // start pulse
26
27   c0 = getcyclecount();
28   while (! *status) {
29     *ctl = (4 << 24);
30     if (*status) break;
31     *ctl = (5 << 24);
32   }
33   c1 = getcyclecount();
34   for (i=0; i<256; i++) {
35     stream[2*i] = *output;
36     *ctl = (4 << 24);
37     stream[2*i+1] = *output;
38     *ctl = (5 << 24);
39   }
40   c2 = getcyclecount();
41
42   for (i=0; i<16; i++) {
43     printf("%8x ", stream[i]);
44     if (!((i+1) % 8))
45       printf("\n");
46   }
47   printf("key_schedule_cycles:",
48          " %lld  stream_cycles: %lld\n",
49          c1 - c0, c2 - c1);
50
51   return 0;
52 }
```

How about the performance of this result? Since the Trivium kernel used in this design is unrolled 32 times (and thus can produce a new word every clock cycle), 512 words in 10,524 clock cycles is not a stellar result. Each word requires around 20 clock cycles. This includes synchronization of software and hardware, transfer of a result, writing that result into memory, and managing the loop counter and address generation (lines 34–39 in Listing 13.9).

Another way to phrase the performance question is: how much better is this result compared to an optimized full-software implementation? To answer this question, we can port an available, optimized implementation to the StrongARM and make a similar profiling. We used the implementation developed by Trivium's authors, C. De Canniere, in this profiling experiment, and found that this implementation takes 3,810 cycles for key schedule and 48,815 cycles for generating 512 words. Thus, each word of the keystream requires close to 100 clock cycles on the ARM. Therefore, we conclude that the hardware coprocessor is still five times faster compared to an optimized software implementation, even though that hardware coprocessor has an overhead factor of 20 times compared to a standalone hardware implementation.

Since we wrote the hardware from scratch, one may wonder if it wouldn't have been easier to try to *port* the Trivium software implementation into hardware. In practice, this may be hard to do, since the optimizations one does for software are very different than the optimizations one does for hardware. As an example, Listing 13.10 shows part of the software-optimized Trivium implementation of De Canniere. This implementation was written with 64-bit execution in mind. Clearly, the efficient translation of this code into hardware is quite difficult, since the specification does not have the same clarity compared to the algorithm definition we discussed at the start of the Chapter.

This completes our discussion of the memory-mapped Trivium coprocessor design. In the next section, we consider a third type of hardware/software interface for the Trivium kernel: the mapping of Trivium into custom instructions on a 32-bit processor.

## 13.3.3 Hardware Platform Using a Custom-Instruction Interface

The integration of a Trivium coprocessor as a custom datapath in a processor requires a processor that supports custom-instruction extensions. As discussed in Chap. 11, this has a strong impact on the tools that come with the processor. In this example, we will make use of the custom-instruction interface of the StrongARM processor discussed in Sect. 11.3.1. Figure 13.10 shows the design of a Trivium Kernel integrated into two custom-instruction interfaces, an OP3X1 and an OP2X2. The former is an instruction that takes three 32-bit operands and produces a single

**Listing 13.10** Optimized software implementation for Trivium

```
// Support Macro's
#define U32TO8_LITTLE(p, v) (((u32*)(p))[0] = U32TO32_LITTLE(v))
#define U8TO32_LITTLE(p) U32TO32_LITTLE(((u32*)(p))[0])
#define U32TO32_LITTLE(v) (v)

#define Z(w) (U32TO8_LITTLE(output + 4 * i, \
               U8TO32_LITTLE(input + 4 * i) ^ w))
#define S(a, n) (s##a##n)
#define T(a) (t##a)

#define S00(a, b) ((S(a, 1)<<( 32-(b))))
#define S32(a, b) ((S(a, 2)<<( 64-(b)))|(S(a, 1)>>((b)-32)))
#define S64(a, b) ((S(a, 3)<<( 96-(b)))|(S(a, 2)>>((b)-64)))
#define S96(a, b) ((S(a, 4)<<(128-(b)))|(S(a, 3)>>((b)-96)))

#define UPDATE()                                               \
  do {                                                         \
    T(1) = S64(1,  66) ^ S64(1,  93);                          \
    T(2) = S64(2,  69) ^ S64(2,  84);                          \
    T(3) = S64(3,  66) ^ S96(3, 111);                          \
                                                               \
    Z(T(1) ^ T(2) ^ T(3));                                     \
                                                               \
    T(1) ^= (S64(1,  91) & S64(1,  92)) ^ S64(2,  78);         \
    T(2) ^= (S64(2,  82) & S64(2,  83)) ^ S64(3,  87);         \
    T(3) ^= (S96(3, 109) & S96(3, 110)) ^ S64(1,  69);         \
  } while (0)

#define ROTATE()                                               \
  do {                                                         \
    S(1, 3) = S(1, 2); S(1, 2) = S(1, 1); S(1, 1) = T(3);      \
    S(2, 3) = S(2, 2); S(2, 2) = S(2, 1); S(2, 1) = T(1);      \
    S(3, 4) = S(3, 3); S(3, 3) = S(3, 2); S(3, 2) = S(3, 1);   \
    S(3, 1) = T(2);                                            \
  } while (0)


// ...

// This is the Trivium keystream generation loop

  for (i = 0; i < msglen / 4; ++i)
    {
      u32 t1, t2, t3;

      UPDATE();
      ROTATE();
    }
```

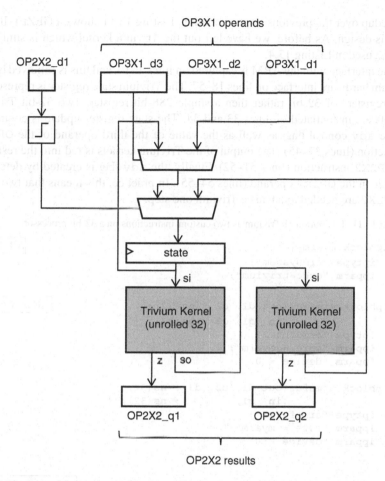

**Fig. 13.10** Custom-instruction integration of Trivium on a 32-bit processor

32-bit result. The latter is an instruction that takes two 32-bit operands and produces two 32-bit results.

During normal operation, the trivium state is fed through two Trivium kernels which each provide 32 bits of keystream. These two words form the results of an OP2x2 instruction. The same OP2x2 instruction also controls the update of the Trivium state. This way, each custom OP2x2 instruction advances the Trivium algorithm for one step, producing 64 bits of keystream. When the Trivium algorithm is not advancing, the state register can be reprogrammed by means of OP3x1 instructions. The third operand of OP3x1 selects which part of the 288-bit state register will be modified. The first and second operands contain 64 bit of state register data. The result of the OP3x1 instruction is ignored.

Thus, both programming and keystream retrieval can be done using a bandwidth of 64 bits, which is larger than the memory-mapped interface. Hence, we can expect a speedup over the previous implementation. Listing 13.11 shows a GEZEL listing for this design. As before, we have left out the Trivium kernel which is similar to the one used in Listing 13.4.

The interface with the ARM is captured on lines 1–16, and this is followed by the Trivium hardware interface on lines 18–57. The Trivium state register is represented as 9 registers of 32 bit rather then a single 288-bit register. Two 32-bit Trivium kernels are instantiated on lines 33 and 34. The state register update is controlled by the adv control flag, as well as the value of the third operand of the OP3X1 instruction (lines 37–45). The output of the Trivium kernels is fed into the result of the OP2X2 instruction (lines 51–52). Finally, the adv flag is created by detecting an *edge* in the OP2x2 operand (lines 54–55). In practice, this means that two calls to OP2X2 are needed to advance Trivium one step.

**Listing 13.11** Integration of Trivium as two custom-instructions on a 32-bit processor

```
1   ipblock myarm {                                              ARM core
2     iptype "armsystem";
3     ipparm "exec=trivium";
4   }
5   ipblock armsfu1(out d1, d2 : ns(32);                      ARM interfaces
6                    in  q1, q2 : ns(32)) {
7     iptype "armsfu2x2";
8     ipparm "core = myarm";
9     ipparm "device = 0";
10  }
11  ipblock armsfu2(out d1, d2, d3 : ns(32);
12                   in  q1          : ns(32)) {
13    iptype "armsfu3x1";
14    ipparm "core = myarm";
15    ipparm "device = 0";
16  }
17
18  dp triviumsfu {                                      Trivium control shell
19    sig o2x2_d1, o2x2_d2, o2x2_q1, o2x2_q2 : ns(32);
20    sig o3x1_d1, o3x1_d2, o3x1_d3, o3x1_q1 : ns(32);
21    use armsfu1( o2x2_d1, o2x2_d2, o2x2_q1, o2x2_q2);
22    use armsfu2( o3x1_d1, o3x1_d2, o3x1_d3, o3x1_q1);
23    use myarm;
24
25    reg w1, w2, w3, w4 : ns(32);
26    reg w5, w6, w7, w8 : ns(32);
27    reg w9             : ns(32);
28    reg tick           : ns(1);
29    sig adv            : ns(1);
30    sig si0, si1       : ns(288);
31    sig so0, so1       : ns(288);
32    sig z0, z1         : ns(32);
33    use trivium320(si0, so0, z0);                        Trivium kernel
34    use trivium321(si1, so1, z1);                        Trivium kernel
35
```

```
36      always {
37        w1 = adv ? so1[  0: 31]  : ((o3x1_d3 == 1) ? o3x1_d1 : w1);
38        w2 = adv ? so1[ 32: 63]  : ((o3x1_d3 == 1) ? o3x1_d2 : w2);
39        w3 = adv ? so1[ 64: 95]  : ((o3x1_d3 == 2) ? o3x1_d1 : w3);
40        w4 = adv ? so1[ 96:127]  : ((o3x1_d3 == 2) ? o3x1_d2 : w4);
41        w5 = adv ? so1[128:159]  : ((o3x1_d3 == 3) ? o3x1_d1 : w5);
42        w6 = adv ? so1[160:191]  : ((o3x1_d3 == 3) ? o3x1_d2 : w6);
43        w7 = adv ? so1[192:223]  : ((o3x1_d3 == 4) ? o3x1_d1 : w7);
44        w8 = adv ? so1[224:255]  : ((o3x1_d3 == 4) ? o3x1_d2 : w8);
45        w9 = adv ? so1[256:287]  : ((o3x1_d3 == 5) ? o3x1_d1 : w9);
46        o3x1_q1 = 0;
47
48        si0 = w9 # w8 # w7 # w6 # w5 # w4 # w3 # w2 # w1;
49        si1 = so0;
50
51        o2x2_q1 = z0;
52        o2x2_q2 = z1;
53
54        tick = o2x2_d1[0];
55        adv  = (tick != o2x2_d1[0]);
56      }
57    }
58
59    system S {
60      triviumsfu;
61    }
```

### 13.3.4   Software Driver for a Custom-Instruction Interface

Listing 13.12 shows a software driver for the Trivium custom-instruction processor that generates a keystream of 512 words in memory. The driver starts by loading key and data (lines 25–30), running the key schedule (lines 34–37), and retrieving the keystream (lines 41–48). At the same time, the getcyclecount system call is used to determine the performance of the key schedule and the keystream generation part.

**Listing 13.12** Custom-instruction software driver for the Trivium ASIP

```
1   #include <stdio.h>
2   #define OP2x2_1(D1,D2,S1,S2) \
3      asm volatile ("smullnv_%0,_%1,_%2,_%3": \
4          "=&r"(D1),"=&r"(D2) : \
5          "r"(S1),"r"(S2));
6
7   #define OP3x1_1(D1, S1, S2, S3) \
8      asm volatile ("mlanv_%0,_%1,_%2,_%3": \
9          "=&r"(D1): "r"(S1), "r"(S2), "r"(S3)); \
10
11  extern unsigned long long getcyclecount();
12
```

```
13  int main() {
14    int z1, z2, i;
15    unsigned int stream[512];
16    unsigned long long c0, c1, c2;
17
18    int key1 = 0x80;
19    int key2 = 0xe0000000;
20
21    // clear 'tick'
22    OP2x2_1(z1, z2, 0, 0);
23
24    // load key = 80 and IV = 0
25    OP3x1_1(z1,key1, 0, 1);
26    OP3x1_1(z1,   0, 0, 2);
27    OP3x1_1(z1,   0, 0, 3);
28    OP3x1_1(z1,   0, 0, 4);
29    OP3x1_1(z1,key2, 0, 5);
30    OP3x1_1(z1,   0, 0, 0);
31
32    // run key schedule
33    c0 = getcyclecount();
34    for (i=0; i<9; i++) {
35      OP2x2_1(z1, z2, 1, 0);
36      OP2x2_1(z1, z2, 0, 0);
37    }
38    c1 = getcyclecount();
39
40    // run keystream
41    for (i=0; i<128; i++) {
42      OP2x2_1(z1, z2, 1, 0);
43      stream[4*i]   = z1;
44      stream[4*i+1] = z2;
45      OP2x2_1(z1, z2, 0, 0);
46      stream[4*i+2] = z1;
47      stream[4*i+3] = z2;
48    }
49    c2 = getcyclecount();
50
51    for (i=0; i<16; i++) {
52      printf("%8x ", stream[i]);
53      if (!((i+1) % 8))
54        printf("\n");
55    }
56    printf("key_schedule_cycles:",
57           "%lld__stream_cycles:_%lld\n",
58           c1 - c0, c2 - c1);
59
60    return 0;
61  }
```

The algorithm can be compiled with the ARM cross-compiler and simulated on top of GEZEL gplatform. This results in the following output.

```
>arm-linux-gcc -static trivium.c cycle.s -o trivium
```

```
>gplatform triviumsfu.fdl
core myarm
armsystem: loading executable [trivium]
ccce757b 99bd7920 9a235a88 1251fc9f aff0a655 7ec8ee4e bfd42128
  86dae608
806ea7eb 58aec102 16fa88f4 c5c3aa3e b1bcc9f2 bb440b3f c4349c9f
  be0a7e3c
key schedule cycles: 289  stream cycles: 8062
Total Cycles: 42688
```

We can verify that, as before, the correct keystream is generated. The cycle count of the algorithm is significantly smaller than before: the key schedule completes in 289 cycles, and the keystream is generated within 8,862 cycles. This implies that each word of keystream required around 17 cycles. If we turn on the O3 flag while compiling the driver code, we obtain 67 and 1,425 clock cycles for key schedule and keystream respectively, implying that each word of the keystream requires less than three cycles! Hence, we conclude that for this design, an ASIP interface is significantly more efficient than a memory-mapped interface.

## 13.4  Summary

In this chapter, we designed a stream cipher coprocessor for three different hosts: a small 8-bit micro-controller, a 32-bit SoC processor, and a 32-bit ASIP. In each of these cases, we created a hardware interface to match the coprocessor to the available hardware-software interface. The stream cipher algorithm was easy to scale over different word-lengths by simply unrolling the algorithm. The performance evaluation results of all these implementations are captured in Table 13.2. These results demonstrate two points. First, it is not easy to exploit the parallelism of hardware. All of the coprocessors are limited by their hardware/software interface or the speed of software on the host, not by the computational limits of the hardware coprocessors. Second, the wide variation of performance results

**Table 13.2** Performance evaluation of Trivium coprocessors on multiple platforms

| Platform | Hardware | 8051 | Hardware | StrongARM | Unit |
|---|---|---|---|---|---|
| Interface | Native | Port-mapped | Native | Memory mapped | |
| Wordlength | 8 | 8 | 32 | 32 | bit |
| Key schedule | 144 | 1,416 | 36 | 435 | cycles |
| Key stream | 4 | 6.7 | 1 | 20.5 | cycles/word |
| Platform | StrongARM | StrongARM | StrongARM | | Unit |
| Interface | SW | ASIP | ASIP ($-$O3) | | |
| Wordlength | 32 | 64 | 64 | | bit |
| Key schedule | 3,810 | 289 | 67 | | cycles |
| Key stream | 95 | 17 | 3 | | cycles/word |

underline the importance of a carefully designed hardware interface, and a careful consideration of the application when selecting a hardware/software interface.


## 13.5   Further Reading

The standard reference of cryptographic algorithms is by Menezes et al. (2001). Of course, cryptography is a fast-moving field. The algorithm described in this section was developed for the eStream Project (ECRYPT 2008) in 2005. The Trivium specifications are by De Canniere and Preneel (2005). The Trivium webpage in the eStream website describes several other hardware implementations of Trivium.


## 13.6   Problems

**Problem 13.1.** Design a hardware interface for the Trivium algorithm on top of a *Fast Simplex Link* interface. Please refer to Sect. 11.2.1 for a description of the FSL timing and the FSL protocol. Assume the following interface for your module.

```
dp trivium_fsl (in   idata   : ns(32);   // input slave interface
                in   exists  : ns(1);
                out  read    : ns(1);
                out  odata   : ns(32);   // output master interface
                in   full    : ns(1);
                out  write   : ns(1))
```

**Problem 13.2.** Consider a simple linear feedback shift register, defined by the following polynomial: $g(x) = x^{35} + x^2 + 1$. A possible hardware implementation of this LFSR is shown in Fig. 13.11. This polynomial is *primitive*, which implies that the LFSR will generate a so-called *m-sequence*: for a given initialization of the registers, the structure will cycle through all possible $2^{35} - 1$ states before returning to the same state.

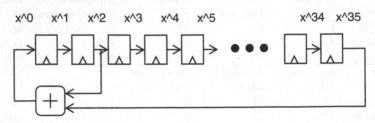

**Fig. 13.11** LFSR for $g(x) = x^{35} + x^2 + 1$

(a) Write an optimized software implementation of an LFSR generator that calculates the first 1,024 states starting from the initialization $x^{32} = x^{33} = x^{34} = x^{35} = 1$ and all other bits 0. For each state you need to store only the first 32 bits.

(b) Write an optimized standalone hardware implementation of an LFSR generator that calculates the first 1,024 states starting from the initialization $x^{32} = x^{33} = x^{34} = x^{35} = 1$ and all other bits 0. You do not need to store the first 32 bits, but can feed them directly to an output port.

(c) Design a control shell for the module you have designed under (b), and use a memory-mapped interface to capture and store the first 1,024 states of the LFSR. You only need to capture the first 32 bits of each state. Compare the resulting performance to the solution of (a).

(d) Design a control shell for the module you have designed under (b), and use a custom-instruction interface to capture and store the first 1,024 states of the LFSR. You only need to capture the first 32 bits of each state. Compare the resulting performance to the solution of (a).

# Chapter 14
# AES Co-processor

## 14.1 AES Encryption and Decryption

Figure 14.1 shows an overview of the AES block cipher encryption and decryption algorithm.

Encryption transforms a block of 128-bits of plaintext, with the help of a 128 bit secret key, into a block of 128 bits of ciphertext. Decryption does the opposite. Using the same key, decryption converts ciphertext into plaintext.

Encryption and decryption are structured as a series of *rounds*. There is an initial round, followed by nine identical regular rounds, followed by a final round. Each round is made out of a combination of the four transformations: addroundkey, subbytes, shiftrows, and mixcolumns.

Each round uses 128 bits of key material, called a roundkey. There is one roundkey for each round, for a total of 11 roundkeys. The roundkeys are derived through a key expansion step that transforms the 128 bit input key into the 11 roundkeys. AES is a symmetric-key algorithm, and encryption and decryption are inverse operations. In Fig. 14.1, you can observe that the steps, as well as the rounds, for encryption and decryption are each other's inverse. Furthermore, the roundkey order for decryption is the opposite of the roundkey order for encryption.

The detailed definition of AES encryption and decryption is defined in a document called *FIPS-197 Advanced Encryption Standard*. It's easy to track it down online, including a reference implementation in software. In addition, the book website includes a reference GEZEL implementation for AES.

## 14.2 Memory-Mapped AES Encryption Coprocessor

Our first implementation is a memory-mapped coprocessor for AES Encryption. We wish to obtain a hardware equivalent for the following function call in software.

P.R. Schaumont, *A Practical Introduction to Hardware/Software Codesign*,                    409
DOI 10.1007/978-1-4614-3737-6_14, © Springer Science+Business Media New York 2013

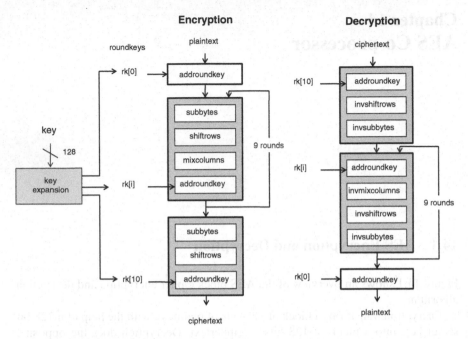

**Fig. 14.1**  AES encryption and decryption

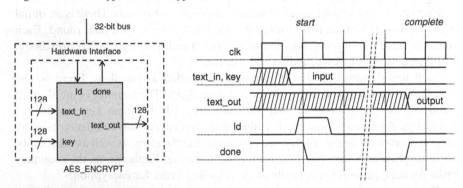

**Fig. 14.2**  AES encryption module interface

```
void encrypt(unsigned plaintext[4],
             unsigned key[4],
             unsigned ciphertext[4]);
```

In this function call, plaintext and key are input arguments, and ciphertext is an output argument. We also assume, in this design, that the key is only programmed occasionally, so that we can treat it as a parameter. The plaintext and ciphertext, on the other hand, are arguments that change for every call to the coprocessor.

Figure 14.2 shows a hardware implementation of the AES algorithm. The module has two data inputs, text_in and key. There is also one data output,

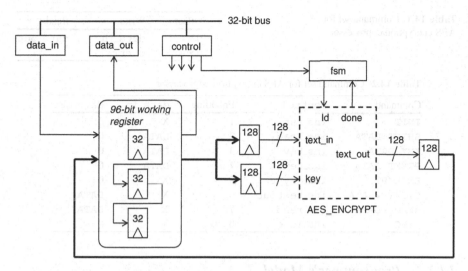

**Fig. 14.3**  Hardware interface for AES encryption

text_out. The timing diagram in Fig. 14.2 shows how the interface works. The encryption starts when both ld and done are high. The hardware reads a key and a plaintext from the 128-bit input ports, and several clock cycles later, a corresponding ciphertext is generated, and the done pin is raised again.

Our objective, in this codesign problem, is to design a hardware interface for this AES module, as well as a software driver, so that the function call encrypt is implemented on the coprocessor hardware.

## 14.2.1  Hardware Interface Operation

In this memory-mapped coprocessor, the hardware interface is operated through a 32-bit bus. We have to design a command set to operate the hardware AES module according to the interface protocol of Fig. 14.2. This includes loading of plaintext and key, retrieving of ciphertext, and controlling the encryption. Furthermore, since the key is a parameter and the plaintext input is an argument, both should be programmable as separate items. Finally, the blocks of data are 128-bit, and they need to be handled using a 32-bit processor. Thus, we will need at least four transfers per block of data.

Figure 14.3 shows the datapath of the control shell. Three 32-bit control shell ports are included: data_in, data_out, and decode. The 128-bit operands are assembled using a 96-bit working register in combination with a data input port and a data output port. The control port steers the update of the working register and the control pins of the encryption module.

**Table 14.1** Command set for
AES encryption coprocessor

| Offset | Write | Read |
|---|---|---|
| $0 \times 0$ | data_in | data_out |
| $0 \times 4$ | control | |

**Table 14.2** Command set for AES encryption coprocessor

| Command | control | Encoding | data_in | data_out |
|---|---|---|---|---|
| INIT | ins_rst | 1 | X | 0 |
| SHIFT DATA | ins_load | 2 | DATA | 0 |
| KEY DATA | ins_key | 3 | DATA | 0 |
| PTEXT DATA | ins_text | 4 | DATA | 0 |
| ENCRYPT | ins_crypt | 5 | X | 0 |
| CTEXT *DATA | ins_textout | 6 | X | DATA |
| READ *DATA | ins_read | 7 | X | DATA |
| SYNC | ins_idle | 0 | X | 0 |

## 14.2.2   Programmer's Model

Next, we consider the software view of the hardware interface, and define the instruction set of the coprocessor. Table 14.1 shows the address map. The three memory-mapped registers are mapped onto two addresses; one for data, and one for control.

An *instruction* for the AES coprocessor is the combination of a value written to the control register in combination with a value written to (or read from) the data register. Table 14.2 describes the command set of the coprocessor. The first column (Command) is a mnemonic representation of each command. The second column (control) shows the value for control port. The third column shows the encoding of control. The fourth and fifth columns (data_in, data_out) show the arguments of each instruction.

These commands have the following meaning.

- INIT is used to initialize the coprocessor.
- SHIFT is used to shift data into the working register of the coprocessor. The argument of this command is DATA.
- KEY is used to copy the working register of the coprocessor to the key register. The argument of this command is DATA.
- PTEXT is used to copy the working register of the coprocessor to the plaintext input register. The argument of this command is DATA.
- ENCRYPT is to initiate the encryption operation on the coprocessor. The encryption for this AES module completes in ten clock cycles.
- CTEXT copies the cipher output register of the coprocessor to the working register. This command returns a result in *DATA.
- READ is used to shift data out of the working register of the coprocessor. This command returns a result in *DATA.

• SYNC is used as part of the high-level synchronization protocol used by the coprocessor. This synchronization needs to make sure that the command written to the control register is consistent with the value on the data_in or data_out ports. This works as follows. Each instruction for the coprocessor is a sequence of two values at the control port, SYNC followed by an active command. To transfer data to the coprocessor, the data_in port needs to be updated between the SYNC command and the active command. The retrieve data from the coprocessor, the data_out port needs to be read after the active command in the sequence SYNC, active command.

Each high level function call in C can now be converted into a sequence of coprocessor commands. The following example illustrates the sequence of commands required to load a key and a plaintext block onto the coprocessor, perform encryption, and retrieve the ciphertext. This command sequence will be generated through software using memory-mapped write and read operations to control, data_in and data_out.

```
// Command Sequence for Encryption
// Input:   plaintext[0..3]    4 words of plaintext
//          key[0..3]          4 words of key
// Output: ciphertext[0..4]    4 words of ciphertext
    SYNC
    SHIFT plaintext[0]
    SYNC
    SHIFT plaintext[1]
    SYNC
    SHIFT plaintext[2]
    SYNC
    PTEXT plaintext[3]

    SYNC
    SHIFT key[0]
    SYNC
    SHIFT key[1]
    SYNC
    SHIFT key[3]
    SYNC
    KEY key[4]

    ENCRYPT

    SYNC
    CTEXT ciphertext[0]
    SYNC
    READ ciphertext[1]
    SYNC
    READ ciphertext[2]
    SYNC
    READ cuphertext[3]
```

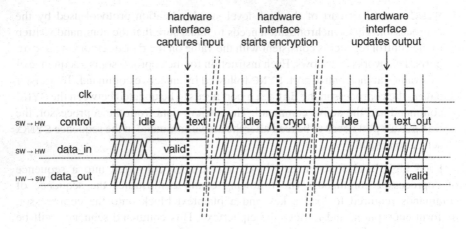

**Fig. 14.4** AES hardware interface operation

Figure 14.4 clarifies the manner in which the software implements a synchronization point with the hardware. The clk signal in this diagram is the hardware clock,
which can be unrelated to the clock of the microprocessor. The signals control,
data_in, and data_out are ports of the hardware interface. They are controlled
by software, and their value can change asynchronously from the hardware. The
interleaved idle/active sequence on the control port enables the hardware to select a
single clock cycle when the data_in must have a known value, when to start the
encryption, and when the data_out must be updated.

### 14.2.3   Software Driver Design

Listing 14.1 shows a software driver for the AES encryption processor. We're
using a memory-mapped hardware/software interface. The three memory-mapped
registers are defined on lines 4–6. The command encoding is captured in an enum
statement on line 1. The pointers to each memory-mapped register are initialized
in the function init. The addresses used by the design depends on the hardware
implementation.

The coprocessor is operated through two functions, set_key and do_encrypt.
The set_key function transfers four words of an 128-bit key using the protocol
described above. First, control is set to ins_idle and the data_in argument is
loaded. Next, the actual command is written into control. Using ins_load, the
first three words of the key are shifted into the working register. Using ins_key,
all 128 bits of the key register are programmed.

The do_encrypt function shows a similar sequence for loading the plaintext.
Next, it will call the encryption command. Note that, in the listing, there is
no test to check if the encryption completes. In this particular design, this is

**Listing 14.1** A C driver for an AES memory-mapped coprocessor

```
1   enum {ins_idle, ins_rst, ins_load, ins_key,
2           ins_text, ins_crypt, ins_textout, ins_read};
3
4   volatile unsigned int *control;   // memory-mapped register
                                      //           for control
5   volatile unsigned int *data_in;   // memory-mapped register
                                      //           for data_in
6   volatile unsigned int *date_out;  // memory-mapped register
                                      //           for data_out
7
8   void init() {
9     control  = (int *) 0x80000000;
10    data_in  = (int *) 0x80000004;
11    data_out = (int *) 0x80000008;
12  }
13
14  void set_key(unsigned key[4]) {
15    unsigned i;
16    for (i=0; i < 4; i++) {
17      *control = ins_idle;
18      *data_in = key[i];
19      *control = (i == 3) ? ins_key : ins_load;
20    }
21  }
22
23  void do_encrypt(unsigned plaintext[4],
24                  unsigned ciphertext[4]) {
25    unsigned i;
26    for (i=0; i < 4; i++) {
27      *control = ins_idle;
28      *data_in = plaintext[i];
29      *control = (i == 3) ? ins_text : ins_load;
30    }
31    *control = ins_idle;
32    *control = ins_crypt;
33    for (i=0; i < 4; i++) {
34      *control = ins_idle;
35      *control = (i == 0) ? ins_textout : ins_read;
36      ciphertext[i] = *data_out;
37    }
38
39  }
```

acceptable because the AES hardware finishes encryption in just ten clock cycles. If a coprocessor would operate for a longer time, or an unknown time, then an additional status completion check must be included.

Finally, in lines 26–40, the do_encrypt function retrieves the result from the coprocessor. This works again using an interleaved ins_idle/command sequence. First, control is set to ins_idle. Next, the actual command retrieves the output argument. Using ins_textout, the working register is initialized with the output encryption result. Using ins_read, this result is gradually shifted out of the working register. In the next section, we discuss an RTL implementation of the hardware interface.

## 14.2.4 Hardware Interface Design

Listing 14.2 shows a GEZEL implementation of a control shell for the AES coprocessor. The AES hardware module is instantiated on line 13, and it is controlled through three ports: control, data_in, data_out. Several registers (key, text_in, text_out) surround the aes module following the arrangement as shown in Fig. 14.3.

The easiest way to understand the operation of this design is to start with the FSM description on lines 45–71. The overall operation of the decoder FSM is an infinite loop that accepts a command from software, and then executes that command. Each state performs a specific step in the command execution.

- In state s1, the FSM tests insreg, which holds the latest value of the control port, against each possible command. Obviously, it must follow the command encoding chosen earlier for the C driver program. Depending on the value of the command, the FSM will transition to state s2, s3, s5, s6.
- State s2 performs the second half of the command handshake protocol, and waits for ins_idle before going back to s1 for the next command. State s2 is used for the SYNC command.
- State s3 is entered after the ENCRYPT command. This state waits for the encryption to complete. Thus, the control shell is unable to accept new instructions while the coprocessor is operational. The software can detect command completion by reading and testing the value of the data_out memory-mapped register. During the ENCRYPT command, the register will reflect the value of the done flag of the coprocessor.
- State s5 is entered when the first word of the output is read back in software.
- State s6 is entered when the next three words of the output are read back in software.

The datapath of the control shell (Listing 14.2, lines 15–43) implements the register transfers that map the control shell ports to the AES module ports and back.

**Listing 14.2** A control shell for the AES coprocessor

```
1  dp aes_decoder (in   control : ns( 8);
2                  in   data_in : ns(32);
3                  out data_out : ns(32)) {
4
5    reg rst, ld, done              : ns(  1);
6    reg key, text_in, text_out    : ns(128);
7    sig sigdone                    : ns(  1);
8    sig sigtext_out                : ns(128);
9    reg wrkreg0, wrkreg1, wrkreg2 : ns( 32);
10   reg insreg                     : ns(  8);
11   reg dinreg                     : ns( 32);
12
13   use aes_top(rst, ld, sigdone, key, text_in, sigtext_out);
14
15   always       { insreg    = control;
16                  dinreg    = data_in;
17                  done      = sigdone;
18                  text_out = sigdone ? sigtext_out: text_out;}
19   sfg dout_d   { data_out = done;                        }
20   sfg dout_t   { data_out = text_out[127:96];            }
21   sfg dout_w   { data_out = wrkreg2;                     }
22   sfg aes_idle { rst = 0;  ld  = 0;                      }
23   sfg aes_rst  { rst = 1;  ld  = 0;                      }
24   sfg aes_ld   { rst = 0;  ld  = 1;                      }
25   sfg putword  { wrkreg0   = dinreg;
26                  wrkreg1   = wrkreg0;
27                  wrkreg2   = wrkreg1;                    }
28   sfg setkey   { key       = (wrkreg2 << 96) |
29                              (wrkreg1 << 64) |
30                              (wrkreg0 << 32) |
31                               dinreg;                    }
32   sfg settext  { text_in  = (wrkreg2 << 96) |
33                              (wrkreg1 << 64) |
34                              (wrkreg0 << 32) |
35                               dinreg;                    }
36   sfg gettext  { data_out = text_out[127:96];
37                  wrkreg2   = text_out[95:64];
38                  wrkreg1   = text_out[63:32];
39                  wrkreg0   = text_out[31:0];             }
40   sfg shiftw   { wrkreg2   = wrkreg1;
41                  wrkreg1   = wrkreg0;                    }
42   sfg getword  { data_out = wrkreg2;                     }
43 }
44
45 fsm faes_decoder(aes_decoder) {
46   initial s0;
47   state s1, s2, s3, s4, s5, s6;
48   @s0 (aes_idle, dout_0) -> s1;
49   @s1 if (insreg == 1)       then (aes_rst,  dout_0)-> s2;
50      else if (insreg == 2) then (aes_idle, putword, dout_0)
          -> s2;
```

```
51          else if (insreg == 3) then (aes_idle, setkey,  dout_0)
                -> s2;
52          else if (insreg == 4) then (aes_idle, settext, dout_0)
                -> s2;
53          else if (insreg == 5) then (aes_ld,   dout_d)      -> s3;
54            else if (insreg == 6) then (aes_idle, gettext) -> s5;
55            else if (insreg == 7) then (aes_idle, getword) -> s6;
56                               else (aes_idle, dout_0)      -> s1;
57    // SYNC
58    @s2 if (insreg == 0)         then (aes_idle, dout_0)      -> s1;
59                                 else (aes_idle, dout_0)      -> s2;
60    // ENCRYPT
61    @s3 if (done == 1)           then (aes_idle, dout_d)      -> s4;
62                                 else (aes_idle, dout_d)      -> s3;
63    @s4 if (insreg == 0)         then (aes_idle, dout_d)      -> s1;
64                                 else (aes_idle, dout_d)      -> s4;
65    // CTEXT
66    @s5 if (insreg == 0)         then (aes_idle, dout_0)      -> s1;
67                                 else (aes_idle, dout_t)      -> s5;
68    // READ
69    @s6 if (insreg == 0)         then (aes_idle, shiftw, dout_0)
                -> s1;
70                                 else (aes_idle, dout_w)      -> s6;
71  }
```

Figure 14.5 illustrates the design process of this codesign. We started from a custom hardware module and integrated that into software. This requires the selection of a hardware/software interface, and the definition of a command set. Once these are defined, the integration of hardware and software follow two independent paths. For software, we created a software driver that provides a smooth transition from a high-level API to the hardware/software interface. For hardware, we encapsulated the hardware module into a control shell that connects directly onto the hardware/software interface.

In the next section, we analyze the performance of the resulting design.

## 14.2.5  System Performance Evaluation

Hardware interface design has substantial impact on the overall system performance of a design. We will evaluate the performance of the AES coprocessor following the scheme of Fig. 14.6. We compare an all-software, optimized AES implementation with an all-hardware standalone AES implementation. We also compare these results against the performance of an integrated coprocessor using the software driver and control shell developed earlier. The experiments were done using GEZEL and the Simit-ARM instruction set simulator. A C version of the AES algorithm

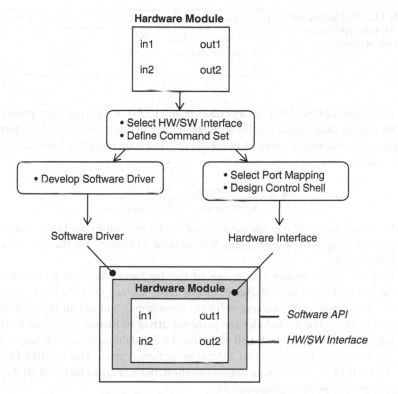

**Fig. 14.5** Development of the hardware interface

**Fig. 14.6** AES coprocessor
performance evaluation

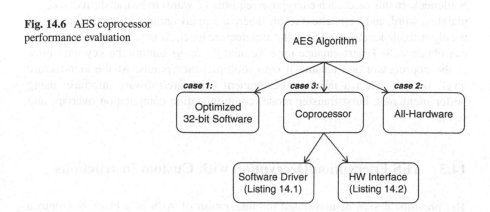

**Table 14.3** Performance of
100 AES encryptions on
different platforms

| Implementation | Cycle count | Speedup over software |
|---|---|---|
| AES software (32-bit) | 362,702 | 1.0 |
| AES custom hardware | 1,100 | 329.7 |
| AES coprocessor | 33,381 | 10.9 |

(derived from the OpenSSH library) takes around 3,600 cycles per encryption on an ARM. On the other hand, a full-hardware implementation of AES which requires 1 clock cycle per round, takes 11 clock cycles. The speedup of the hardware design is now defined as:

$$S = \frac{cycles_{software}}{cycles_{hardware}} \times \frac{T_{clock,software}}{T_{clock,hardware}} \tag{14.1}$$

If we assume that the microprocessor and the custom-hardware run at the same clock frequency, the speedup $S$ is around 327 times for the full hardware implementation.

Next, we also compare the design of the hardware AES integrated onto the hardware interface discussed above. In this case, we use the GEZEL cosimulator to obtain combine hardware and software simulations and obtain the overall performance. The software includes the software driver of Listing 14.1. The hardware includes the hardware control shell of Listing 14.2 and the custom hardware module. The system cycle count is around 334 cycles per encryption. This is still a factor of 10.9 faster than the all-software implementation, but is it also a factor of 30.2 *slower* than the all-hardware implementation.

Table 14.3 shows the performance summary of the AES design. The analysis of the speedup factors shows that a hardware/software interface can easily become a bottleneck. In this case, each encryption requires 12 words to be transferred: a key, a plaintext word, and a ciphertext result. There are many optimization possibilities, at the algorithmic level as well as at the architecture level. At the algorithmic level, we can obtain a 30 % performance improvement by programming the key only once in the coprocessor and reusing it over multiple encryptions. At the architecture level, we can select a faster, more efficient hardware/software interface: using buffer memories, burst transfer mode, communication-computation overlap, and so on.

## 14.3   AES Encryption/Decryption with Custom Instructions

The previous design demonstrated the integration of AES as a black-box onto a memory bus. The integration in software was implemented 'bottom-up'. In this section, we will design an AES coprocessor in a 'top-down' fashion. Starting from a reference implementation in C, we will identify a suitable set of custom-instructions to accelerate the software. The custom-instructions will support both encryption and decryption, and will use the NiosII custom-instruction interface (See Sect. 11.3.3).

### 14.3.1 AES T-box Reference Implementation

Listing 14.3 demonstrates a so-called T-box implementation of AES Encryption. The T-box design is a version of AES specifically optimized for 32-bit processors. The name T-box stems from the fact that it is heavily based on the use of lookup tables.

This implementation organizes the 128-bit intermediate state of AES in four 32-bit variables. The initial round, lines 10–13, adds the first roundkey to the plaintext data and produces the intermediate state s0, s1, s2 and s3. The loop on lines 16–61 computes two AES rounds per loop iteration. The first round (lines 17–36) reads the intermediate state s0 through s3 and produces t0 through t3. The second round (lines 41–60) reads the intermediate state t0 through t3 and produces s0 through s3. This particular arrangement minimizes the amount of data copying. The final round, lines 63–82, produces the ciphertext.

**Listing 14.3** AES T-box reference implementation for encryption

```
1   void AES_encrypt(const unsigned char *in,
2               unsigned char *out,
3               const AES_KEY *key) {
4       const u32 *rk;
5       u32 s0, s1, s2, s3, t0, t1, t2, t3;
6       int r;
7
8       rk = key->rd_key;
9
10      s0 = GETU32(in      ) ^ rk[0];              Initial Round
11      s1 = GETU32(in +  4) ^ rk[1];
12      s2 = GETU32(in +  8) ^ rk[2];
13      s3 = GETU32(in + 12) ^ rk[3];
14
15      r = key->rounds >> 1;
16      for (;;) {
17          t0 = Te0[(s0 >> 24)       ] ^          Regular Round
18               Te1[(s1 >> 16) & 0xff] ^
19               Te2[(s2 >>  8) & 0xff] ^
20               Te3[(s3      ) & 0xff] ^
21               rk[4];
22          t1 = Te0[(s1 >> 24)       ] ^
23               Te1[(s2 >> 16) & 0xff] ^
24               Te2[(s3 >>  8) & 0xff] ^
25               Te3[(s0      ) & 0xff] ^
26               rk[5];
27          t2 = Te0[(s2 >> 24)       ] ^
```

```
28                    Te1[(s3 >> 16) & 0xff] ^
29                    Te2[(s0 >>  8) & 0xff] ^
30                    Te3[(s1      ) & 0xff] ^
31                    rk[6];
32            t3 = Te0[(s3 >> 24)       ] ^
33                    Te1[(s0 >> 16) & 0xff] ^
34                    Te2[(s1 >>  8) & 0xff] ^
35                    Te3[(s2      ) & 0xff] ^
36                    rk[7];
37          rk += 8;
38          if (--r == 0)
39              break;
40
41          s0 = Te0[(t0 >> 24)       ] ^
42                    Te1[(t1 >> 16) & 0xff] ^
43                    Te2[(t2 >>  8) & 0xff] ^
44                    Te3[(t3      ) & 0xff] ^
45                    rk[0];
46          s1 = Te0[(t1 >> 24)       ] ^
47                    Te1[(t2 >> 16) & 0xff] ^
48                    Te2[(t3 >>  8) & 0xff] ^
49                    Te3[(t0      ) & 0xff] ^
50                    rk[1];
51          s2 = Te0[(t2 >> 24)       ] ^
52                    Te1[(t3 >> 16) & 0xff] ^
53                    Te2[(t0 >>  8) & 0xff] ^
54                    Te3[(t1      ) & 0xff] ^
55                    rk[2];
56          s3 = Te0[(t3 >> 24)       ] ^
57                    Te1[(t0 >> 16) & 0xff] ^
58                    Te2[(t1 >>  8) & 0xff] ^
59                    Te3[(t2      ) & 0xff] ^
60                    rk[3];
61      }
62
63      s0 = (Te4[(t0 >> 24)       ] & 0xff000000) ^
64            (Te4[(t1 >> 16) & 0xff] & 0x00ff0000) ^
65            (Te4[(t2 >>  8) & 0xff] & 0x0000ff00) ^
66            (Te4[(t3      ) & 0xff] & 0x000000ff) ^
67            rk[0];
68      s1 = (Te4[(t1 >> 24)       ] & 0xff000000) ^
69            (Te4[(t2 >> 16) & 0xff] & 0x00ff0000) ^
70            (Te4[(t3 >>  8) & 0xff] & 0x0000ff00) ^
71            (Te4[(t0      ) & 0xff] & 0x000000ff) ^
72            rk[1];
73      s2 = (Te4[(t2 >> 24)       ] & 0xff000000) ^
74            (Te4[(t3 >> 16) & 0xff] & 0x00ff0000) ^
75            (Te4[(t0 >>  8) & 0xff] & 0x0000ff00) ^
76            (Te4[(t1      ) & 0xff] & 0x000000ff) ^
77            rk[2];
78      s3 = (Te4[(t3 >> 24)       ] & 0xff000000) ^
79            (Te4[(t0 >> 16) & 0xff] & 0x00ff0000) ^
```

Regular Round

Final Round

```
80                  (Te4[(t1 >>  8) & 0xff] & 0x0000ff00) ^
81                  (Te4[(t2     ) & 0xff] & 0x000000ff) ^
82              rk[3];
83
84          PUTU32(out      , s0);
85          PUTU32(out +  4, s1);
86          PUTU32(out +  8, s2);
87          PUTU32(out + 12, s3);
88     }
```

The round computations make use of five different lookup tables. The regular rounds use Te0, Te1, Te2 and Te3. The final round uses Te4. Each of these tables has 256 entries of 32 bits each, and it is indexed by means of 1 byte from the state. For example, consider the following expression, part of a regular round (Listing 14.3, line 27).

```
t2 = Te0[(s2 >> 24)         ] ^
     Te1[(s3 >> 16) & 0xff] ^
     Te2[(s0 >>  8) & 0xff] ^
     Te3[(s1      ) & 0xff] ^
     rk[6];
```

This expression extracts byte 3 from s2, and uses it as an index into Te0. Similarly it uses byte 2 from s3 as an index into Te1, byte 1 from s0 as an index into Te2, and byte 0 from s1 as an index into Te3. The output of all of these table lookups are combined with xor. In addition, 32-bits of a roundkey is used. This produces the resulting word t2. Thus, this expression involves five memory lookups, four xor operations, and four byte-select operations, implemented with bitwise-and and shift operations. Four such expressions need to be computed for every round of AES.

The last round of AES is computed in a slightly different manner. Only a single lookup table is used. For example, the following expression computes s1.

```
s1 = (Te4[(t1 >> 24)         ] & 0xff000000) ^
     (Te4[(t2 >> 16) & 0xff] & 0x00ff0000) ^
     (Te4[(t3 >>  8) & 0xff] & 0x0000ff00) ^
     (Te4[(t0      ) & 0xff] & 0x000000ff) ^
     rk[1];
```

Notice how the output of each table lookup is masked off. Each table lookup contributes only a single byte, and each time it's a different byte. The xor operations merely serve to merge all bytes resulting from the table lookup into a single word. Hence, the last round can be thought of as four independent table lookups, each of them working as a byte-to-byte table lookup.

Decryption looks very similar to encryption, but uses a different set of T-box tables. We will therefore focus our attention on implementing encryption, and later expand the implementation to cover encryption as well as decryption.

**Table 14.4** Performance of AES T-box on different NiosII configurations

| Operation | 512/1 K,O0[a] | 4 K/4 K,O0[b] | 512/1 K,O3[c] | 4 K/4 K,O3[d] | Unit |
|---|---|---|---|---|---|
| Encryption | 15,452 | 4,890 | 9,692 | 2,988 | Cycles |
| Encrypt key expansion | 3,960 | 2,464 | 4,924 | 1,006 | Cycles |
| Decryption | 16,550 | 5,208 | 9,744 | 3,266 | Cycles |
| Decrypt key expansion | 30,548 | 12,074 | 17,548 | 5,922 | Cycles |

[a] 512 byte of data cache, 1 KB of instruction cache, and compiler optimization level 0
[b] 4 KB of data cache, 4 KB of instruction cache, and compiler optimization level 0
[c] 512 byte of data cache, 1 KB of instruction cache, and compiler optimization level 3
[d] 4 KB of data cache, 4 KB of instruction cache, and compiler optimization level 3

The performance of the T-box version of AES in software on a 32-bit NiosII processor is shown in Table 14.4. The table illustrates the performance for four different functions (encryption and decryption, as well as key expansion for each). These functions execute on a Nios/f processor, a seven-stage pipelined RISC processor. In each case, four different processor configurations are used, which vary in cache size and compiler optimization level. In the most optimized case, encryption uses 186.75 cycles per byte, and decryption uses 204.1 cycles per byte. In comparison, a full hardware implementation of AES (with a cycle budget of 11 cycles), performs encryption at 0.68 cycles per byte.

## 14.3.2   AES T-box Custom Instruction Design

In this section, we will develop a custom instruction to execute AES more efficiently. From observing Listing 14.3, it's clear that this design will fit well into hardware. The T-box lookup tables can be moved from main-memory (where they use precious memory bandwidth) to dedicated lookup tables in hardware. In addition, a hardware implementation does not require the bit-wise anding and shifting used in software to mask off a single byte from a word.

In the design of the memory-mapped coprocessor, the entire AES round was calculated in a single clock cycle. Fast, but also resource-hungry. In the design of the T-box, we would need 16 parallel T-box custom instruction operations per round, each using a table of 256 32-bit words. To reduce the resource cost, we will build a 4-cycle-per-round design. Such a design uses four T-box tables, and computes one output word per clock cycle. Close inspection of Listing 14.3 reveals several useful symmetries:

- The byte index into Te0 is always byte 3 of a state word.
- The byte index into Te1 is always byte 2 of a state word.
- The byte index into Te2 is always byte 1 of a state word.
- The byte index into Te3 is always byte 0 of a state word.
- From the computation of one expression to the next, the state words are rotated. For example, t0 is computed with s0, s1, s2, and s3. Next, t1 is computed with the same state, rotated one word: s1, s2, s3, and s0. The same goes for t2 and t3.

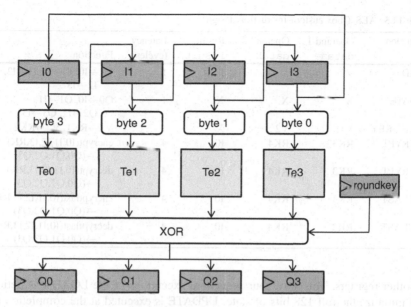

**Fig. 14.7**  Datapath for a four cycle/round T-box calculation

Figure 14.7 shows a datapath for the computation of an AES round in four clock cycles, using T-box tables. The inputs of the datapath include four words of state, I0 through I3, and a round key. Each cycle, an output word is generated in one of the output registers Q0 through Q3.

An output word is computed by taking 1 byte from each input register (simple wiring), accessing a T-box lookup table, merging all outputs with xor, and writing one of the output registers (Q0, Q1, Q2 or Q3). Each cycle, the input registers I0 through I3 are rotated to align the proper inputs to the T-box lookup tables for the next cycle. After four clock cycles, all output registers are computed. To proceed to the next round, we can now swap the positions of the I registers and the Q registers, and repeat the same process.

Next, we will integrate the coprocessor datapath into a custom-instruction. We will use the Nios-II custom-instruction interface, as described in Sect. 11.3.3. We recall the main characteristics of this interface.

- The interface has three data-inputs, including two 32-bit operands dataa and datab, and an 8-bit opcode field n.
- The interface has one data-output, a 32-bit result.
- The interface has a start control input to initiate execution of a custom instruction, and a done flag to indicate completion of a custom instruction.

Table 14.5 shows a design for an encryption/decryption instruction set. The data-input/output instructions are LOAD and UPDATE. LOAD initializes the input registers I0 through I3. This is done by storing the data operand in I0 and rotating

**Table 14.5** AES T-box custom-instruction design

| Instruction n | Operand 1 dataa | Operand 2 datab | Result result | Latency (cycles) | Function |
|---|---|---|---|---|---|
| LOAD | V1 | X | I0 | 1 | V1→I0, I0→I1, I1→I2, I2→I3 |
| UPDATE | X | X | X | 1 | Q0→I0, Q1→I1, Q2→I2, Q3→I3 |
| ROUNDKEY | V1 | V2 | X | 1 | V1→RK1, V2→RK2 |
| ENCRYPT | RK3 | RK4 | I0 | 4 | encrypt(I0,I1,I2,I3,RK) →(Q0,Q1,Q2,Q3) |
| DECRYPT | RK3 | RK4 | I0 | 4 | decrypt(I0,I1,I2,I3,RK) →(Q0,Q1,Q2,Q3) |
| ENCLAST | RK3 | RK4 | I0 | 4 | encryptlast(I0,I1,I2,I3,RK) →(Q0,Q1,Q2,Q3) |
| DECLAST | RK3 | RK4 | I0 | 4 | decryptlast(I0,I1,I2,I3,RK) →(Q0,Q1,Q2,Q3) |

the other registers. This way, four sequential executions of the LOAD instruction will initialize the full 128 bits of state. UPDATE is executed at the completion of encryption or decryption. It copies the output state register Q0 through Q3 to I0 through I3. Extracting output data from the custom instruction is done in overlap with LOAD: for each word loaded into the custom datapath, a result from the previous instruction is returned.

The ROUNDKEY instruction is used to load two words of the roundkey into the custom-instruction datapath. These two words are stored in two 32-bit registers RK1 and RK2. Of course, a complete roundkey is 128 bit. The second half of the roundkey needed for an encryption or decryption is included as an operand to the encryption and decryption instructions.

There are four instructions for encryption and decryption: ENCRYPT, DE-CRYPT, ENCLAST and DECLAST. Each of these instructions has a latency of four clock cycles, as explained previously in the design of the custom-instruction datapath. We are now ready to implement the custom instruction in hardware, and to develop the software driver for AES encryption/decryption.

### 14.3.3  AES T-box Custom Instruction in GEZEL

In this section, we discuss the custom-instruction design implementation using GEZEL.

We start with a remark on the design of T-box tables in hardware. For AES encryption and decryption, we need 16 tables: four tables (256 entries of 32 bit) for each of the regular encryption round and regular decryption round, and four tables (256 entries of 8 bit) for each of the last encryption and decryption round. Coding this in GEZEL is rather tedious, in particular since we already have a reference implementation of AES T-box encryption/decryption in software. It is easy to write a short C program that generates these tables in a file. Listing 14.4 shows a function to do this. A T-box file generated with this function look as follows (some lines have been skipped).

```
dp tbox0(in a : ns(8); out q : ns(32)) {
  reg qr : ns(32);
  lookup T : ns(32) = {
    0xc66363a5, 0xf87c7c84, 0xee777799, 0xf67b7b8d,
    ...
    0x7bb0b0cb, 0xa85454fc, 0x6dbbbbd6, 0x2c16163a};

  always {
    q  = qr;
    qr = T(a);
  }
}
```

The T-box is implemented in hardware as a lookup table with a register at the output. The register is included because we intend to map this design on an FPGA implementation, and make use of the on-chip memory blocks to implement the T-boxes. The memory blocks of modern FPGA technologies (Xilinx, Altera) have a similar register at their output. Of course, adding a register to a T-box means that the latency of a T-box table lookup now equals one clock cycle. We need to keep this in mind while implementing the schedule of encryption and decryption.

Listing 14.4 Function to convert T-boxes from C to GEZEL

```
1   void writetablegezel(char *n, char *m, const u32 T[256]) {
2     FILE *f = fopen(n, "w");
3     unsigned i;
4
5     fprintf(f, "dp%s(ina:ns(8);outq:ns(32)){\n", m);
6     fprintf(f, "regqr:ns(32);\n");
7
8     fprintf(f, "lookupT:ns(32)={\n");
9     fprintf(f, "");
10    for (i=0; i<255; i++) {
11      fprintf(f,"0x%08x,", T[i]);
12      if (!((i+1)%4)) fprintf(f, "\n");
13    }
14    fprintf(f,"0x%08x};\n", T[255]);
15    fprintf(f, "\n");
16
17    fprintf(f,"always{\n");
18    fprintf(f,"q=qr;\n");
19    fprintf(f,"qr=T(a);\n");
20    fprintf(f,"}\n");
21    fprintf(f,"}\n");
22    fclose(f);
23  }
```

Listing 14.5 shows the GEZEL implementation of the custom datapath. For brevity, this listing does not include the definition of T-box lookup tables. The tables are instantiated directly in this datapath, starting at line 13 for the encryption tables, and line 21 for the decryption tables. The always block, at line 35, wires bytes from the input state register the T-box inputs.

The datapath also includes several sfg, which implement the custom instructions. The best place to start is the control FSM, starting at line 87. This FSM is the custom-instruction decoder. State sdec implements the instruction decoding. Single-cycle instructions, such as LOAD, UPDATE and ROUNDKEY are implemented with a single clock cycle. Multi-cycle instructions, such as ENCRYPT, DECRYPT, ENCLAST and DECLAST, have a schedule of their own. In this particular implementation, each multi-cycle instruction takes five rather than four clock cycles of latency. This is caused by the registers in the T-box implementation.

The design can be simulated with a testbench and converted into VHDL for implementation on FPGA (see Problem Section).

**Listing 14.5**  AES T-box custom instruction implementation in GEZEL

```
1   dp tboxtop(in   start  : ns(1);
2              out  done   : ns(1);
3              in   n      : ns(8);
4              in   dataa  : ns(32);
5              in   datab  : ns(32);
6              out  result : ns(32)) {
7       sig e0_a, e1_a, e2_a, e3_a, d0_a, d1_a, d2_a, d3_a :ns(8)
        ;
8       sig e0_q, e1_q, e2_q, e3_q, d0_q, d1_q, d2_q, d3_q:ns(32)
        ;
9       sig e41_q, d41_q : ns(32);
10      sig e42_q, d42_q : ns(32);
11      sig e43_q, d43_q : ns(32);
12      sig e44_q, d44_q : ns(32);
13      use tboxe0 (e0_a, e0_q);               Encryption T-box
14      use tboxe1 (e1_a, e1_q);
15      use tboxe2 (e2_a, e2_q);
16      use tboxe3 (e3_a, e3_q);
17      use tboxe4 (e0_a, e41_q);
18      use tboxe42 (e1_a, e42_q);
19      use tboxe43 (e2_a, e43_q);
20      use tboxe44 (e3_a, e44_q);
21      use tboxd0 (d0_a, d0_q);               Decryption T-box
22      use tboxd1 (d1_a, d1_q);
23      use tboxd2 (d2_a, d2_q);
24      use tboxd3 (d3_a, d3_q);
25      use tboxd4 (d0_a, d41_q);
26      use tboxd42 (d1_a, d42_q);
27      use tboxd43 (d2_a, d43_q);
28      use tboxd44 (d3_a, d44_q);
29
30      reg i0, i1, i2, i3 : ns(32);
```

```
31    reg q0, q1, q2, q3 : ns(32);
32    reg rk0, rk1     : ns(32);
33    reg ndecode      : ns(7);
34
35    always       { result  = i0;
36                    e0_a = i0[31:24];
37                    e1_a = i1[23:16];
38                    e2_a = i2[15: 8];
39                    e3_a = i3[ 7: 0];
40                    d0_a = i0[31:24];
41                    d1_a = i3[23:16];
42                    d2_a = i2[15: 8];
43                    d3_a = i1[ 7: 0];
44                  }
45    sfg decode   { ndecode = start ? n : 0;
46                    done = 1;
47                  }
48    sfg loadins  { i3  = dataa;
49                    i2  = i3;
50                    i1  = i2;
51                    i0  = i1;
52                  }
53    sfg updateins {i0  = q0;
54                    i1  = q1;
55                    i2  = q2;
56                    i3  = q3;
57                    rk0 = dataa;
58                    rk1 = datab;
59                  }
60    sfg rkins    { rk0 = dataa;   // 1/4 round key
61                    rk1 = datab;   // 1/4 round key
62                  }
63    sfg irotate  { i0  = i1;
64                    i1  = i2;
65                    i2  = i3;
66                    i3  = i0;
67                    done = 0;
68                  }
69    sfg enc1     { q0 = e0_q ^ e1_q ^ e2_q ^ e3_q ^ rk0;
                    }
70    sfg enc2     { q1 = e0_q ^ e1_q ^ e2_q ^ e3_q ^ rk1;
                    }
71    sfg enc3     { q2 = e0_q ^ e1_q ^ e2_q ^ e3_q ^ dataa;
                    }
72    sfg enc4     { q3 = e0_q ^ e1_q ^ e2_q ^ e3_q ^ datab;
                    }
73    sfg enclast1 { q0 = (e41_q # e42_q # e43_q # e44_q) ^ rk0
      ; }
74    sfg enclast2 { q1 = (e41_q # e42_q # e43_q # e44_q) ^ rk1
      ; }
75    sfg enclast3 { q2 = (e41_q # e42_q # e43_q #e44_q)^dataa;
                    }
```

```
76    sfg enclast4 { q3 = (e41_q # e42_q # e43_q #e44_q)^datab;
           }
77    sfg dec1       { q0 = d0_q ^ d1_q ^ d2_q ^ d3_q ^ rk0;
           }
78    sfg dec2       { q1 = d0_q ^ d1_q ^ d2_q ^ d3_q ^ rk1;
           }
79    sfg dec3       { q2 = d0_q ^ d1_q ^ d2_q ^ d3_q ^dataa;
           }
80    sfg dec4       { q3 = d0_q ^ d1_q ^ d2_q ^ d3_q ^datab;
           }
81    sfg declast1 { q0 = (d41_q # d42_q # d43_q # d44_q) ^ rk0
         ; }
82    sfg declast2 { q1 = (d41_q # d42_q # d43_q # d44_q) ^ rk1
         ; }
83    sfg declast3 { q2 = (d41_q # d42_q # d43_q # d44_q)^dataa
         ; }
84    sfg declast4 { q3 = (d41_q # d42_q # d43_q # d44_q)^datab
         ; }
85  }
86
87  fsm c_tboxtop(tboxtop) {
88    initial sdec;
89    state   se0,      se1,      se2,      se3;
90    state   sd0,      sd1,      sd2,      sd3;
91    state   se0last, se1last, se2last, se3last;
92    state   sd0last, sd1last, sd2last, sd3last;
93    @sdec if       (ndecode == 1) then (loadins, decode)   ->
         sdec;
94          else if (ndecode == 2) then (updateins, decode) ->
            sdec;
95          else if (ndecode == 3) then (rkins, decode)     ->
            sdec;
96          else if (ndecode == 4) then (irotate)           ->
            se0;
97          else if (ndecode == 5) then (irotate)           ->
            sd0;
98          else if (ndecode == 6) then (irotate)           ->
            se0last;
99          else if (ndecode == 7) then (irotate)           ->
            sd0last;
100         else (decode)                                   ->
            sdec;
101   @se0     (enc1,    irotate) -> se1;
102   @se1     (enc2,    irotate) -> se2;
103   @se2     (enc3,    irotate) -> se3;
104   @se3     (enc4,    decode)  -> sdec;
105   @sd0     (dec1,    irotate) -> sd1;
106   @sd1     (dec2,    irotate) -> sd2;
107   @sd2     (dec3,    irotate) -> sd3;
108   @sd3     (dec4,    decode)  -> sdec;
109   @se0last (enclast1, irotate) -> se1last;
110   @se1last (enclast2, irotate) -> se2last;
111   @se2last (enclast3, irotate) -> se3last;
```

**Table 14.6** Performance of AES T-box custom-instruction design on different NiosII configurations

| Operation | 512/1 K,O0[a] | 4 K/4 K,O0[b] | 512/1 K,O3[c] | 4 K/4 K,O3[d] | Unit |
|---|---|---|---|---|---|
| Encryption | 1,622 | 1,132 | 894 | 454 | Cycles |
| Speedup over Software | 9.5x | 4.3x | 10.8x | 6.6x | |
| Decryption | 1,648 | 1,132 | 566 | 454 | Cycles |
| Speedup over Software | 10.0x | 4.6x | 17.2x | 7.2x | |

[a]512 byte of data cache, 1 KB of instruction cache, and compiler optimization level 0
[b]4 KB of data cache, 4 KB of instruction cache, and compiler optimization level 0
[c]512 byte of data cache, 1 KB of instruction cache, and compiler optimization level 3
[d]4 KB of data cache, 4 KB of instruction cache, and compiler optimization level 3

```
112    @se3last (enclast4, decode)  -> sdec;
113    @sd0last (declast1, irotate) -> sd1last;
114    @sd1last (declast2, irotate) -> sd2last;
115    @sd2last (declast3, irotate) -> sd3last;
116    @sd3last (declast4, decode)  -> sdec;
117  }
```

### *14.3.4 AES T-box Software Integration and Performance*

After integration of the custom-instruction in a Nios processor, we can use the custom instructions in a software driver. Listing 14.6 shows the implementation of an AES encryption. It uses five custom instructions; it is helpful to compare this code with the earlier, software-only implementation in Listing 14.3. In particular, note that the first round of the AES encryption still executes on the Nios processor. For decryption, a similar implementation can be made.

Table 14.6 shows the performance of the accelerated encryption and decryption using custom instructions. The speedup factors listed in the table make a comparison with the performance of the software implementation shown in Table 14.4. The design delivers substantial speedup, even when a fully optimized software implementation is used. Let's evaluate how good these results are. Given the amount of lookup table hardware, this design needs at least 4 cycles per round, and thus at least 40 cycles per encryption. The custom-instruction implementation gives, in the most optimal case, 454 cycles per encryption, a factor 11.3 times slower than the ideal case (40 cycles per encryption). This overhead represents the cost of integrating the custom hardware with software. This overhead is large, but it is still much better than the case of the memory-mapped coprocessor, which had an overhead factor of more then 30 times.

**Listing 14.6**   A C driver for the AES T-box design

```
1   void AES_encrypt_CI(const unsigned char *in,
2                       unsigned char *out,
3                       const AES_KEY *key) {
4     const u32 *rk;
5     u32 s0, s1, s2, s3;
6     int r;
7
8     rk = key->rd_key;
9
10    s0 = GETU32(in     ) ^ rk[0];              [First Round]
11    s1 = GETU32(in +  4) ^ rk[1];
12    s2 = GETU32(in +  8) ^ rk[2];
13    s3 = GETU32(in + 12) ^ rk[3];
14
15    ALT_CI_TBOXTOP_INST(LOADINS, s0, 0);       [Load state]
16    ALT_CI_TBOXTOP_INST(LOADINS, s1, 0);
17    ALT_CI_TBOXTOP_INST(LOADINS, s2, 0);
18    ALT_CI_TBOXTOP_INST(LOADINS, s3, 0);
19
20    rk += 4;
21    ALT_CI_TBOXTOP_INST(RKINS,  rk[0], rk[1]);  [Second Round]
22    ALT_CI_TBOXTOP_INST(ENCINS, rk[2], rk[3]);
23
24    for (r=0; r<8; r++) {
25      rk += 4;
26      ALT_CI_TBOXTOP_INST(UPDATEINS, rk[0], rk[1]); [Regular Round]
27      ALT_CI_TBOXTOP_INST(ENCINS,    rk[2], rk[3]);
28    }
29
30    rk += 4;
31    ALT_CI_TBOXTOP_INST(UPDATEINS,  rk[0], rk[1]);  [Last Round]
32    ALT_CI_TBOXTOP_INST(ENCLASTINS, rk[2], rk[3]);
33
34    s0 = ALT_CI_TBOXTOP_INST(UPDATEINS, 0,    0);
35    s1 = ALT_CI_TBOXTOP_INST(LOADINS,   0,    0);
36    s2 = ALT_CI_TBOXTOP_INST(LOADINS,   0,    0);
37    s3 = ALT_CI_TBOXTOP_INST(LOADINS,   0,    0);
38
39    PUTU32(out     , s0);
40    PUTU32(out +  4, s1);
41    PUTU32(out +  8, s2);
42    PUTU32(out + 12, s3);
43  }
```

## 14.4   Summary

In this chapter, we discussed hardware acceleration of the Advanced Encryption Standard. We developed a memory-mapped coprocessor design and a custom-instruction design. In both cases, we obtained a large speedup over software – roughly an order of magnitude. However, we also found that the overhead of custom-hardware integration is quite large – at least an order of magnitude as well. This underlines the importance of developing a system perspective when designing a specialized hardware module.

## 14.5   Further Reading

The Advanced Encryption Standard was published by NIST in Federal Information Processing Standard 197 (FIPS 197), which can be consulted online (NIST 2001). Gaj has published a detailed treatment of AES hardware design, covering various strategies in (Gaj et al. 2009). Furthermore, Hodjat discusses the integration issues of AES as a hardware accelerator in (Hodjat and Verbauwhede 2004).

## 14.6   Problems

**Problem 14.1.** Implement the AES memory-mapped coprocessor on a Xilinx FPGA, interfacing a Microblaze processor. Consult the book webpage for the design files.

**Problem 14.2.** Implement the AES custom-instruction coprocessor on an Altera FPGA, interfacing a NiosII processor. Consult the book webpage for the design files.

# Chapter 15
# CORDIC Co-processor

## 15.1 The Coordinate Rotation Digital Computer Algorithm

In this section we introduce the CORDIC algorithm, including a reference implementation in C.

### 15.1.1 The Algorithm

The CORDIC algorithm calculates the rotation of a two-dimensional vector $x_0, y_0$ over an arbitrary angle $\alpha$. Figure 15.1a describes the problem of coordinate rotation. Given $(x_0, y_0)$ and a rotation angle $\alpha$, the coordinates $(x_T, y_T)$ are given by

$$\begin{bmatrix} x_T \\ y_T \end{bmatrix} = \begin{bmatrix} cos\ \alpha & -sin\ \alpha \\ sin\ \alpha & cos\ \alpha \end{bmatrix} \begin{bmatrix} x_0 \\ y_0 \end{bmatrix} \tag{15.1}$$

This rotation can be written in terms of a single function $tan\ \alpha$ by using

$$cos\ \alpha = \frac{1}{\sqrt{1 + tan^2\ \alpha}} \tag{15.2}$$

$$sin\ \alpha = \frac{tan\ \alpha}{\sqrt{1 + tan^2\ \alpha}} \tag{15.3}$$

The resulting coordinate rotation now becomes

$$\begin{bmatrix} x_T \\ y_T \end{bmatrix} = \frac{1}{\sqrt{1 + tan^2\ \alpha}} \begin{bmatrix} 1 & -tan\ \alpha \\ tan\ \alpha & 1 \end{bmatrix} \begin{bmatrix} x_0 \\ y_0 \end{bmatrix} \tag{15.4}$$

P.R. Schaumont, *A Practical Introduction to Hardware/Software Codesign*,
DOI 10.1007/978-1-4614-3737-6_15, © Springer Science+Business Media New York 2013

**a**                                                **b**

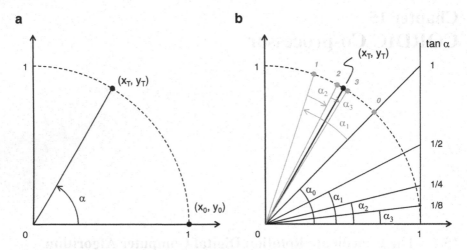

**Fig. 15.1** (**a**) Coordinate rotation over $\alpha$. (**b**) Decomposition of the rotation angle $\alpha = \alpha_0 + \alpha_1 - \alpha_2 - \alpha_3$

The clever part of the CORDIC algorithm is that the rotation over the angle *alpha* can be expressed in terms of rotations over smaller angles. The CORDIC algorithm chooses a decomposition in angles whose tangent is a power of 2, as illustrated in Fig. 15.1b. Thus, we choose a set of angles $\alpha_i$ so that

$$tan\ \alpha_i = \frac{1}{2^i} \tag{15.5}$$

From the Figure, we can see that $\alpha$ can be reasonably approximated as $\alpha_0 + \alpha_1 - \alpha_2 - \alpha_3$. Because of the particular property of these angles, Formula (15.4) becomes easy to evaluate: $(x_{i+1}, y_{i+1})$ can be found using addition, subtraction, and shifting of $(x_i, y_i)$. For example, suppose that we want to rotate *clockwise* over $alpha_i$, then we compute

$$x_{i+1} = K_i \left\{ x_i + \frac{y_i}{2^i} \right\} \tag{15.6}$$

$$y_{i+1} = K_i \left\{ \frac{-x_i}{2^i} + y_i \right\} \tag{15.7}$$

If instead we want to rotate *counter-clockwise* over $alpha_i$, then we use

$$x_{i+1} = K_i \left\{ x_i - \frac{y_i}{2^i} \right\} \tag{15.8}$$

$$y_{i+1} = K_i \left\{ \frac{x_i}{2^i} + y_i \right\} \tag{15.9}$$

In these formulas, $K_i$ is a constant.

$$K_i = \frac{1}{\sqrt{1 + 2^{-2i}}} \qquad (15.10)$$

We can approximate an arbitrary, but unknown, angle $\beta$ by means of a binary-search process as follows. We precalculate the set of angles $\alpha_i = arctan\ 2^{-i}$ and store them in a lookup table. Assume our current approximation of $\beta$ is $\beta_i$. If $\beta_i > \beta$, we rotate clockwise and $\beta_{i+1} = \beta_i - alpha_i$. If $\beta_i < \beta$, we rotate counterclockwise and $\beta_{i+1} = \beta_i + alpha_i$. We continue this process iteratively until $\beta_n \simeq \beta$. We gain around 1 bit of precision per iteration. For example, after 20 iterations, the precision on the angle is around one part in one million (six significant digits).

## 15.1.2   Reference Implementation in C

A distinctive property of the CORDIC algorithm is that it only needs additions, subtractions and shift operations. It maps well to integer arithmetic, even though the numbers being handled are still fractional. We will discuss a CORDIC implementation in C that uses scaled `int` types.

Fractional arithmetic can be implemented using integer numbers, by scaling each number by a power of 2. The resulting representation is called a <M,N> *fixed point representation*. M represents the integer wordlength, and N the fractional wordlength. For example, a <32,28> fixed-point number has a wordlength of 32 bits, and has 28 fractional bits. Thus, the weight of the least significant bit is $2^{-28}$. Fixed-point numbers adjust the weight of the bits in a binary number. They work just like integers in all other respects – you can add, subtract, compare, and shift them. For example, a 32-bit unsigned number has the value 8,834,773. As a <32,28> number, it has the value $8,834,773/2^{28} = 0.3291209...$

**Listing 15.1**   Reference implementation of a fixed-point CORDIC algorithm

```
 1  #include <stdio.h>
 2  #define K_CONST 163008218 /* 0.60725293510314      */
 3  #define PI 843314856        /* 3.141593.. in <32,28> */
 4  typedef int fixed;           /* <32,28>              */
 5
 6  static const int angles[] = {
 7     210828714,  124459457,   65760959,   33381289,
 8      16755421,    8385878,    4193962,    2097109,
 9       1048570,     524287,     262143,     131071,
10         65535,      32767,      16383,       8191,
11          4095,       2047,       1024,        511 };
12
13  void cordic(int target, int *rX, int *rY) {
14      fixed X, Y, T, current;
```

```
15      unsigned step;
16      X           = K_CONST;
17      Y           = 0;
18      current = 0;
19      for(step=0; step < 20; step++) {
20          if (target > current) {
21              T           = X - (Y >> step);
22              Y           = (X >> step) + Y;
23              X           = T;
24              current   += angles[step];
25          } else {
26              T           = X + (Y >> step);
27              Y           = -(X >> step) + Y;
28              X           = T;
29              current   -= angles[step];
30          }
31      }
32      *rX = X;
33      *rY = Y;
34  }
35
36  int main(void) {
37      fixed X, Y, target;
38      fixed accsw, accfsl;
39
40      target = PI / 17;
41      cordic(target, &X, &Y);
42
43      printf("Target_%d:_(X,Y)_=_(%d,%d)\n", target, X, Y);
44      return(0);
45  }
```

Listing 15.1 shows a fixed-point version of a 32-bit CORDIC algorithm, using `<32,28>` fixed point arithmetic. The CORDIC is evaluated using 20 iterations, which means that it can approximate angles with a precision of $arctan\ 2^{-20}$, or around one-millionth of a radian. At the start, the program defines a few relevant constants.

- PI, the well-known mathematical constant, equals $\pi * 2^{28}$ or 843,314,856.
- K_CONST is the product of the 21st $K_i$ according to Eq. 15.10. This constant factor needs to be evaluated once.
- angles[] is an array of constants that holds the angles $\alpha_i$ defined by Eq. 15.5. For example, the first element is 210,828,714, which is a `<32,28>` number corresponding to $atan(1) = 0.78540$.

The cordic function, on lines 13–34, first initializes the angle accumulator current, and the initial vector (X,Y). Next, it goes through 20 iterations. Every iteration, the angle accumulator is compared with the target angle, and the vector is rotated clockwise or counterclockwise.

The main function, on lines 36–45, demonstrates the operation of the function with a simple testcase, a rotation of (1,0) over $\pi/17$. We can compile and run this program on a PC, or for the SimIT-ARM simulator. The program generates the following output.

```
Target 49606756:  (X,Y) = (263864846, 49324815)
```

Indeed, after scaling everything by $2^{28}$, we can verify that for the target $\pi/17$, $(X,Y)$ equals $(0.98297, 0.18375)$, or $(cos(\pi/17), sin(\pi/17))$.

To evaluate the performance of this function on an embedded processor, a similar technique as in Sect. 13.3.2 can be used. Measurement of the execution time for cordic on Simit-ARM yields 485 cycles (-O3 compiler optimization) per call. In the next section, we will develop a hardware implementation of the CORDIC algorithm. Next, we will integrate this hardware design as a coprocessor to the software.

## 15.2   A Hardware Coprocessor for CORDIC

We'll develop a hardware implementation of the CORDIC design presented in the previous section. The objective is to create a coprocessor, and the first step is to create a hardware kernel to implement CORDIC. Next, we convert the kernel into a coprocessor design. As before, the selection of the hardware/software interface is a crucial design decision. In this case, we intend to map the design onto an FPGA, and the selection is constrained by what is available in the FPGA design environment. We will be making use of the Fast Simplex Link interface discussed in Sect. 11.2.1.

### 15.2.1   A CORDIC Kernel in Hardware

Listing 15.2 illustrates a CORDIC hardware kernel. In anticipation of using the FSL-based interface, the input/output protocol of the algorithm uses two-way handshake interfaces. The input uses a slave interface, while the output implements a master interface. The computational part of the algorithm is in sfg iterate, lines 30–38. This iterate instruction is very close to the inner-loop of the cordic function in Listing 15.1, including the use of a lookup table angles to store the rotation angles. Note that while GEZEL supports lookup tables, it does not support read/write arrays.

**Listing 15.2** A Standalone hardware implementation of the CORDIC algorithm

```
1  dp cordic_fsmd    (in   rdata   : tc(32);  // interface to
                                                        slave
2                     in  exists   : ns(1);
3                     out read     : ns(1);
4                     out wdata    : tc(32);  // interface to
                                                        master
5                     in  full     : ns(1);
6                     out write    : ns(1)) {
7     lookup angles : tc(32) = {
8        210828714,    124459457,    65760959,    33381289,
```

```
 9          16755421,       8385878,       4193962,        2097109,
10           1048570,        524287,        262143,         131071,
11             65535,         32767,         16383,            8191,
12              4095,          2047,          1024,             511 };
13   reg X, Y, target, current: tc(32);
14   reg step                  : ns( 5);
15   reg done, rexists, rfull : ns( 1);
16   sig cmp : ns(1);
17   always      { rexists = exists;
18                 rfull   = full; }
19   sfg dowrite   { write    = 1; }
20   sfg dontwrite { write    = 0;
21                   wdata    = 0; }
22   sfg doread    { read     = 1; }
23   sfg dontread  { read     = 0; }
24   sfg capture   { step     = 0;
25                   done     = 0;
26                   current  = 0;
27                   X        = 163008218;  // K
28                   Y        = 0;
29                   target   = rdata;      }
30   sfg iterate   { step     = step + 1;
31                   done     = (step == 19);
32                   cmp      = (target > current);
33                   X        = cmp ? X - (Y >> step):
34                                    X + (Y >> step);
35                   Y        = cmp ? Y + (X >> step):
36                                    Y - (X >> step);
37                   current  = cmp ? current + angles(step):
38                                    current - angles(step); }
39   sfg writeX    { wdata = X; }
40   sfg writeY    { wdata = Y; }
41
42 }
43 fsm fsm_cordic_fsmd(cordic_fsmd) {
44    initial s0;
45    state s1, s2, s22;
46    state c1;
47
48    // wait for SW to write slave
49    @s0 if (rexists) then (capture , doread, dontwrite)   -> c1;
50                     else (dontread, dontwrite)           -> s0;
51
52    // calculate result
53    @c1 if (done)    then (dontread, dontwrite)           -> s1;
54                     else (iterate, dontread, dontwrite) -> c1;
55
56    // after read op completes, do a write to the master
57    @s1 if (rfull)   then (dontread, dontwrite)           -> s1;
58                     else (dowrite , writeX, dontread )   -> s2;
59    @s2 if (rfull)   then (dontread, dontwrite)           -> s2;
60                     else (dowrite , writeY, dontread )   -> s0;
61 }
```

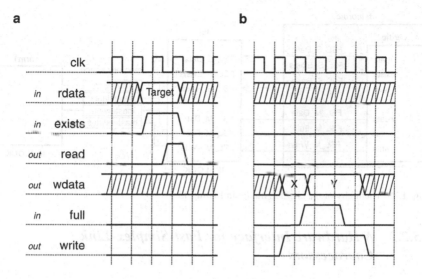

**Fig. 15.2** (a) Input slave handshake. (b) Output master handshake

The remaining datapath instructions in Listing 15.2 support the implementation of the input/output operations of the algorithm, and they are most easily understood by studying the controller description on lines 45–63. The four states of the finite state machine correspond to the following activities:

- State s0: Reading the target angle.
- State c1: Perform the rotation.
- State s1: Produce output X.
- State s2: Produce output Y.

Figure 15.2 shows a sample input operation and a sample output operation. The CORDIC coprocessor goes through an infinite loop consisting of the operations: read target, calculate, write X, write Y. Each time it needs a new target angle, the coprocessor will wait for exists to be raised. The coprocessor will acknowledge the request through read and grab a target angle. Next, the coprocessor proceeds to evaluate the output coordinates X and Y. When they are available, the write output is raised and, as long as the full input remains low, X and Y are passed to the output in a single clock cycle. In Fig. 15.2, four clock cycles are required for the complete output operation because the full input is raised for two clock cycles.

The hardware testbench for this design is left as an exercise for the reader (See Problem 15.1).

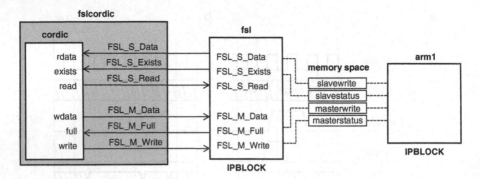

**Fig. 15.3** Hierarchy of the GEZEL model in Listing 15.3

## 15.2.2   A Hardware Interface for Fast-Simplex-Link Coprocessors

We will now integrate the hardware kernel into a control shell with FSL interfaces. The FSL interface is natively supported on the MicroBlaze processor. GEZEL has no built-in MicroBlaze instruction-set simulator, but it emulates FSL interfaces through memory-mapped operations on the ARM, and an ipblock with the outline of a real Fast Simplex Link interface. Figure 15.3 demonstrates this approach. The CORDIC kernel is encapsulated in a module, fslcordic, which defines the proper FSL interface. The FSL interface is driven from an ARM simulator, which drives the value of the interface signals through memory-mapped read and write operations.

Listing 15.3 shows the hardware platform of the complete design. The FSL interface is on lines 6–24. The pinout of this interface follows the specifications of the Xilinx FSL interface, although this GEZEL model does not use all features of the interface and will not use all signals on the interface pinout. In particular, the Xilinx FSL supports asynchronous operation and control information in conjunction with data, while GEZEL sticks to synchronous operation and data-only FSL transfers. The operation of the FSL interface is emulated with read and write operations on memory addresses in the ARM simulation model (lines 19–23). The control shell module, fslcordic, is very simple. In fact, the CORDIC kernel can be directly instantiated (line 40) and wired to the FSL ports (lines 43–53). Finally, the top-level module interconnects the system simulation.

**Listing 15.3** The CORDIC coprocessor attached to a Fast Simplex Link

```
1  ipblock arm1 {
2    iptype "armsystem";
3    ipparm "exec = cordic_fixp";
4  }
5
6  ipblock fsl(in  FSL_S_Clk    : ns(1);   // notused
```

```
7              in  FSL_S_Read    : ns(1);   // hshk for slave
                                                        side
8              out FSL_S_Data    : ns(32);  // data for slave
                                                        side
9              out FSL_S_Control : ns(1);   // control for slave
                                                        side
10             out FSL_S_Exists  : ns(1);   // hshk for slave
                                                        side
11             in  FSL_M_Clk     : ns(1);   // notused
12.            in  FSL_M_Write   : ns(1);   // hshk for master
                                                        side
13             in  FSL_M_Data    . ns(32);  // data for master
                                                        side
14             in  FSL_M_Control : ns(1);   // control for
                                                      master side
15             out FSL_M_Full    : ns(1)) { // hshk for master
                                                        side
16      iptype "xilinx_fsl";
17      ipparm "core=arm1";                  // strongarm core
18
19      ipparm "slavewrite  = 0x80000000"; // write slave data
20      ipparm "slavestatus = 0x80000004"; // poll slave status
21
22      ipparm "masterread  = 0x80000008"; // read master data
23      ipparm "masterstatus= 0x8000000C"; // poll master status
24   }
25
26   dp fslcordic( out FSL_S_Clk     : ns(1);   // notused
27                 out FSL_S_Read    : ns(1);   // hshk for slave
                                                        side
28                 in  FSL_S_Data    : ns(32);  // data for slave
                                                        side
29                 in  FSL_S_Control : ns(1);   // control for
                                                      slave side
30                 in  FSL_S_Exists  : ns(1);   // hshk for
                                                      slave side
31                 out FSL_M_Clk     : ns(1);   // notused
32                 out FSL_M_Write   : ns(1);   // hshk for
                                                      master side
33                 out FSL_M_Data    : ns(32);  // data for
                                                      master side
34                 out FSL_M_Control : ns(1);   // control for
                                                      master side
35                 in  FSL_M_Full    : ns(1)) { // hshk for
                                                      master side
36      sig rdata, wdata    : ns(32);
37      sig write, read     : ns(1);
38      sig exists, full    : ns(1);
39
40      use cordic_fsmd    (rdata, exists, read,
41                          wdata, full, write);
42
43      always {
```

```
44        rdata         = FSL_S_Data;
45        exists        = FSL_S_Exists;
46        FSL_S_Read    = read;
47
48        FSL_M_Data    = wdata;
49        FSL_M_Control = 0;
50        FSL_M_Write   = write;
51        full          = FSL_M_Full;
52
53        FSL_S_Clk     = 0;
54        FSL_M_Clk     = 0;
55    }
56  }
57
58  dp top {
59    sig FSL_Clk, FSL_Rst, FSL_S_Clk, FSL_M_Clk : ns( 1);
60    sig FSL_S_Read, FSL_S_Control, FSL_S_Exists : ns( 1);
61    sig FSL_M_Write, FSL_M_Control, FSL_M_Full : ns( 1);
62    sig FSL_S_Data, FSL_M_Data : ns(32);
63
64    use arm1;
65
66    use fslcordic (FSL_S_Clk, FSL_S_Read, FSL_S_Data,
67                   FSL_S_Control, FSL_S_Exists, FSL_M_Clk,
68                   FSL_M_Write, FSL_M_Data, FSL_M_Control,
                       FSL_M_Full);
69
70    use fsl (FSL_S_Clk, FSL_S_Read, FSL_S_Data, FSL_S_Control,
71             FSL_S_Exists, FSL_M_Clk, FSL_M_Write, FSL_M_Data,
72             FSL_M_Control, FSL_M_Full);
73  }
74
75  system S {
76    top;
77  }
```

In order to verify the design, we also need a software driver. Listing 15.4 shows an example driver to compare the reference software implementation (Listing 15.1) with the FSL coprocessor. The driver evaluates 4096 rotations from 0 to $\frac{\pi}{2}$ and accumulates the coordinates. Since the CORDIC design is in fixed point, the results of the hardware coprocessor must be exactly the same as the results from the software reference implementation. Of particular interest in the software driver is the emulation of the FSL interface signals through memory-mapped operations. The driver first transfers a token to the coprocessor (lines 8–9), and then reads two coordinates from the coprocessor (lines 11–15).

**Listing 15.4** Driver for the CORDIC coprocessor on the emulated FSL interface

```
1  void cordic_driver(int target, int *rX, int *rY) {
2    volatile unsigned int *wchannel_data   = (int *)
         0x80000000;
3    volatile unsigned int *wchannel_status = (int *)
         0x80000004;
```

```
4    volatile unsigned int *rchannel_data    = (int *)
        0x80000008;
5    volatile unsigned int *rchannel_status = (int *)
        0x8000000C;
6    int i;
7
8    while (*wchannel_status == 1) ;
9    *wchannel_data = target;
10
11   while (*rchannel_status != 1) ;
12   *rX = *rchannel_data;
13
14   while (*rchannel_status != 1) ;
15   *rY = *rchannel_data;
16 }
17
18 // Reference implementation
19 extern void cordic(int target, int *rX, int *rY);
20
21 int main(void) {
22   fixed X, Y, target;
23   fixed accsw, accfsl;
24
25   accsw = 0;
26   for (target = 0; target < PI/2; target += (1 <<
        (UNIT -   12))) {
27     cordic(target, &X, &Y);
28     accsw += (X + Y);
29   }
30
31   accfsl = 0;
32   for (target = 0; target < PI/2; target += (1 <<
        (UNIT -   12))) {
33     cordic_driver(target, &X, &Y);
34     accfsl += (X + Y);
35   }
36
37   printf("Checksum_SW_%x_FSL_%x\n", accsw, accfsl);
38   return(0);
39 }
```

The cosimulation of this model, consisting of the hardware design in Listing 15.3 and the software design in Listing 15.4, confirms that the reference implementation and the hardware coprocessor behave identically.

In the next section, we will port this coprocessor to the FPGA for a detailed performance analysis.

```
> /usr/local/arm/bin/arm-linux-gcc -static -O3 cordic_fixp.c
  -o cordic_fixp
> gplatform fsl.fdl
core arm1
armsystem: loading executable [cordic_fixp]
Checksum SW 4ae1ee FSL 4ae1ee
Total Cycles: 3467162
```

## 15.3   An FPGA Prototype of the CORDIC Coprocessor

Our next step is to map the complete system – processor and coprocessor – to an FPGA prototype. The prototyping environment we are using contains the following components.

- Spartan-3E Starter Kit, including a Spartan 3ES500 Xilinx FPGA and various peripherals. We will be making use of one peripheral besides the FPGA: a 64 MB DDR SDRAM Module.
- FPGA Design Software, consisting of Xilinx Platform Studio (EDK 9.2) and associated hardware synthesis tools (ISE 9.2.04).

Figure 15.4 shows the system architecture of the CORDIC design. The coprocessor connects through a Fast Simplex Link (FSL) to a Microblaze processor. The processor includes an instruction-cache and a data-cache of two 2 KB. Besides the processor, several other components are included on the platform: an 8 KB local memory, a debug unit, a timer, and a Dual Data Rate (DDR) DRAM Controller. The interconnect architecture of the system is quite sophisticated, if we consider how these components interact.

- The 8 KB local memory is intended as a local store for the Microblaze, for example to hold the stack or the heap segment of the embedded software. The local memory uses a dedicated Local Memory Bus (LMB) so that local memory can be accessed with a fixed latency (two cycles).

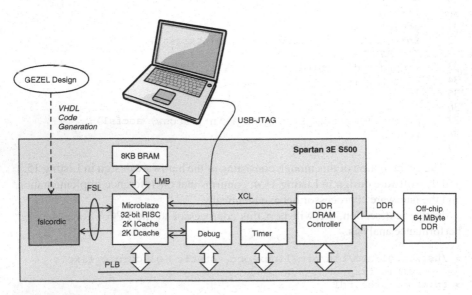

**Fig. 15.4**  FPGA prototype

- The DDR Controller provides access to a large off-chip 64 MB DRAM. The DDR Controller has two connections to the Microblaze processor. The first uses the common Processor Local Bus (PLB), and is used for control operations on the DDR Controller. The second uses the Xilinx Cache Link (XCL), a fast point-to-point bus similar to FSL, and is used for data transfer from the DDR Controller to the cache memories.
- The debug unit allows the system to be controlled using debug software on a laptop. The debug unit has two connections to the Microblaze processor: a PLB connection for control operations on the debug unit, and a dedicated connection to the debug port on the Microblaze processor. The debug port provides access to all internal processor registers, and it supports low level control operations such as single-stepping the processor.
- The timer unit is used to obtain accurate performance estimations, by counting the number of elapsed cycles between two positions in the program. The timer unit is controlled through the PLB.

The platform design and – implementation flow relies on Xilinx Platform Studio (XPS) and will not be discussed in detail here. We will clarify the implementation path from GEZEL to FPGA using XPS. Once we have a working GEZEL system simulation, we can convert GEZEL code into synthesizable VHDL. The code generator is called fdlvhd, and a sample run of the tool on Listing 15.3 is as follows.

```
> fdlvhd -c FSL_Clk FSL_Rst fsl.fdl
Pre-processing System ...
Output VHDL source ...
----------------------------
Generate file: arm1.vhd
Generate file: fsl.vhd
Generate file: cordic_fsmd.vhd
Generate file: fslcordic.vhd
Generate file: top.vhd
Generate file: system.vhd
Generate file: std_logic_arithext.vhd
```

The code generator creates a separate file for each module in the system. In addition, one extra library file is generated (std_logic_arithext.vhd), which is needed for synthesis of GEZEL-generated VHDL code. The code generator also uses a command line parameter, -c FSL_Clk FSL_Rst, which enables a user to choose the name of the clock and reset signal on the top-level module. This makes the resulting code pin-compatible with VHDL modules expected by XPS. If we consider the system hierarchy of Fig. 15.3 once more, we conclude that not all of the generated code is of interest. Only cordic and fslcordic constitute the actual coprocessor. The other modules are 'simulation stubs'. Note that GEZEL does not generate VHDL code for an ARM or and fsl interface; the ipblock modules in Listing 15.3 translate to black-box views. The transfer of the code to XPS relies on the standard design flow to create new peripherals. By designing pin-

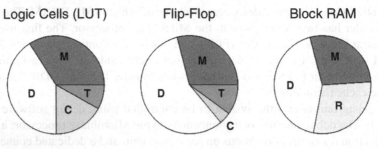

M = Microblaze, D = DDR Controller, C = CORDIC, T = Timer, R = Local RAM

**Fig. 15.5**  Relative resource cost of platform components for Fig. 15.3

compatible modules, we can convert GEZEL modules to XPS peripherals without writing a single line of VHDL.

A detailed discussion of the synthesis results for the platform is beyond the scope of this example. However, it is useful to make a brief summary of the results, in particular because it illustrates the relative hardware cost of a coprocessor in a system platform such as in Fig. 15.4. All components of the platform run at 50 MHz. We partition the implementation cost of the system into logic cells (lookup-tables for FPGA), flip-flops, and BlockRAM cells. The entire system requires 4,842 logic cells, 3,411 flip-flops and 14 BlockRAM cells. Figure 15.4 shows the relative resource cost for each of the major components in the platform. There are several important conclusions to make from this Figure. First, the processor occupies around one quarter of the resources on the platform. The most expensive component in terms of resources is the DDR controller. The coprocessor cost is relatively minor, although still half of the Microblaze in logic area (Fig. 15.5).

## 15.4   Handling Large Amounts of Rotations

In this section, we investigate the performance of the FPGA prototype of the CORDIC coprocessor, and we consider performance optimizations on the overall system throughput.

A coprocessor is not very useful if you use it only once. Therefore, we will consider a scenario that includes a large number of rotations. We use a large table of target angles, stored in off-chip memory. The objective is to convert this table into an equivalent table of $(X, Y)$ coordinates, also stored in off-chip memory. Refering to Fig. 15.4, the table will be stored in the 64 MB DDR memory, and the elements of that table need to be fetched by the Microblaze and processed in the attached CORDIC coprocessor. The outputs of the CORDIC need to be written back to the 64 MB DDR memory.

Let's start with a simple software driver for this coprocessor. Listing 15.5 illustrates a Microblaze program that performs CORDIC transformations on an array of 8,192 elements. The coprocessor driver, cordic_driver, is implemented on lines 11–18. The Microblaze processor uses dedicated instructions to write to, and read from, the FSL interface: (putfsl) and (getfsl). One function call to cordic_driver will complete a single CORDIC rotation. While the coprocessor is active, the Microblaze processor will stall and wait for the result. As discussed above, each rotation takes 20 clock cycles in the hardware implementation. The function compare, lines 17–45, compares the performance of 8192 CORDIC rotations in software versus 8192 CORDIC rotations in hardware. Performance measurements are obtained from a timer module: XTmrCtr_Start, XTmrCtr_Stop, and XTmrCtr_GetValue will start, stop and query the timer module respectively. The resulting rotations are also accumulated (in accsw and accfsl) as a simple checksum to verify that the software and the hardware obtain the same result.

**Listing 15.5** Driver for the CORDIC coprocessor on Microblaze

```
 1  #include "fsl.h"
 2  #include "xparameters.h"
 3  #include "xtmrctr.h"
 4  #define N 8196
 5  int arrayT[N];
 6  int arrayX[N];
 7  int arrayY[N];
 8
 9  void cordic_driver(int target, int *rX, int *rY) {
10    int r;
11    putfslx(target,0,FSL_ATOMIC);
12    getfslx(r,0,FSL_ATOMIC);
13    *rX = r;
14    getfslx(r,0,FSL_ATOMIC);
15    *rY = r;
16  }
17
18  int compare(void) {
19    unsigned i;
20    int accsw = 0, accfsl = 0;
21    int timesw, timefsl;
22    XTmrCtr T;
23
24    XTmrCtr_Start(&T, 0);
25    for (i=0; i<N; i++) {
26      cordic(arrayT[i], &arrayX[i], &arrayY[i]);
27      accsw += (arrayX[i] + arrayY[i]);
28    }
29    XTmrCtr_Stop(&T, 0);
30    timesw = XTmrCtr_GetValue(&T, 0);
31
32    XTmrCtr_Start(&T, 0);
33    for (i=0; i<N; i++) {
34      cordic_driver(arrayT[i], &arrayX[i], &arrayY[i]);
```

```
35      accfsl += (arrayX[i] + arrayY[i]);
36    }
37    XTmrCtr_Stop(&T, 0);
38    timefsl = XTmrCtr_GetValue(&T, 0);
39
40    xil_printf("Checksum_SW_%x_FSL_%x\n", accsw, accfsl);
41    xil_printf("Cycles___SW_%d_FSL_%d\n", timesw, timefsl);
42
43    return(0);
44 }
```

We compile Listing 15.5 while allocating all sections in off-chip memory. The compilation command line selects medium optimization (-O2), as well as several options specific to the Microblaze processor hardware. This includes a hardware integer multiplier (-mno-xl-soft-mul) and a hardware pattern comparator (-mxl-pattern-compare). The compiler command line also shows the use of a *linker script*, which allows the allocation of sections to specific regions of memory (See Sect. 7.3).

```
> mb-gcc -O2 \
      cordiclarge.c         \
      -o executable.elf \
      -mno-xl-soft-mul  \
      -mxl-pattern-compare \
      -mcpu=v7.00.b         \
      -Wl,-T -Wl,cordiclarge_linker_script.ld \
      -g \
      -I./microblaze_0/include/ \
      -L./microblaze_0/lib/
> mb-size executable.elf
   text      data      bss       dec       hex   filename
   7032       416   100440    107888     1a570   executable.elf
```

The actual sizes of the program sections are shown using the mb-size command. Recall that text contains instructions, data contains initialized data, and bss contains uninitialized data. The large bss section is occupied by 3 arrays of 8192 elements each, which require 98,304 bytes. The remainder of that section is required for other global data, such as global variables in the C library.

The resulting performance of the program is shown in scenarios 1 and 2 of Table 15.1. The software CORDIC requires 358 million cycles, while the hardware-accelerated CORDIC requires 4.8 million cycles, giving a speedup of 74.5 times. While this is an excellent improvement, it also involves significant overhead. Indeed, 8192 CORDIC rotations take 163,840 cycles in the FSL coprocessor. Over the total runtime of 4.8 million cycles, the coprocessor thus has only 3 % utilization! Considering the program in Listing 15.5, this is also a peak utilization, since the coprocessor is called in a tight loop with virtually no other software activities. Clearly, there is still another bottleneck in the system.

That bottleneck is the off-chip memory, in combination with the PLB memory bus leading to the microprocessor. Since all program segments are stored in off-chip memory, the MicroBlaze will fetch not only all CORDIC data elements, but also all

**Table 15.1** Performance evaluation over 8,192 CORDIC rotations

| Scenario | CORDIC | Cache | text segment | data segment | bss segment | Performance (cycles) | Speedup |
|---|---|---|---|---|---|---|---|
| 1 | SW | No | DDR | DDR | DDR | 358,024,365 | 1 |
| 2 | FSL | No | DDR | DDR | DDR | 4,801,716 | 74.5 |
| 3 | SW | No | On-chip | On-chip | DDR | 16,409,224 | 21.8 |
| 4 | FSL | No | On-chip | On-chip | DDR | 1,173,651 | 305 |
| 5 | SW | Yes | DDR | DDR | DDR | 16,057,950 | 22.3 |
| 6 | FSL | Yes | DDR | DDR | DDR | 594,655 | 602 |
| 7 | FSL (prefetch) | Yes | DDR | DDR | DDR | 405,840 | 882 |
| 8 | FSL (prefetch) | Yes/8 | On-chip | On-chip | DDR | 387,744 | 923 |

instructions from the off-chip memory. Worse, the cache memory on a MicroBlaze is not enabled by default until instructed so by software, so that the program does not benefit from on-chip memory at all.

There are two possible solutions: local on-chip memories, and cache memory. We will show that the effect of both of them is similar.

- To enable the use of the on-chip memory (Fig. 15.4), we need to modify the linker script and re-allocate sections to on-chip memory. In this case, we need to move the text segments as well as the constant data segment to on-chip memory. In addition, we can also allocate the stack and heap to on-chip memory, which will ensure that local variables and dynamic variables will remain on-chip.
- To enable the use of cache memory, we need to include

```
microblaze_enable_icache();
microblaze_enable_dcache();
```

at the start of the program. The data- and instruction cache of a microblaze is a direct-mapped, four-word-per-line cache architecture.

The result of each of these two optimizations is shown in scenarios 3–6 in Table 15.1. For the software implementations, the use of on-chip local memory, and the use of a cache each provide a speedup of approximately 22 times. For the hardware-accelerated implementations, the use of on-chip local memory provides a speedup of 305 times, while the use of a cache provides a speedup of 602 times. These results confirm that off-chip memory clearly was a major bottleneck in system performance. In general the effect of adding a cache is larger than the effect of moving the text segment/local data into on-chip memory. This is because of two reasons: (a) the cache improves memory-access time, and (b) the cache improves the *off-chip* memory-access time. Indeed, the 'XCL' connections, shown in Fig. 15.4, enable burst-access to the off-chip memory, while the same burst-access effect cannot be achieved through the 'PLB' connection.

The impact of cache on the hardware coprocessor is much more dramatic (600 times speedup instead of 300 times speedup) than its impact on the software CORDIC (22.3 speedup instead of 21.8 speedup). This can be understood by looking at the absolute performance numbers. For the case of software, the cache

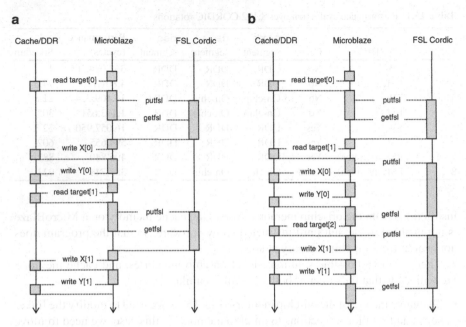

**Fig. 15.6** (a) Sequential memory access and coprocessor execution. (b) Folded memory access and coprocessor execution

provides an advantage of 3.5 million cycles over local-memory (scenarios 3 vs. 5). For the case of hardware, the cache provides an advantage of only 500,000 cycles over local memory (scenarios 4 vs. 6). However, the hardware-accelerated system is already heavily optimized, and hence very sensitive to inefficiencies.

How can we improve this design even further? By close inspection of the loop that drives the FSL coprocessor, we find that the memory accesses and the coprocessor execution are strictly sequential. This is in particular the case for memory-writes, since the write-through cache of the Microblaze forces all of them to be an off-chip access. Indeed, the code first accesses arrayT, then runs the coprocessor through putfsl and getfsl, and finally writes back the results into arrayX and arrayY. This is illustrated in Fig. 15.6a.

```
for (i=0; i<N; i++) {
    cordic_driver(arrayT[i], &arrayX[i], &arrayY[i]);
    accfsl += (arrayX[i] + arrayY[i]);
}
```

The key optimization is to exploit parallelism between the memory accesses and the coprocessor execution. Specifically, instead of waiting for the result of the coprocessor (using getfsl), the Microblaze processor may use that time to read from/ write to memory. A solution that uses this overlap for memory writes is shown in Fig. 15.6b. An equivalent C code fragment that achieves this behavior looks as follows.

```
cordic_put(arrayT[0]);
for (i=1; i<N; i++) {
  cordic_get(&tmpX, &tmpY);
  cordic_put(arrayT[i]);
  arrayX[i-1]  = tmpX;
  arrayY[i-1]  = tmpY;
  accfsl      += (tmpX + tmpY);
}
cordic_get(&arrayX[N-1], &arrayY[N-1]);
accfsl += ( arrayX[N-1] + arrayY[N-1]);
```

The effect of this optimization is illustrated in scenario 7 of Table 15.1. An additional 200,000 clock cycles are gained, and the overall execution time is around 400,000 clock cycles, or a speedup of 882. At this point, the coprocessor utilization has increased to 41 %, which is an improvement of more than ten times over the original case.

Further improvements are still possible. For example, we know that the accesses to arrayT are strictly sequential. Hence, it makes sense to increase the line size of the cache as much as possible (the line size is the number of consecutive elements that are read after a cache miss). In addition, we can use an on-chip memory for the program, but reserve all the cache memory for data accesses. The result of these optimizations is an additional 18,000 cycles, as shown in Table 15.1, scenario 8. The overall speedup is now 923 times, and the coprocessor utilization is 42 %. As long as the utilization is not 100 %, the system-level bottleneck is *not* in hardware but rather between the MicroBlaze and the off-chip memory. The next step in the optimization is to investigate the assembly code of the loop, and to profile the behavior of the loop in detail.

In conclusion, this section has demonstrated that the design of a hardware module is only the first step in an efficient hardware/software codesign. System integration and system performance optimization is the next step.

## 15.5 Summary

In this chapter we discussed the implementation of the Coordinate Rotation Digital Computer (CORDIC) algorithm as a hardware coprocessor. The CORDIC algorithm rotates, iteratively, a vector $(X, Y)$ over a given angle $\alpha$. The algorithm uses only integer operations, which makes it very well suited for embedded system implementation. We discussed a coprocessor design based on the Fast Simple Link coprocessor interface, and we created a simulation model of the CORDIC in GEZEL, first as a standalone module, and next as a coprocessor module. After functional verification of the coprocessor at high abstraction level, we ported the design to an FPGA prototype using a Spartan 3E chip. The resulting embedded system architecture consists roughly of one-third microprocessor, one-third memory-controller, and one-third peripherals (with the coprocessor being included in the peripherals). Early implementation results showed that the coprocessor provided a speedup of 74 over the CORDIC software implementation. Through careful tuning, in particular by optimizing off-chip accesses, that speedup can be further improved to 923 times.

## 15.6  Further Reading

The CORDIC algorithm is 50 years old, and was developed for 'real-time airborne computation', in other words, for a military application. The original CORDIC proposal, by Volder, is a good example of a paper that truly stands the test of time (Volder 1959). More recently Meher has provided a comprehensive overview (Maharatna et al. 2009). Valls discusses CORDIC applications in digital radio receivers (Valls et al. 2006).

## 15.7  Problems

**Problem 15.1.** Design a GEZEL testbench for the standalone CORDIC design shown in Listing 15.2. Verify the sine and cosine values shown in Table 15.2.

**Problem 15.2.** The CORDIC algorithm in this chapter is working in the so-called rotation mode. In rotation mode, the CORDIC iterations aim to drive the value of the angle accumulator to 0 (refer to Listing 15.1). CORDIC can also be used in vector mode. In this mode, the CORDIC rotations aim to drive the value of the Y coordinate to 0. In this case, the input of the algorithm consists of the vector $(x_0, y_0)$, and the angle accumulator is initialized to zero.

(a)  Show that, in the vector mode, the final values of $(x_T, y_T)$ are given by

$$\begin{bmatrix} x_T \\ y_T \end{bmatrix} = \begin{bmatrix} K \cdot \sqrt{(x_0{}^2 + y_0{}^2)} \\ 0 \end{bmatrix} \tag{15.11}$$

with $K$ a similar magnitude constant as used in the rotation mode.

**Table 15.2** Test cases for Problem 15.1

| Angle | cos(angle) | sin(angle) |
|-------|------------|------------|
| 0 | 1 | 0 |
| $\pi/6$ | $\sqrt{3}/2$ | $1/2$ |
| $\pi/4$ | $1/\sqrt{2}$ | $1/\sqrt{2}$ |
| $\pi/3$ | $1/2$ | $\sqrt{3}/2$ |
| $\pi/2$ | 0 | 1 |

(b) Show that, in the vector mode, the final value of the angle accumulator is given by

$$\alpha = arctan\left(\frac{y_0}{x_0}\right) \quad\quad\quad (15.12)$$

(c) Adjust the hardware design in Listing 15.2 so that it implements CORDIC in vector mode. Verify your design with some of the tuples shown in Table 15.2.

**Problem 15.3.** Develop a CORDIC coprocessor in rotation mode using the custom-instruction interface discussed in Sect. 13.3.3. The recommended approach is to build a coprocessor that does a single CORDIC iteration per custom-instruction call. Hence, you will need an OP2X2 instruction for each iteration. You will also need a mechanism to program the rotation angle. For example, the software driver could look like:

```
int target, X, Y;
unsigned i;

// argument 1: target angle
// argument 2: 10 iterations
OP2x2_1(target, 10, 0, 0);

for (i=0; i<10; i++)
  OP2x2_1(X, Y, X, Y);
```

# Appendix A
# Hands-on Experiments in GEZEL

## A.1 Overview of the GEZEL Tools

GEZEL is a set of open-source tools for hardware/software codesign. The GEZEL website is at http://rijndael.ece.vt.edu/gezel2. This website distributes source code, pre-compiled versions (for Ubuntu), examples, and an online manual with installation instructions.

Figure A.1 shows an overview of the GEZEL tools. GEZEL is constructed as a C++ library with several components: a parser for the GEZEL language, a cycle-accurate simulation kernel, a VHDL code-generator, an interface to four different instruction-set simulators, and an interface to user-defined simulator extensions. The GEZEL tools are created on top of the GEZEL library. The examples throughout the book make use of these tools.

- fdlsim is the stand-alone cycle-accurate simulator. It uses the parser component, the cycle-accurate simulation kernel, and optional user-defined simulator extensions.
- gplatform is the co-simulator. It uses all of the components of fdlsim, in addition to several instruction-set simulators.
- fdlvhd is the code-generator. It uses the parser component, the code-generator, and optional user-defined code-generation extensions.

For the reader interested in simulation tools, it is useful to study the user-defined simulation extension interface. This interface supports the creation of new ipblock types. All of the cosimulation environments were generated using this model.

P.R. Schaumont, *A Practical Introduction to Hardware/Software Codesign*,
DOI 10.1007/978-1-4614-3737-6, © Springer Science+Business Media New York 2013

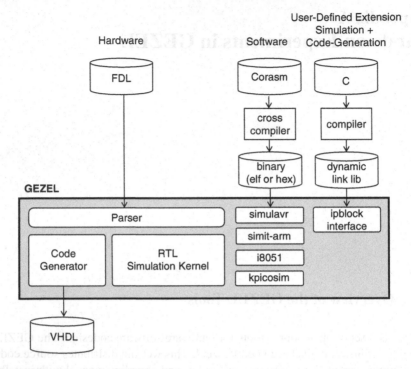

**Fig. A.1**  Overview of the GEZEL tools

## A.2   Installing the GEZEL Tools

In this section, we briefly review the installation procedure for GEZEL. There are two different methods: installing pre-compiled packages, and recompiling from scratch. Installation of pre-compiled packages is the preferred method, in particular in a classroom environment. For example, the author has relied on the use of a Ubuntu Virtual Machine in his class, so that all students in the class can install a cosimulation environment on their own machine.

### A.2.1   Installation on a Ubuntu System

Ubuntu uses the Debian packaging system. This provides an easy installation procedure. GEZEL is distributed as six packages. In the following package names, the dist suffix should be replaced with the name of the Ubuntu distribution. For example, on Ubuntu Precise (12.04), you would use gezel-base-precise.

- gezel-base-dist provides basic GEZEL capabilities, including simulation, cosimulation and code generation.

- `gezel-sources-dist` provides GEZEL source code modules, which you need if you want to recompile GEZEL from source.
- `gezel-debug-dist` provides a compiled version of the GEZEL tools with full debug info. This package is useful for GEZEL debugging purposes.
- `gezel-simulavr-dist` is a version of simulavr with GEZEL cosimulation interfaces. You need this package if you want to do AVR-based cosimulations.
- `gezel-simitarm-dist` is a version of Simit-ARM with GEZEL cosimulation interfaces. You need this package if you want to do ARM-based cosimulation.
- `gezel-examples-dist` contains demonstration examples of GEZEL, including many of the examples described in this book.

GEZEL packages are stored in a repository, a web server that provides easy access to the distribution. The URL of the GEZEL repository is http://rijndael. ece.vt.edu/gezel2repo. Installation of GEZEL packages now proceeds through four steps.

1. Configure the packaging system on your Ubuntu machine to read the GEZEL repository. The easiest way is to add the following line to the file `/etc/apt/sources.list`:

   ```
   deb http://rijndael.ece.vt.edu/gezel2repo precise
   main
   ```

   This line includes the name of the Ubuntu release. If you are working on Ubuntu Lucid (10.04), for example, you have to replace 'precise' with 'lucid'. Also, you need superuser privileges to edit the file `/etc/apt/sources.list`. Use `sudo vi /etc/apt/sources.list`.
2. Add the author's public key to your system. This will enable you to verify the authenticity of the packages. Adding a public key will require superuser privileges as well.

   ```
   sudo apt-key adv --keyserver pgp.mit.edu \
                    --recv-keys 092EF91B
   ```

3. Refresh the directory of available packages

   ```
   sudo apt-get update
   ```

4. Decide what GEZEL packages to install and proceed. For example, to install `gezel-base-precise` you would use

   ```
   sudo apt-get install gezel-base-precise
   ```

Once you have configured the packaging system to recognize the GEZEL repository, you can easily remove and add packages with additional `apt-get` commands. Furthermore, if GEZEL is upgraded, you will automatically be notified by the Ubuntu system that an upgrade is available. The current version (Summer 2012) of GEZEL is 2.5.13. There are approximately two new releases of GEZEL per year.

The installation directory of the GEZEL tools is /opt/gezel. This directory
is not included in the standard PATH of a Ubuntu system. To run, for example,
fdlsim on the file aes.fdl, you should use the full path as follows.

```
/opt/gezel/bin/fdlsim aes.fdl 100
```

Alternately, you can adjust your path to include the GEZEL directory by default.
This allows you to use just the GEZEL executable name in a command line.

```
export PATH=$PATH:/opt/gezel/bin
fdlsim aes.fdl 100
```

## A.2.2 Installation of Cross-Compiler Tools

The instruction-set simulators in GEZEL need a cross-compiler to generate binaries
for simulation. Depending on the cores you wish to use, you need to install one of
the following compiler toolchains.

- Simit-ARM uses an arm-linux-gcc compiler. The repository for GEZEL
  includes such a compiler (package arm-linux-gcc). Executables for Simit-ARM
  need to be created with the -static flag.

```
/usr/local/arm/bin/arm-linux-gcc -static \
                                 -o myprogram \
                                 myprogram.c
```

- Dalton 8051 uses the Small Devices C Compiler (package sdcc). sdcc is
  included in the 'universe' part of the Ubuntu Debian repository. Executables for
  the 8051 Dalton simulator are created as ihx files.

```
sdcc myprogram.c
```

- SimulAVR uses the avr-gcc compiler (package gcc-avr). Additional packages include binutils-avr and avr-libc. gcc-avr is included in the
  'universe' part of the Ubuntu Debian repository. Executables for SimulAVR are
  created as ELF files:

```
avr-gcc -mmcu=atmega128 myprog.c -o myproc
```

## A.2.3 Compiling GEZEL from Source Code on a 32-bit System

Compiling GEZEL from source requires multiple compilation steps, and it requires
additional packages on your system. The following steps illustrate compilation on a
clean Ubuntu Precise (12.04) system.

Start by downloading the gezel-sources-precise package. You can either follow the package configuration steps above, or else directly download the package from the repository. Direct installation of a Debian package can be done with the dpkg command.

```
sudo dpkg -i gezel-sources-precise_2.5.13_i386.deb
```

To compile all source, you need to install several additional packages: auto conf, libtool, g++, bison, texinfo, flex, libgmp3-dev, binutils-dev. Use the following command.

```
sudo apt-get install autoconf      \
                     libtool       \
                     g++           \
                     bison         \
                     flex          \
                     texinfo       \
                     libgmp3-dev   \
                     binutils-dev
```

### A.2.3.1 Compiling the Stand-Alone Simulation Tool fdlsim

Extract the source code of the gezel simulation tools (gplatform, fdlsim)

```
tar zxfv /opt/gezel-sources/gezel-sim.tgz
cd gezel-sim
./bootstrap
```

To compile the stand-alone simulator, and install it in gezel-sim/build, use the following command:

```
./configure --enable-standalone
make install
```

### A.2.3.2 Compiling the Instruction-Set Simulator Simit-ARM

Before you can compile the GEZEL cosimulator, you will need to compile the instruction-set simulators you would like to use. The Dalton 8051 and Picoblaze ISS are integrated at source-code level. The Simit-ARM and simulavr simulators, however, are integrated at the library level. Compile Simit-ARM as follows. First, extract the source code of the instruction-set simulator.

```
tar zxfv /opt/gezel-sources/simit-arm-sfu.tgz
```

To compile `simit-arm-sfu`, and install it in `/opt/simit-arm-sfu`, use the following command. Note the use of the pre-processor directive `CPPFLAGS`, which is needed to enable cosimulation stubs in `simit-arm`.

```
cd simit-arm-sfu
./configure CPPFLAGS='-DCOSIM_STUB' \
            CXXFLAGS='-fpermissive' \
            --prefix=/opt/simit-arm-sfu
make
sudo make install
```

### A.2.3.3   Compiling the Instruction-Set Simulator simulavr

First, extract the source code of the instruction-set simulator.

```
tar zxfv /opt/gezel-sources/simulavr.tgz
cd simulavr
./bootstrap
```

To compile `simulavr`, and install it in `/opt/simulavr`, use the following command.

```
./configure --prefix=/opt/simulavr
make
sudo make install
```

### A.2.3.4   Compiling the Cosimulation Tool gplatform

To compile the cosimulator, and install it in `gezel-sim/build`, first install `gezel-simitarm-dist` and `gezel-simulavr-dist`, or compile them from source as indicated above. Then, use the following command:

```
./configure --enable-gplatform \
            --enable-simitarm --enable-simitsfu \
            --with-simit=/opt/simit-arm-sfu \
            --with-simulavr=/opt/simulavr
make install
```

### A.2.3.5   Compiling the Code Generation Tool fdlvhd

Extract the source code of the gezel code generation tools (`fdlvhd`, `igc`)

```
tar zxfv /opt/gezel-sources/gezel-cg.tgz
cd gezel-cg
./bootstrap
```

To compile the code generation tools, and install them in gezel-cg/build, use the following command:

```
./configure --enable-vhdl \
            --enable-igc \
            --with-gezel=/opt/gezel
make install
```

## A.2.4   Compiling GEZEL from Source Code on a 64-bit System

The compilation for a 64-bit platform is largely identical to the compilation for a 32-bit platform. This section only explains the steps needed for 64-bit compilation.

The main difference between the 32-bit and 64-bit version of GEZEL is that the Simit-ARM instruction-set simulator is not available as part of the cosimulation.

To compile a 32-bit (standalone) version of Simit-ARM, first install multilib support for g++:

```
sudo apt-get install g++-multilib
```

### A.2.4.1   Compiling the Instruction-Set Simulator Simitarm (on 64-bit Platform)

Extract the source code of the instruction-set simulator.

```
tar zxfv /opt/gezel-sources/simit-arm-sfu.tgz
```

To compile simit-arm-sfu, and install it in /opt/simit-arm-sfu, use the following command. Note the use of the pre-processor directive, which selects a 32-bit compile.

```
cd simit-arm-sfu
./configure CPPFLAGS=-m32 LDFLAGS=-m32 --prefix=/opt/
simit-arm-sfu make
sudo make install
```

**A.2.4.2   Compiling the Cosimulation Tool gplatform (on 64-bit Platform)**

To compile the cosimulator, and install it in `gezel-sim/build`, first install
`gezel-simulavr`, or compile it from source as indicated above. Then, use the
following command:

```
./configure --enable-gplatform \
            --with-simulavr=/opt/simulavr
make install
```

## A.3   Running the Examples

The package `gezel-examples-dist` includes many examples from the book.
Their source code can be found on `/opt/gezel-examples/bookex`. The
examples are ready-to-run, assuming a complete GEZEL installation is available.
This section briefly describes how to compile and run the examples.

### A.3.1   Examples from FSMD Chapter

The `C05_FSMD` directory includes two subdirectories: `gcd` and `median`. The `gcd`
directory contains the source code of a Greatest Common Divisor Algorithm in
several Hardware Description Languages. The `median` directory includes three
versions of the Median computation example: a version in C, a fully parallel
hardware version, and a sequentialized hardware version. The following sequence
of commands demonstrates their execution.

- Reference implementation in C.

  ```
  > make median
  gcc -o median median.c
  > ./median
  The median of 4, 56, 2, 10, 32 is 10
  ```

- Fully parallel version in GEZEL.

  ```
  > make sim1
  /opt/gezel/bin/fdlsim m1.fdl 1
  The median is 10
  ```

- Sequential version in GEZEL.

  ```
  > make sim2
  /opt/gezel/bin/fdlsim m2.fdl 200
  12 a1 1234/1234 q1 0
  ```

```
25 a1 91a/91a q1 0
38 a1 848d/848d q1 91a
51 a1 c246/c246 q1 1234
64 a1 e123/e123 q1 848d
77 a1 7091/7091 q1 848d
90 a1 b848/b848 q1 b848
103 a1 dc24/dc24 q1 c246
116 a1 6e12/6e12 q1 b848
129 a1 3709/3709 q1 7091
142 a1 1b84/1b84 q1 6e12
155 a1 8dc2/8dc2 q1 6e12
168 a1 46e1/46e1 q1 46e1
181 a1 2370/2370 q1 3709
194 a1 91b8/91b8 q1 46e1
```

## A.3.2 Examples from Microprogrammed Architectures Chapter

The C06_MICRO directory includes three examples: an implementation of the Hypothetical Microprogrammed Machine (HMM) in GEZEL, the Bresenham Algorithm written for an 8051 microcontroller, and the Bresenham Algorithm written for an 8051-based microprogrammed machine. The following sequence of commands demonstrates their execution.

- Hypothetical Microprogrammed Machine

```
> make
cpp -P hmm.fdl | /opt/gezel/bin/fdlsim  200
0 IO 1 0 14 14
1 IO 1 0 32 32
2 IO 0 0 87 14
3 IO 0 0 87 14
4 IO 0 0 87 14
5 IO 0 0 87 14
6 IO 0 0 87 14
7 IO 0 0 87 14
8 IO 0 0 87 14
9 IO 0 0 87 14
...
```

- Bresenham Algorithm on an 8051 Microcontroller

```
> make
sdcc bresen.c
> make sim
```

```
/opt/gezel/bin/gplatform bresen.fdl
i8051system: loading executable [bresen.ihx]
0x00 0xFF 0xFF 0xFF
0x00 0x00 0xFF 0xFF
0x00 0x00 0x00 0xFF
0x00 0x00 0x00 0x00
0x00 0x00 0x00 0xFF
0x01 0x00 0x00 0xFF
0x01 0x01 0x00 0xFF
...
```

- Bresenham Algorithm on an 8051-based Microprogrammed Machine

```
> make
sdcc bresen.c
> make sim
/opt/gezel/bin/gplatform bresen.fdl
i8051system: loading executable [bresen.ihx]
0 x 0/0 y 0/0 e 0/0 x2 0/0 y2 0/0 e2 0/0 xs 0/0 ...
0xFF 0xD7 0xFF 0xFF
0 x 0/0 y 0/0 e 0/0 x2 0/0 y2 0/0 e2 0/0 xs 0/0 ...
0x80 0xD7 0xFF 0xFF
80 x 0/0 y 0/0 e 0/0 x2 0/0 y2 0/0 e2 0/0 xs 0/0 ...
80 x 0/0 y 0/0 e 0/0 x2 0/0 y2 0/0 e2 0/0 xs 0/0 ...
0x00 0xD7 0xFF 0xFF
0 x 0/0 y 0/0 e 0/0 x2 0/0 y2 0/0 e2 0/0 xs 0/0 ...
0x00 0x17 0xFF 0xFF
...
```

## A.3.3  Examples from System on Chip Chapter

The C08_SOC directory includes two examples. The first, pingpong, illustrates a pingpong buffer communication scheme between an 8051 and hardware. The second, uart, shows simple UART analyzer (written in GEZEL), attached to an AVR microcontroller. The following sequence of commands demonstrates their execution.

- Ping Pong buffer in 8051

```
> make
sdcc ramrw.c
> make sim
/opt/gezel/bin/gplatform -c 50000 pingpong.fdl
i8051system: loading executable [ramrw.ihx]
0x00 0x00 0xFF 0xFF
```

```
0x01 0x00 0xFF 0xFF
28984 ram radr 0/1 data 40
28985 ram radr 1/2 data 3f
28986 ram radr 2/3 data 3e
28987 ram radr 3/4 data 3d
28988 ram radr 4/5 data 3c
0x00 0x01 0xFF 0xFF
38464 ram radr 20/21 data df
38465 ram radr 21/22 data de
38466 ram radr 22/23 data dd
38467 ram radr 23/24 data dc
38468 ram radr 24/25 data db
0x01 0x00 0xFF 0xFF
47944 ram radr 0/1 data 80
47945 ram radr 1/2 data 7f
47946 ram radr 2/3 data 7e
47947 ram radr 3/4 data 7d
47948 ram radr 4/5 data 7c
Total Cycles: 50000
```

- AVR UART Analyzer

```
> make
avr-gcc -mmcu=atmega128 avruart.c -o avruart.elf
> make sim
/opt/gezel/bin/gplatform -c 8000 uart.fdl
atm128core: Load program avruart.elf
atm128core: Set clock frequency 8 MHz
@237: ->1
@361: ->0
@569: ->1
@1401: ->0
@1609: ->1
@1817: ->0
@2233: ->1
@2649: ->0
@2857: ->1
@3689: ->0
@3897: ->1
@4105: ->0
@4521: ->1
Total Cycles: 8000
```

## A.3.4   Examples from Microprocessor Interfaces Chapter

The C11_ITF directory includes two examples. The first, gcd, demonstrates a
memory-mapped GCD algorithm. The second, endianess, demonstrates four
different implementations of an Endianess conversion module. The four implementations are a software implementation, a memory-mapped coprocessor, and two
different ASIP implementations. The following sequence of commands demonstrates their execution.

- Memory-mapped Greatest Common Divisor

```
> make
/usr/local/arm/bin/arm-linux-gcc -static \
                                 gcddrive.c \
                                 -o gcddrive
> make sim
gplatform gcdmm.fdl
core my_arm
armsystem: loading executable [gcddrive]
armsystemsink: set address 2147483652
armsystemsink: set address 2147483660
gcd(80,12) = 4
gcd(80,13) = 1
Total Cycles: 14764
```

- Endianness Conversion: all software design

```
> make
/usr/local/arm/bin/arm-linux-gcc -O3 \
                                 -static \
                                 -o endian.elf \
                                 endian.c cycle.s
> make sim
/opt/gezel/bin/gplatform endian.fdl
core myarm
armsystem: loading executable [endian.elf]
4K conversions take 53786 cycles
   Per conversion: 13 cycles
Total Cycles: 70035
```

- Endianess Conversion: memory-mapped coprocessor design

```
> make
/usr/local/arm/bin/arm-linux-gcc -O3 \
                                 -static \
                                 -o endian.elf \
                                 endian.c cycle.s
```

```
> make sim
/opt/gezel/bin/gplatform endian.fdl
core myarm
armsystem: loading executable [endian.elf]
armsystemsink: set address 2147483652
4K conversions take 41502 cycles
    Per conversion: 10 cycles
Total Cycles: 57746
```

- Endianess Conversion: ASIP design with single-argument instructions

```
> make
/usr/local/arm/bin/arm-linux-gcc -O3 \
                                 -static \
                                 -o endian.elf \
                                 endian.c cycle.s
> make sim
/opt/gezel/bin/gplatform endian.fdl
core myarm
armsystem: loading executable [endian.elf]
4K conversions take 37401 cycles
    Per conversion: 9 cycles
Total Cycles: 57702
```

- Endianess Conversion: ASIP design with double-argument instruction

```
> make
/usr/local/arm/bin/arm-linux-gcc -O3 \
                                 -static \
                                 -o endian.elf \
                                 endian.c cycle.s
> make sim
/opt/gezel/bin/gplatform endian.fdl
core myarm
armsystem: loading executable [endian.elf]
4K conversions take 29209 cycles
    Per conversion: 7 cycles
Total Cycles: 47459
```

### A.3.5  Examples from Trivium Chapter

The C13_Trivium directory includes four examples, all of them based on the Trivium coprocessor. The first, trivium_hw, includes three different implementations of the Trivium design as a stand-alone module. The second, trivium_arm_sw, shows a reference implementation of Trivium in software, for ARM. The third, `trivium_arm_t32` demonstrates a memory-mapped coprocessor for ARM. The last one, `trivium_arm_sfu` shows a custom-instruction design of Trivium, emulated on an ARM. The following sequence of commands demonstrates their execution.

- Trivium standalone module. Use make `sim1`, make `sim2` or make `sim3` to run a 1 bit-per-cycle, 8-bit-per-cycle or 32-bit-per-cycle implementation. The output of the 32-bit-per-cycle simulation is shown below.

```
> make sim3
/opt/gezel/bin/fdlsim trivium32.fdl 50
39 11001100110011100111010101111011 ccce757b
40 10011001101111010111100100100000 99bd7920
41 10011010001000110101101010001000 9a235a88
42 00010010010100011111110010011111 1251fc9f
43 10101111111100001010011001010101 aff0a655
44 01111110110010001110111001001110 7ec8ee4e
45 10111111110101000010000100101000 bfd42128
46 10000110110110101110011000001000 86dae608
47 10000000011011101010011111101011 806ea7eb
48 01011000101011101100000100000010 58aec102
49 00010110111110101000100011110100 16fa88f4
```

- Trivium reference implementation in software

```
> make
/usr/local/arm/bin/arm-linux-gcc -static \
                                 -O3 \
                                 trivium.c cycle.s \
                                 -o trivium
> make sim
/opt/gezel/bin/gplatform trivium32.fdl
core myarm
armsystem: loading executable [trivium]
7b
75
ce
cc
...
51
12
key schedule cycles: 3810   stream cycles: 48815
Total Cycles: 81953
```

- Trivium memory-mapped coprocessor

```
> make
/usr/local/arm/bin/arm-linux-gcc -static \
```

```
                                     trivium.c cycle.s\
                                     -o trivium
> make sim
/opt/gezel/bin/gplatform trivium32.fdl
core myarm
armsystem: loading executable [trivium]
armsystemsink: set address 2147483648
armsystemsink: set address 2147483656
ccce757b ccce757b 99bd7920 9a235a88 ...
86dae608 806ea7eb 58aec102 16fa88f4 ...
...
c2cecf02 o18e5chc 533dbb8f 4faf90ef ...
key schedule cycles: 435   stream cycles. 10524
Total Cycles: 269120
```

- Trivium custom-instruction design

```
> make
/usr/local/arm/bin/arm-linux-gcc -static \
                                     trivium.c cycle.s\
                                     -o trivium
> make sim
/opt/gezel/bin/gplatform triviumsfu.fdl
core myarm
armsystem: loading executable [trivium]
ccce757b 99bd7920 9a235a88 1251fc9f ...
806ea7eb 58aec102 16fa88f4 c5c3aa3e ...
key schedule cycles: 289   stream cycles: 8862
Total Cycles: 39219
```

## A.3.6 Examples from AES Chapter

The C14_AES subdirectory includes the design of a memory-mapped coprocessor for ARM. The following sequence of commands demonstrates its execution.

- AES memory-mapped coprocessor

```
> make
/usr/local/arm/bin/arm-linux-gcc -static  \
   aes_coproc_armdriver.c \
   -o aes_coproc_armdriver
> make sim
/opt/gezel/bin/gplatform aes_coproc_arm.fdl
core myarm
armsystem: loading executable [aes_coproc_armdriver]
```

```
armsystemsink: set address 2147483652
cycle 10164: set key0/102030405060708090a0b0c0d0e0f
cycle 10222: set text_in 0/112233445566778899aabbcc
ddeeff
cycle 10235: start encryption
cycle 10255: get text_out 69c4e0d86a7b0430d8cdb7807
0b4c55...
text_out 69c4e0d8 6a7b0430 d8cdb780 70b4c55a
Total Cycles: 15188
```

### A.3.7   Examples from CORDIC Chapter

The C15_CORDIC subdirectory includes the design of an FSL-mapped CORDIC
design. The FSL link is emulated on an ARM processor, as explained in Chap. 15.
The following sequence of commands demonstrates its execution.

- CORDIC FSL-mapped accelerator

```
> make
/usr/local/arm/bin/arm-linux-gcc -static \
                                 -O3 \
                                 cordic.c \
                                 -o cordic
> make sim
/opt/gezel/bin/gplatform cordic.fdl
core arm1
armsystem: loading executable [cordic]
Checksum SW 55affcee FSL 55affcee
Total Cycles: 23228
```

# References

Aeroflex G (2009) Leon-3/grlib intellectual property cores. Technical report, http://www.gaisler.com

Altera (2011) Avalon interface specifications. http://www.altera.com/literature/manual/mnl_avalon_spec.pdf

Appel AW (1997) Modern compiler implementation in C: basic techniques. Cambridge University Press, New York, NY, USA

Atmel (2008) AT91SAM7l128 preliminary technical information. http://www.atmel.com/dyn/products/product_card.asp?part_id=4293

Berry G (2000) The foundations of esterel. In: Milner R (ed) Proof, language, and interaction. MIT, Cambridge, pp 425–454

Bogdanov A, Knudsen L, Leander G, Paar C, Poschmann A, Robshaw M, Seurin Y, Vikkelsoe C (2007) Present: an ultra-lightweight block cipher. In: Proceedings of the cryptographic hardware and embedded systems 2007, Vienna, Springer, Heidelberg, pp 450–466

Butenhof D (1997) Programming with POSIC Threads. Addison-Wesley Professional, 1997. ISBN 978-0201633924.

Claasen T (1999) High speed: not the only way to exploit the intrinsic computational power of silicon. In: Solid-state circuits conference, 1999. Digest of technical papers, ISSCC. IEEE International, Piscataway, Piscataway, NJ, USA, pp 22–25

Claasen T (2006) An industry perspective on current and future state of the art in system-on-chip (soc) technology. Proc IEEE 94(6):1121–1137

Committee T (1995) Tool interface standard executable and linkable format (elf) specification, version 1.2. Technical report, http://refspecs.freestandards.org/elf/elf.pdf

Cytron R, Ferrante J, Rosen BK, Wegman MN, Zadeck FK (1991) Efficiently computing static single assignment form and the control dependence graph. ACM Trans Program Lang Syst 13(4):451–490

Davio M, Deschamps JP, Thayse A (1983) Digital systems with algorithm implementation. Wiley, New York

De Canniere C, Preneel B (2005) Trivium specifications. Technical report, ESAT/SCD-COSIC, K.U.Leuven, http://www.ecrypt.eu.org/stream/p3ciphers/trivium/trivium_p3.pdf

Dennis J (2007) A dataflow retrospective – how it all began. http://csg.csail.mit.edu/Dataflow/talks/DennisTalk.pdf

D'Errico J, Qin W (2006) Constructing portable compiled instruction-set simulators: an adl-driven approach. In: DATE '06: proceedings of the conference on design, automation and test in Europe, Munich, pp 112–117

Dijkstra EW (2009) The E.W. Dijkstra Archive. Technical report, http://www.cs.utexas.edu/users/EWD/

P.R. Schaumont, *A Practical Introduction to Hardware/Software Codesign*,
DOI 10.1007/978-1-4614-3737-6, © Springer Science+Business Media New York 2013

ECRYPT (2008) The estream project. Technical report, http://www.ecrypt.eu.org/stream/technical.
html

Edwards SA (2006) The challenges of synthesizing hardware from c-like languages. IEEE Des
Test Comput 23(5):375–386

Eker J, Janneck J, Lee E, Liu J, Liu X, Ludvig J, Neuendorffer S, Sachs S, Xiong Y (2003) Taming
heterogeneity – the ptolemy approach. Proc IEEE 91(1):127–144

Gaj K, Chodowiec P (2009) FPGA and ASIC implementations of AES. In: Koc C (ed) Cryptographic engineering. Springer, New York. ISBN 978-0-387-71817-0.

Gajski DD, Abdi S, Gerstlauere A, Schirner G (2009) Embedded system design: modeling,
synthesis, verification. Springer, Boston

Ganesan P, Venugopalan R, Peddabachagari P, Dean A, Mueller F, Sichitiu M (2003) Analyzing
and modeling encryption overhead for sensor network nodes. In: WSNA '03: proceedings of the
2nd ACM international conference on wireless sensor networks and applications. ACM, New
York, pp 151–159. doi:http://doi.acm.org/10.1145/941350.941372

Good T, Benaissa M (2007) Hardware results for selected stream cipher candidates. Technical
report, eSTREAM project, http://www.ecrypt.eu.org/stream/hw.html

Gupta S, Gupta R, Dutt N, Nicolau A (2004) SPARK: a parallelizing approach to the high-level
synthesis of digital circuits. Springer, Boston

Harel D (1987) Statecharts: a visual formulation for complex systems. Sci Comput Program
8(3):231–274

Hennessy JL, Patterson DA (2006) Computer architecture: a quantitative approach, 4th edn.
Morgan Kaufmann, Boston

Hillis WD, Steele GL Jr (1986) Data parallel algorithms. Commun ACM 29(12):1170–1183

Hodjat A, Verbauwhede I (2004) High-throughput programmable cryptocoprocessor. IEEE Micro
24(3):34–45

Hoe JC (2000) Operation-centric hardware description and synthesis. Ph.D. thesis, MIT

IBM (2009) Coreconnect bus architecture. Technical report, https://www-01.ibm.com/chips/
techlib/techlib.nsf/productfamilies/CoreConnect_Bus_Architecture

Ivanov A, De Micheli G (2005) Guest editors' introduction: The network-on-chip paradigm in
practice and research. IEEE Des Test Comput 22(5):399–403

Kaps JP (2008) Chai-tea, cryptographic hardware implementations of xtea. In: INDOCRYPT.
Springer, New York, pp 363–375

Karlof C, Sastry N, Wagner D (2004) Tinysec: a link layer security architecture for wireless
sensor networks. In: SenSys '04: proceedings of the 2nd international conference on embedded
networked sensor systems. ACM, New York, pp 162–175. doi:http://doi.acm.org/10.1145/
1031495.1031515

Kastner R, Kaplan A, Sarrafzadeh M (2003) Synthesis techniques and optimizations for reconfigurable systems. Kluwer, Boston

Keutzer K, Newton A, Rabaey J, Sangiovanni-Vincentelli A (2000) System-level design: orthogo-
nalization of concerns and platform-based design. IEEE Trans Comput Aided Des Integr Circuit
Syst 19(12):1523–1543

Keppel D (1994), Tools and Techniques for building fast portable thread packages. http://www.cs.
washington.edu/research/compiler/papers.d/quickthreads.html

Kogge PM (1981) The architecture of pipelined computers. McGraw-Hill, New York

Leander G, Paar C, Poschmann A, Schramm K (2007) New lightweight des variants. In: Biryukov
A (ed) Fast software encryption. Lecture notes on computer science, vol 4593. Springer, New
York, pp 196–200

Lee EA, Messerschmitt DG (1987) Static scheduling of synchronous data flow programs for digital
signal processing. IEEE Trans Comput 36(1):24–35

Lee EA, Seshia SA (2011) Introduction to embedded systems, a cyber-physical systems approach.
http://LeeSeshia.org, ISBN 978-0-557-70857-4.

Leupers R, Ienne P (2006) Customizable embedded processors: design technologies and applications. Morgan Kaufmann, San Francisco

Ltd A (2009a) The amba system architecture. Technical report, http://www.arm.com/products/solutions/AMBAHomePage.html

Ltd A (2009b) Arm infocenter. Technical report, http://infocenter.arm.com/help/index.jsp

Lynch M (1993) Micro-programmed state machine design, CRC, Boca Raton

Madsen J, Steensgaard-Madsen J, Christensen L (2002) A sophomore course in codesign. Computer 35(11):108–110. doi:http://dx.doi.org/10.1109/MC.2002.1046983

Maharatna K, Valls J, Juang TB, Sridharan K, Meher P (2009) 50 years of cordic: algorithms, architectures, and applications. IEEE Trans Circuit Syst I Regul Pap 56(9):1893–1907

McKee S (2004) Reflections on the memory wall. In: Conference on computing frontiers. ACM, New York, pp 162–168

Meiser G, Eisenbarth T, Lemke-Rust K, Paar C (2007) Software implementation of estream profile i ciphers on embedded 8-bit avr microcontrollers. Technical report, eSTREAM project. http://www.ecrypt.eu.org/stream/sw.html

Menezes A, van Oorschot P, Vanstone S (2001) Handbook of applied cryptography. CRC, Boca Raton

Micheli GD, Benini L (2006) Networks on chips: technology and tools (Systems on silicon). Morgan Kaufmann, San Francisco

Micheli GD, Wolf W, Ernst R (2001) Readings in hardware/software co-design. Morgan Kaufmann, San Francisco

Moderchai BA (2006) Principles of concurrent and distributed programming, 2nd edn. Addison Wesley, Boston

Muchnick SS (1997) Advanced compiler design and implementation. Morgan Kaufmann, San Francisco

NIST (2001) Federal information processing standards publication 197: announcing the advanced encryption standard (aes). Technical report, http://csrc.nist.gov/publications/fips/fips197/fips-197.pdf

Panda PR, Catthoor F, Dutt ND, Danckaert K, Brockmeyer E, Kulkarni C, Vandecappelle A, Kjeldsberg PG (2001) Data and memory optimization techniques for embedded systems. ACM Trans Des Autom Electron Syst 6(2):149–206

Parhi KK (1999) VLSI digital signal processing: design and implementation. Wiley, New York. ISBN 978-0471241867.

Parhi KK, Messerschmitt DG (1991) Static rate-optimal scheduling of iterative data-flow programs via optimum unfolding. Computers, IEEE Transactions on 40(2):178–195.

Pasricha S, Dutt N (2008) On-chip communication architectures: system on chip interconnect. Morgan Kaufmann, Amsterdam

Potop-Butucaru D, Edwards SA, Berry G (2007) Compiling esterel. Springer, New York

Qin W (2004) Modeling and description of embedded processors for the development of software tools. Ph.D. thesis, Princeton University

Qin W, Malik S (2003) Flexible and formal modeling of microprocessors with application to retargetable simulation. In: DATE '03: proceedings of the conference on design, automation and test in Europe, Munich, p 10556

Rabaey JM (2009) Low power design essentials. Springer, New York

Rowen C (2004) Engineering the complex SOC: fast, flexible design with configurable processors. Prentice Hall, Upper Saddle River

Saleh R, Wilton S, Mirabbasi S, Hu A, Greenstreet M, Lemieux G, Pande P, Grecu C, Ivanov A (2006) System-on-chip: reuse and integration. Proc IEEE 94(6):1050–1069

Satoh A, Morioka S (2003) Hardware-focused performance comparison for the standard block ciphers aes, camellia, and triple-des. In: ISC, no. 2851. Lecture notes on computer science. Springer, New York, pp 252–266

Schaumont P, Shukla S, Verbauwhede I (2006) Design with race-free hardware semantics. In: DATE'06: Proceedings on design, automation and test in Europe, IEEE 1, vol. 1, pp 6

Smotherman M (2009) A brief history of microprogramming. Technical report, Clemson University. http://www.cs.clemson.edu/~mark/uprog.html

Stanford Graphics Lab (2003) Brook language. http://graphics.stanford.edu/projects/brookgpu/lang.html

Talla D, Hung CY, Talluri R, Brill F, Smith D, Brier D, Xiong B, Huynh D (2004) Anatomy of a portable digital mediaprocessor. IEEE Micro 24(2):32–39

Taubenfeld G (2006) Synchronization algorithms and concurrent programming. Pearson/Prentice Hall, Harlow

Thies W (2008) Language and compiler support for stream programs. Ph.D. thesis, MIT. http://groups.csail.mit.edu/cag/streamit/shtml/documentation.shtml

Vahid F (2003) The softening of hardware. Computer 36(4):27–34

Vahid F (2007a) Digital design. Wiley, Hoboken

Vahid F (2007b) It's time to stop calling circuits "hardware". Computer 40(9):106–108

Vahid F (2009) Dalton project. Technical report, http://www.cs.ucr.edu/~dalton/

Valls J, Sansaloni T, Perez-Pascual A, Torres V, Almenar V (2006) The use of cordic in software defined radios: a tutorial. IEEE Commun Mag 44(9):46–50

Volder JE (1959) The cordic trigonometric computing technique. IEEE Trans Electron Comput EC-8(3):330–334

Wolf W (2003) A decade of hardware/software codesign. Computer 36(4):38–43

Wulf W, McKee S (1995) Hitting the memory wall: implications of the obvious. In: ACM SIGARCH computer architecture news, 23, http://www.cs.virginia.edu/papers/Hitting_Memory_Wall-wulf94.pdf

Xilinx I (2009a) Xilinx embedded development toolkit. Technical report, http://www.xilinx.com/support/documentation/dt_edk.htm

Yaghmour K, Masters J, Ben-Yossef G, Gerum P (2008) Building embedded Linux systems, 2nd edn. O'Reilly, Sebastopol

# Index

P.R. Schaumont, *A Practical Introduction to Hardware/Software Codesign*,
DOI 10.1007/978-1-4614-3737-6, © Springer Science+Business Media New York 2013

477